Frontiers in Applied Dynamical Systems: Reviews and Tutorials

Volume 8

More information about this series at http://www.springer.com/series/13763

Frontiers in Applied Dynamical Systems: Reviews and Tutorials

The Frontiers in Applied Dynamical Systems (FIADS) covers emerging topics and significant developments in the field of applied dynamical systems. It is a collection of invited review articles by leading researchers in dynamical systems, their applications and related areas. Contributions in this series should be seen as a portal for a broad audience of researchers in dynamical systems at all levels and can serve as advanced teaching aids for graduate students. Each contribution provides an informal outline of a specific area, an interesting application, a recent technique, or a "how-to" for analytical methods and for computational algorithms, and a list of key references. All articles will be refereed.

Stefanie Winkelmann • Christof Schütte

Stochastic Dynamics in Computational Biology

 Springer

Stefanie Winkelmann (iD)
Zuse Institute Berlin
Berlin, Germany

Christof Schütte (iD)
Institute of Mathematics
Freie Universität Berlin
Berlin, Germany

Zuse Institute Berlin
Berlin, Germany

ISSN 2364-4532   ISSN 2364-4931   (electronic)
Frontiers in Applied Dynamical Systems: Reviews and Tutorials
ISBN 978-3-030-62386-9   ISBN 978-3-030-62387-6   (eBook)
https://doi.org/10.1007/978-3-030-62387-6

Mathematics Subject Classification: 92C45, 60J28, 82C22, 65C20, 34A38

This Springer imprint is published by the registered company Springer Nature Switzerland AG
The registered company address is: Gewerbestrasse 11, 6330 Cham, Switzerland

Introduction

Computational biology is the science of developing mathematical models and numerical simulation techniques to understand biological and biochemical systems. Since the early 1970s, researchers have used analytical methods and numerical algorithms to evaluate and interpret biological data. Many biochemical or biological processes can be considered as systems of individual entities (molecules, cells, particles, etc.), which move within a spatial environment and interact with each other through different types of chemical reactions or other forms of interplay. Such cellular reaction systems are the focus of this book; examples include gene regulatory systems, signaling cascades, viral infection kinetics, or enzyme cascades.

The book is predominantly addressed to graduate students and researchers wanting to get a sound theoretical background of modeling biochemical reaction systems as well as an overview of appropriate algorithmic concepts and simulation methods. Therefore, the mathematical technicalities have been limited to a level that the authors believe to be accessible for theory-interested non-mathematicians. The aim of this book is to provide a well-structured and coherent overview of the existing mathematical modeling approaches; relations between the models are explained, and several types of model recombinations (hybrid models) are investigated. In the main text, we circumvent rather technical parts of the mathematical theory for the purpose of readability while clarifying the respective mathematical details in the Appendix. Here, also a background for non-mathematicians is provided, as we explain some fundamental mathematical concepts (such as Markov processes and stochastic differential equations), which are repeatedly used throughout the book but might not be familiar to the overall readership.

Another main objective is the extensive illustration and comparison of diverse numerical simulation schemes and their computational effort. To this end, the text includes many examples and helpful illustrations. Everyone interested in deeper insights into the matter will find comprehensive references to the relevant literature. The authors hope that the present book may help to bridge the still existing gap between mathematical research in computational biology and its practical use for realistic biological, biochemical, or biomedical systems by providing readers with tools for performing more in-depth analysis of modeling, simulation, and data analysis methods.

Traditionally, biochemical reaction networks are modeled by deterministic dynamics of continuous variables. The system's state is defined by the mean concentrations of all involved chemical species and evolves in time according to a coupled set of ordinary differential equations called *reaction rate equations*. The numerical simulation of these reaction rate equations is well established and possible even for very large systems with rather small computational effort (at least compared to simulations based on alternative models in computational biology). However, this macroscopic modeling approach is only appropriate for systems with large molecular populations where stochastic effects in the dynamics are negligible. In case that some species appear in low copy numbers, random fluctuations are of crucial importance and the deterministic model may fail to capture the characteristic behavior of the reaction kinetics. For example, problems do naturally occur when using concentration-based approaches in order to compute statistics of extinction times or for modeling bimodal reaction systems.

There exist many applications, especially in the biochemical context, which call for a stochastic description in terms of natural numbers instead of real-valued, continuous concentrations. A summary of the most early attempts to mathematically address the intrinsic stochastic nature of chemical reaction systems can be found in [178], with the first applications of stochastic concepts going back to Kramers [153] and Delbrück [47] in the 1940s. The standard stochastic framework, which was more rigorously derived by Gillespie in the 1990s [102], treats the system's state as a set of non-negative integers defining the number of particles for each species. Chemical reactions appear as separate, instantaneous physical events causing a jump in the population state. The temporal evolution of the system is modeled as a continuous-time, discrete-state Markov process, called *reaction jump process*.

The central evolution equation for the time-dependent probability distribution of the reaction jump process is called *chemical master equation*. This is a countable system of ordinary differential equations, which has a rather simple structure. Nevertheless, due to its high dimensionality, it can rarely be solved analytically. As a consequence, there is a broad interest in efficient numerical solution techniques or indirect approaches based on Monte Carlo methods, which generate a statistically significant ensemble of realizations of the underlying reaction jump process in order to estimate its distribution. The most prominent sampling scheme is given by Gillespie's *stochastic simulation algorithm*, which allows to draw statistically exact trajectories of the process. For highly reactive systems, however, these exact simulations become computationally expensive, such that approximate simulation schemes or alternative modeling approaches are required.

The stochastic simulation becomes inefficient and sometimes even infeasible if the respective system exhibits high reactivity. High reactivity may be induced either by a large number of molecules or by large firing rates of some reactions contained in the system. In case of a large molecular population, the classical model of deterministic dynamics given by the reaction rate equation delivers a suitable approximation of the stochastic reaction jump process [156, 157]. Augmenting the reaction rate equation by a stochastic noise term leads to a higher-order approximation given by the so-called *chemical Langevin equation* [103], a set of stochastic differential equations that is meant to capture both mean and variance of the stochastic dynamics. For systems with multiple population scales, where only part of the molecular population is highly abundant while some species appear in low copy numbers, there exist *hybrid models*, which combine stochastic jump dynamics with reaction rate equations or chemical Langevin equations. This leads to the theory of *piecewise-deterministic Markov processes* [85] or *piecewise chemical Langevin equations* [41].

If, instead, the system's high reactivity stems from various levels of reaction rate constants giving rise to multiple timescales in the dynamics, completely different model reduction techniques are needed. Here, a reformulation of the dynamics in terms of *reaction extents* may be helpful [109, 116], or averaging methods can be applied in order to find effective dynamics of reduced complexity [244].

These different types of modeling and approximation methods have one thing in common: They assume the reaction–diffusion dynamics to be *well mixed* in space, meaning that the particles move comparatively fast in space and meet very often without undergoing a chemical reaction. For such a well-mixed scenario, spatial information becomes redundant and a description in terms of particle numbers or spatially homogeneous concentrations is justified. Systems that breach the well-mixed condition, on the contrary, require a more detailed description that captures certain information about the particles' positions in space. Existing modeling approaches reach from particle-based models, which track the spatial movement of each individual particle and its interaction with the surrounding population [56, 231], to concentration-based models given by partial differential equations [57]. Alternatively, the spatial environment may also be discretized, in order to approximate the continuous diffusion dynamics by spatial jump processes. This leads to extended chemical master equations, called *reaction–diffusion master equations* [130] or *spatiotemporal master equations* [246], containing not only biochemical reactions but also spatial jump events for the particles' movement.

In the last decades, much research has been performed on each of the described topics. A structured, self-contained survey thereof is contained in this book. In contrast to most review papers, which mainly focus on simulation-based approaches [61, 106, 167, 176, 188, 223, 230], this book, in particular, incorporates comprehensive mathematical derivations of the underlying theoretical models. The main motivation for the techniques discussed in this book comes from biochemical contexts, but the theory is also applicable to other scientific fields like epidemiology, ecology, social science, or neuroscience.

Outline   The general structure of the book is as follows: Chaps. 1–3 consider well-mixed systems, while Chap. 4 presents models and algorithms for systems requiring spatial resolution.

We start in Chap. 1 with the fundamental stochastic model of biochemical reaction kinetics. The reaction jump process is introduced, the chemical master equation is formulated, and we give a summary of methods to compute system-dependent expectations. This chapter is concluded by an illustrative example that motivates why (and when) considering stochastic modeling approaches instead of deterministic dynamics is advantageous. The case of a large molecular population, where the reaction jump process can be well approximated by reaction rate equations or chemical Langevin equations, is considered in Chap. 2. Here, also hybrid approaches for systems with multiple population scales are presented. The corresponding *hybrid master equation* is formulated, and a system of gene expression is considered as a multiscale biochemical application. In Chap. 3, systems with multiple temporal scales induced by different levels of reaction rate constants are considered. Our survey comprises both approaches: the first building on the formulation of an alternative master equation in terms of reaction extents, and the second referring to a separation of the multiscale dynamics in terms of observables. Finally, Chap. 4 covers non-well-mixed systems requiring spatial resolution, presenting particle-based reaction–diffusion approaches, models using deterministic and stochastic partial differential equations, as well as spatially extended chemical master equations.

Throughout all chapters, different types of exemplary reaction networks will be considered for illustration, some of them serving as abstract paragons (e.g., simple binding–unbinding systems or birth–death processes), while others referring to concrete biological applications like enzyme kinetics or gene expression dynamics.

Acknowledgments   Finally, it is our pleasure to thank our coworkers and colleagues for fruitful discussions and kind help in connection with the preparation of this book. In particular, we would like to express our gratitude to Rainald Ehrig and Marian Moldenhauer for reading the manuscript carefully and for making valuable suggestions, to Vivian Köneke and Nathalie Unger for useful assistance, and to Martin Weiser for helpful advice. We also wish to acknowledge the work of the anonymous referees who gave helpful comments on improving the quality of the manuscript.

The work on this book has been funded by the Deutsche Forschungsgemeinschaft (DFG, German Research Foundation) under Germany's Excellence Strategy—MATH + : The Berlin Mathematics Research Center, EXC-2046/1—project ID: 390685689, and through DFG grant CRC 1114.

Contents

Abbreviations

a.s.	Almost surely
CFPE	Chemical Fokker–Planck equation 51
CLE	Chemical Langevin equation 47
CME	Chemical master equation 12
DAE	Differential algebraic equation 124
FHD	Fluctuating hydrodynamics 152
hME	Hybrid master equation 83
hP	Hybrid process 83
LNA	Linear noise approximation 55
MSM	Markov state model 170
ODE	Ordinary differential equation 37
PDE	Partial differential equation 132
PBRD	Particle-based reaction–diffusion 159
PCLE	Piecewise chemical Langevin equation 70
PDMP	Piecewise-deterministic Markov process 65
PDRP	Piecewise-deterministic reaction process 61
QEA	Quasi-equilibrium approximation 122
RDME	Reaction–diffusion master equation 161
RJP	Reaction jump process 31
RME	Reaction master equation 107
RRE	Reaction rate equation 31
SDE	Stochastic differential equation 38
SPDE	Stochastic partial differential equation 152
SSA	Stochastic simulation algorithm 24
ST-CME	Spatiotemporal chemical master equation 161

List of Figures

Chapter 1

Well-Mixed Stochastic Reaction Kinetics

Let us start by considering a system of particles/molecules that belong to different biochemical species. The particles move in a given environment, the (spatial) *domain* $\mathbb{D}$ (typically $\mathbb{D} \subset \mathbb{R}^d$ for $d \in \{1, 2, 3\}$), and undergo biochemical reactions which affect the population. As for the relation between spatial movement and reactivity, we make the following central assumption.

Assumption 1.1 (Well-Mixed Assumption). Spatial movement of particles is fast compared to reactions, i.e., the majority of close encounters of particles (where molecules come close enough so that reactions are possible) are nonreactive and particle positions are uniformly distributed throughout the space $\mathbb{D}$ at any time.

Given the well-mixed assumption, the spatial positions of particles become insignificant for the system's dynamics, such that modeling approaches without spatial resolution are justified. The system's state may be specified by counting the number of particles of each species, regardless of their positions in space. As time evolves, the particle numbers change due to chemical reactions that take place. All investigations of Chaps. 1–3 build upon this well-mixed assumption. Modeling approaches for dynamics which do not comply with the well-mixed assumption are topic of Chap. 4.

S. Winkelmann, C. Schütte, *Stochastic Dynamics in Computational Biology*, Frontiers in Applied Dynamical Systems: Reviews and Tutorials 8, https://doi.org/10.1007/978-3-030-62387-6_1

Especially in biochemical systems with comparatively small molecular populations, stochastic effects play an important role for the system's dynamics, which motivates to describe the dynamics by means of stochastic processes. In this chapter, we introduce Markov jump processes as a modeling approach for well-mixed stochastic reaction dynamics. The basic components of a chemical reaction network are introduced in Sect. 1.1. The Markov jump process for the system's stochastic reaction dynamics is defined and characterized in Sect. 1.2. We introduce the path-wise formulation of the process and formulate the characteristic chemical master equation (CME) as well as the corresponding moment equations. In Sect. 1.3 different computation methods for solving the CME and calculating expectations of the process are presented, including numerical approximation methods in terms of stochastic simulation.

1.1 The Chemical Reaction Network

A chemical reaction network comprises a set of *chemical species*, a set of *chemical reactions*, and a set of *propensity functions*. The set of species consists of the *reactants* and *products* of the chemical reactions (each species can be both reactant and product), and the propensity functions define the likelihood for each reaction to occur depending on the system's population state.

More precisely, we consider a system of $L \in \mathbb{N}$ chemical species $\mathcal{S}_1, \ldots, \mathcal{S}_L$ which interact through $K \in \mathbb{N}$ reactions $\mathcal{R}_1, \ldots, \mathcal{R}_K$. Each reaction is represented by a stoichiometric equation of the form

$$\mathcal{R}_k : \quad s_{1k}\mathcal{S}_1 + \ldots + s_{Lk}\mathcal{S}_L \xrightarrow{\gamma_k} s'_{1k}\mathcal{S}_1 + \ldots + s'_{Lk}\mathcal{S}_L, \tag{1.1}$$

with the *stoichiometric coefficients* $s_{lk}, s'_{lk} \in \mathbb{N}_0$ denoting the numbers of reactant and product molecules, respectively. The associated *stoichiometric vector* $\boldsymbol{\nu}_k = (\nu_{1k}, \ldots, \nu_{Lk})^\mathsf{T} \in \mathbb{Z}^L$, also called *state-change vector*, is defined as

$$\nu_{lk} := s'_{lk} - s_{lk} \tag{1.2}$$

and describes the *net change* in the number of molecules of each species $\mathcal{S}_l$ due to reaction $\mathcal{R}_k$ [108]. The constant $\gamma_k > 0$ is the *reaction rate constant* which quantifies the rate for the reaction to take place such that

$\gamma_k \, dt =$ probability, to first order in dt, that a randomly selected
           combination of $\mathcal{R}_k$-reactant molecules will react according to $\mathcal{R}_k$
           within the next infinitesimal time interval $[t, t + dt)$,

cf. [99, 102]. For a reaction of the form $\mathcal{R}_k : \mathcal{S}_1 + \mathcal{S}_2 \xrightarrow{\gamma_k} \ldots$, for example, this means that the probability for any randomly selected $\mathcal{S}_1$-$\mathcal{S}_2$-pair of particles to react according to $\mathcal{R}_k$ within the next time interval of length dt is equal to $\gamma_k dt + o(dt)$.[1] Supposing that the selected $\mathcal{R}_k$-reactants are not affected by any other reaction, the waiting time for them to react follows an exponential distribution with mean $1/\gamma_k$. In general, however, reactants can anytime "appear or disappear" due to the occurrence of other reactions, such that statements about reaction probabilities are only possible on infinitesimal short time intervals. Note that the dimension of γ_k is always the inverse of time, as it characterizes the number of reaction events per time unit.

A reaction $\mathcal{R}_k$ given by (1.1) is called *of order* $m \in \mathbb{N}_0$ if it involves m reactant molecules, i.e., if $\sum_{l=1}^{L} s_{lk} = m$. This number of reacting molecules is also called the *molecularity* of the reaction. The elemental types are first-order reactions ($m = 1$) of the form $\mathcal{S}_l \to \ldots$, also called *unimolecular reactions*; second-order reactions ($m = 2$) of the form $\mathcal{S}_l + \mathcal{S}_{l'} \to \ldots$, called *bimolecular reactions*; and reactions of order $m = 0$ given by $\emptyset \to \ldots$ which describe particle formation happening independently of the molecular population. Reactions of order $m \geq 3$ usually play a minor role because they appear to be very unlikely and can always be expressed by a sequence of two or more elemental reactions [105]. The central basic reactions repeatedly to appear in this work are

- production $\emptyset \to \mathcal{S}_l$ and degradation $\mathcal{S}_l \to \emptyset$,

- conversion $\mathcal{S}_l \to \mathcal{S}_{l'}$,

- binding/association $\mathcal{S}_{l'} + \mathcal{S}_{l''} \to \mathcal{S}_l$ and unbinding/dissociation $\mathcal{S}_l \to \mathcal{S}_{l'} + \mathcal{S}_{l''}$,

- (auto-)catalytic reactions $\mathcal{S}_l \to \mathcal{S}_l + \mathcal{S}_{l'}$.

A reaction network is called *linear* if it holds $m \leq 1$ for all involved reactions.

In the well-mixed scenario, see Assumption 1.1; the *state* of the reaction network is given by a vector of the form

$$\boldsymbol{x} = (x_1, \ldots, x_L)^\mathsf{T} \in \mathbb{N}_0^L,$$

where $x_l \in \mathbb{N}_0$ refers to the number of particles of species $\mathcal{S}_l$, $l = 1, \ldots, L$. Each reaction induces instantaneous changes of the state by transitions of the form

$$\mathcal{R}_k : \quad \boldsymbol{x} \to \boldsymbol{x} + \boldsymbol{\nu}_k$$

[1] The asymptotic notation $o(dt)$ represents some function decreasing faster than dt such that $\lim_{dt \to 0} \left| \frac{o(dt)}{dt} \right| = 0$.

with $\boldsymbol{\nu}_k$ given in (1.2). The probability for such a transition to occur depends on the reaction rate constant γ_k and on the state $\boldsymbol{x}$ of the system. It is defined by the *propensity function* $\alpha_k : \mathbb{N}_0^L \to [0, \infty)$ with

$\alpha_k(\boldsymbol{x})\, dt =$ probability, to first order in dt, that reaction $\mathcal{R}_k$ will occur once within the next infinitesimal time interval $[t, t + dt)$ given that the system is in state $\boldsymbol{x}$ at time t.

Moreover, the probability for the reaction to occur *more than once* within $[t, t + dt)$ is assumed to be of order $o(dt)$, i.e., to be negligible.

As proposed by Gillespie in 1976 [99], the propensity function is of the general form

$$\alpha_k(\boldsymbol{x}) = \gamma_k h_k(\boldsymbol{x})$$

for the *combinatorial function* $h_k : \mathbb{N}_0^L \to \mathbb{N}_0$, where $h_k(\boldsymbol{x})$ defines the *number of distinct combinations* of reactant molecules for reaction $\mathcal{R}_k$ given the system's state $\boldsymbol{x}$. Basic combinatoric arguments give

$$h_k(\boldsymbol{x}) = \prod_{l=1}^{L} \binom{x_l}{s_{lk}} = \prod_{l=1}^{L} \frac{x_l!}{s_{lk}!(x_l - s_{lk})!}$$

whenever $x_l \geq s_{lk}$ for all $l = 1, \ldots, L$, where $x_l \in \mathbb{N}_0$ is the number of molecules of species $\mathcal{S}_l$ and $0! := 1$. This leads to the fundamental propensity functions of *stochastic mass-action kinetics* given by

$$\alpha_k(\boldsymbol{x}) = \begin{cases} \gamma_k \displaystyle\prod_{l=1}^{L} \frac{x_l!}{s_{lk}!(x_l - s_{lk})!} & \text{if } x_l \geq s_{lk} \text{ for all } l = 1, \ldots, L, \\ 0 & \text{otherwise.} \end{cases} \tag{1.3}$$

Table 1.1 gives an overview of these mass-action propensity functions for reactions up to order two. The dimension of $\alpha_k(\boldsymbol{x})$ is always the inverse of time.

Non-mass-Action Propensities

Besides the propensity functions of mass-action type given in (1.3), there are other types of propensity functions commonly used in literature. These non-mass-action-type propensities typically occur for reduced models where an effective reaction replaces a sequence of several mass-action-type reactions. One of the most popular non-mass-action propensity functions is

Table 1.1. Mass-action propensity functions of elementary chemical reactions in terms of particle numbers

Order	Reaction	Propensity
0th	$\emptyset \xrightarrow{\gamma_0} \dots$	$\alpha_0(\boldsymbol{x}) = \gamma_0$
1st	$\mathcal{S}_l \xrightarrow{\gamma_1} \dots$	$\alpha_1(\boldsymbol{x}) = \gamma_1 x_l$
2nd	$\mathcal{S}_l + \mathcal{S}_{l'} \xrightarrow{\gamma_2} \dots$	$\alpha_2(\boldsymbol{x}) = \gamma_2 x_l x_{l'}$
2nd	$2\mathcal{S}_l \xrightarrow{\gamma_{2'}} \dots$	$\alpha_{2'}(\boldsymbol{x}) = \frac{\gamma_{2'}}{2} x_l (x_l - 1)$

the Michaelis-Menten propensity which is used, e.g., for modeling nonlinear degradation processes or enzyme kinetics. It is of the form

$$\alpha(\boldsymbol{x}) = \frac{\gamma x_1}{\gamma_d + x_1}$$

for some constants $\gamma, \gamma_d > 0$ and refers to a reaction of type $\mathcal{S}_1 \to \mathcal{S}_2$. A motivation for this type of reaction propensity is given in Chap. 3, where the Michaelis-Menten system of enzyme kinetics is investigated in more detail (see Example 3.1).

The following natural assumption for general propensity functions guarantees that the system never reaches any negative state. For mass-action propensities, this is always satisfied.

Assumption 1.2. The propensity functions satisfy $\alpha_k(\boldsymbol{x}) \geq 0$ for all $\boldsymbol{x} \in \mathbb{N}_0^L$ and $\alpha_k(\boldsymbol{x}) = 0$ if $\boldsymbol{x} + \boldsymbol{\nu}_k \in \mathbb{Z}^L \setminus \mathbb{N}_0^L$.

Example 1.3 (Binding and Unbinding). As a simple general example of nonlinear reaction networks, we consider the system of binding and unbinding (also called *association* and *dissociation*) of three species $\mathcal{S}_1, \mathcal{S}_2, \mathcal{S}_3$ given by the stoichiometric equations

$$\mathcal{R}_1 : \mathcal{S}_1 + \mathcal{S}_2 \xrightarrow{\gamma_1} \mathcal{S}_3, \quad \mathcal{R}_2 : \mathcal{S}_3 \xrightarrow{\gamma_2} \mathcal{S}_1 + \mathcal{S}_2.$$

see Fig. 1.1 for an illustration.
In case of well-mixed dynamics, the system's state is given by a vector $\boldsymbol{x} \in \mathbb{N}_0^3$ containing the number of particles of each species. The related state-change vectors are

$$\boldsymbol{\nu}_1 = \begin{pmatrix} -1 \\ -1 \\ 1 \end{pmatrix}, \quad \boldsymbol{\nu}_2 = \begin{pmatrix} 1 \\ 1 \\ -1 \end{pmatrix}.$$

and for the mass-action propensity functions, it holds $\alpha_1(\boldsymbol{x}) = \gamma_1 x_1 x_2$ and $\alpha_2(\boldsymbol{x}) = \gamma_2 x_3$ (see Table 1.1). $\diamondsuit$

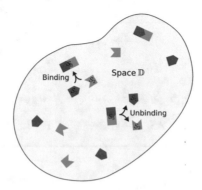

Figure 1.1. System of binding and unbinding. The space $\mathbb{D}$ is given by a biological cell containing particles of three chemical species $\mathcal{S}_1$, $\mathcal{S}_2$, $\mathcal{S}_3$ that interact by binding $\mathcal{S}_1 + \mathcal{S}_2 \to \mathcal{S}_3$ and unbinding $\mathcal{S}_3 \to \mathcal{S}_1 + \mathcal{S}_2$. The well-mixed Assumption 1.1 means that spatial diffusion is comparatively fast and the vast majority of particle crossings is nonreactive

1.2 The Reaction Jump Process

In order to describe the temporal evolution of a system, we consider the continuous-time stochastic process $\boldsymbol{X} = (\boldsymbol{X}(t))_{t \geq 0}$,

$$\boldsymbol{X}(t) = (X_1(t), \ldots, X_L(t))^{\mathsf{T}} \in \mathbb{N}_0^L,$$

with $X_l(t)$ denoting the number of particles of species $\mathcal{S}_l$ at time $t \geq 0$ for $l = 1, \ldots, L$. Let $\boldsymbol{X}(0) = \boldsymbol{x}_0$ be the initial state of the system and denote by $R_k(t)$ the number of times that reaction $\mathcal{R}_k$ has occurred by time t given the initial state $\boldsymbol{x}_0$. That is, it holds

$$R_k(t) = r \in \mathbb{N}_0$$

if reaction $\mathcal{R}_k$ has occurred r times within the time interval $[0, t]$. Each of the processes $R_k = (R_k(t))_{t \geq 0}$ is a monotonically increasing process on $\mathbb{N}_0$. Each time that reaction $\mathcal{R}_k$ occurs, the process R_k increases by one. Due to the special structure of the dynamics with jumps in $\boldsymbol{X}$ always having the form $\boldsymbol{x} \to \boldsymbol{x} + \boldsymbol{\nu}_k$, the state of the system at time $t > 0$ is given by

$$\boldsymbol{X}(t) = \boldsymbol{x}_0 + \sum_{k=1}^{K} R_k(t) \boldsymbol{\nu}_k. \tag{1.4}$$

In the stochastic setting, the counting processes R_k are represented in terms of *Poisson processes*. These are special types of *Markov jump processes*, i.e., continuous-time, discrete-state Markov processes (see Sect. A.1 in the Appendix for a summary of the theoretical background). In this case, the resulting process $X = (X(t))_{t \geq 0}$ defined by (1.4) is a Markov jump process itself and will subsequently be called *reaction jump process*. Depending on the reaction network, the state space $\mathbb{X} \subset \mathbb{N}_0^L$ of X can be finite or infinite, with the former case referring to dynamics that own a natural upper bound for the total population size (as in Example 1.3 of binding and unbinding). An infinite state space $\mathbb{X}$, on the other hand, is given for systems which can produce arbitrarily large numbers of particles, such as a simple birth-death process (see Example 1.7 in Sect. 1.3).

There exist different ways to characterize the reaction jump process X. First, there is a *path-wise representation* in terms of Poisson processes, which is derived in Sect. 1.2.1. Second, the dynamics may be characterized by the Kolmogorov forward equation for the probability distribution of the process, which leads to the *chemical master equation* presented in Sect. 1.2.2. Finally, the characteristic equations for the system's moments are considered in Sect. 1.2.3.

1.2.1 Path-Wise Representation in Terms of Poisson Processes

The fundamental hypothesis of stochastic chemical kinetics states that for each time t the rate for reaction $\mathcal{R}_k$ to occur and for the process R_k to increase by one is a time-dependent random number given by $\alpha_k(X(t))$. That is, the probability for reaction $\mathcal{R}_k$ to take place once within the infinitesimal time interval $[t, t+dt)$ is given by $\alpha_k(X(t)) + o(dt)$, while the probability for several reactions happening within this time span is of order $o(dt)$ (remember the assumptions for α_k on page 4). As for the counting processes R_k, $k = 1, \ldots, K$, these assumptions can be summarized as follows.

Assumption 1.4 (Stochastic Reaction Kinetics). For all $k = 1, \ldots, K$ and $t \geq 0$ it holds

$$\mathbb{P}(R_k(t+dt) - R_k(t) = 0) = 1 - \alpha_k(X(t))dt + o(dt),$$
$$\mathbb{P}(R_k(t+dt) - R_k(t) = 1) = \alpha_k(X(t))dt + o(dt),$$
$$\mathbb{P}(R_k(t+dt) - R_k(t) > 1) = o(dt)$$

for $dt \to 0$.

These characteristics are satisfied by the processes R_k given as

$$R_k(t) = \mathcal{U}_k \left(\int_0^t \alpha_k(\boldsymbol{X}(s)) \, ds \right), \quad k = 1, \ldots, K, \tag{1.5}$$

where $\mathcal{U}_k$ for $k = 1, \ldots, K$ denote independent, unit-rate Poisson processes [7, 10] (see also Sect. A.1.4 in Appendix). That is, for each k it is $\mathcal{U}_k = (\mathcal{U}_k(t'))_{t' \geq 0}$, a monotonically increasing Markov jump process on $\mathbb{N}_0$ which jumps from n to $n+1$ at constant rate $v = 1$. Each Poisson process $\mathcal{U}_k$ brings its own "internal" time t' as well as internal jump times $t'_1, t'_2, \ldots$ where $\mathcal{U}_k(t'_i) \to \mathcal{U}_k(t'_i) + 1$. In (1.5), $\mathcal{U}_k$ is "evaluated" at $t' = \int_0^t \alpha_k(\boldsymbol{X}(s)) \, ds$, meaning that the value of the cumulative propensity $\int_0^t \alpha_k(\boldsymbol{X}(s)) \, ds$ determines the amount of internal time advance in $\mathcal{U}_k$.[2] For example, whenever $\alpha_k(\boldsymbol{X}(t)) = 0$ holds for some "real" time t, the time advance of the internal time t' of $\mathcal{U}_k$ is stopped and the process R_k stagnates. In contrast, large values of α_k lead to a fast time advance in $\mathcal{U}_k$ and consequently more jumps (on average) in R_k. The "real" jump times t_i of the process R_k therefore deviate from the internal jump times t'_i of the Poisson process $\mathcal{U}_k$. In fact, time is rescaled by the propensities α_k, and together with the "real" time t of the overall process $\boldsymbol{X}$ there are in total $K+1$ different time frames.

Since the propensity functions $\alpha_k(\boldsymbol{X}(t))$ are all constant until the next reaction takes place, it follows from the properties of Poisson processes (cf. Eq. (A.13) in Appendix) that

$$\mathbb{P}\left(\mathcal{U}_k \left(\int_0^{t+dt} \alpha_k(\boldsymbol{X}(s)) ds \right) - \mathcal{U}_k \left(\int_0^t \alpha_k(\boldsymbol{X}(s)) \right) = 1 \right)$$
$$= \alpha_k(\boldsymbol{X}(t)) dt + o(dt).$$

Likewise, the probability for each Poisson process $\mathcal{U}_k$ to stay constant within $[t, t+dt)$ is given by $1 - \alpha_k(\boldsymbol{X}(t)) dt + o(dt)$, and the probability for more than one jump of $\mathcal{U}_k$ within $[t, t+dt)$ is of order $o(dt)$ for $dt \to 0$, which is all in consistency with Assumption 1.4.

In combination, Eqs. (1.4) and (1.5) relate the different counting processes R_k, $k = 1, \ldots, K$, to each other. Hence, they are in general not independent of each other, although, fixing a state $\boldsymbol{x}$ of the system, the individual reaction events happen independently of each other. With the reaction propensities depending on the state $\boldsymbol{X}(t) \in \mathbb{X}$ of the system, the processes R_k do normally not have independent increments and are therefore *not* Poisson processes themselves.

---

[2]Note that the internal time of the Poisson process is dimensionless, just like the integral $\int_0^t \alpha_k(\boldsymbol{X}(t)) \, ds$ is.

Inserting (1.5) into (1.4) finally gives the following path-wise formulation of the reaction jump process in terms of independent, unit-rate Poisson processes $\mathcal{U}_k$ [9, 128]:

Path-wise representation of the reaction jump process:

$$X(t) = x_0 + \sum_{k=1}^{K} \mathcal{U}_k \left(\int_0^t \alpha_k(X(s))\,ds \right) \nu_k \qquad (1.6)$$

This means that the process starts at time $t = 0$ in state x_0 and then, in the course of time, switches repeatedly to new states by performing jumps of the form $x \mapsto x + \nu_k$ whenever one of the Poisson processes $\mathcal{U}_k$ indicates such a jump. The waiting times for the jumps depend on the propensity functions α_k.

Equation (1.6) has a unique solution provided that $\sum_k \alpha_k(x) < \infty$ for all x, which is trivially fulfilled if we consider a finite set of reactions. Yet, the solution might only exist for a finite time interval if a "blow up" happens, meaning that at least one of the process components hits infinity in finite time. Additional assumptions are necessary to exclude this (cf. [10]). The path-wise representation (1.6) will be of fundamental relevance for studying approximative dynamics in a continuous state space, which is the topic of Chap. 2.

Note that although $X = (X(t)_{t\geq 0}$ is uniquely determined by the process $(R(t))_{t\geq 0} = (R_1(t), \ldots, R_K(t))_{t\geq 0}^\mathsf{T}$ of reaction extents, the reverse need not be true, because different combinations of reaction extents can lead to the same molecular population. In this sense, the counting process R contains more information than the reaction jump process X. In some situations, a direct characterization and analysis of the dynamics in terms of the reaction extents can be fruitful, as we will see in Sect. 3.1. For most questions of interest, however, the process X optimally captures the relevant information and is thus used as the standard modeling approach.

Example 1.3 (Continued). For the system of binding and unbinding given by the reactions $\mathcal{R}_1 : \mathcal{S}_1 + \mathcal{S}_2 \xrightarrow{\gamma_1} \mathcal{S}_3$ and $\mathcal{R}_2 : \mathcal{S}_3 \xrightarrow{\gamma_2} \mathcal{S}_1 + \mathcal{S}_2$, the reaction jump process is of the form $(X(t))_{t\geq 0} = (X_1(t), X_2(t), X_3(t))_{t\geq 0}^\mathsf{T}$. It is related to the process of reactions extents $(R(t))_{t\geq 0} = (R_1(t), R_2(t))_{t\geq 0}^\mathsf{T}$ via

$$X_1(t) = X_1(0) - R_1(t) + R_2(t),$$
$$X_2(t) = X_2(0) - R_1(t) + R_2(t),$$
$$X_3(t) = X_3(0) + R_1(t) - R_2(t),$$

with $R_1(t)$ and $R_2(t)$ referring to the number of binding and unbinding reactions that occurred by time t, respectively. Both processes are illustrated in Fig. 1.2. Fixing the initial state $\boldsymbol{X}(0) = \boldsymbol{x}_0 = (x_0^1, x_0^2, x_0^3) \in \mathbb{N}_0^3$, the state space $\mathbb{X}$ of the process is naturally bounded, as it holds $X_l(t) \le x_0^1 + x_0^2 + x_0^3$ for all $t \ge 0$ and $l = 1, 2, 3$. $\diamond$

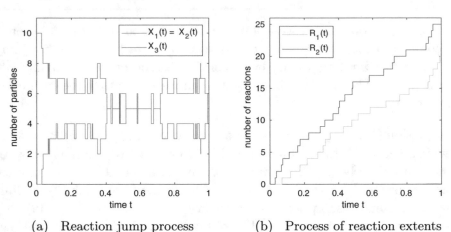

(a) Reaction jump process   (b) Process of reaction extents

Figure 1.2. Process of binding and unbinding. (**a**) One realization of the reaction jump process $(\boldsymbol{X}(t))_{t \ge 0}$ for the system of binding and unbinding (see Example 1.3) with rate constants $\gamma_1 = 1$, $\gamma_2 = 4$ and initial state $\boldsymbol{X}(0) = (0, 0, 10)^\mathsf{T}$. (**b**) Corresponding realization of the process of reaction extents $(\boldsymbol{R}(t))_{t \ge 0} = (R_1(t), R_2(t))_{t \ge 0}^\mathsf{T}$ with initial state $\boldsymbol{R}(0) = (0, 0)^\mathsf{T}$

1.2.2 The Chemical Master Equation

As an alternative to the path-wise representation (1.6) of the reaction jump process $\boldsymbol{X}$, the reaction dynamics can also be characterized by the *Kolmogorov forward equation* for the distribution of $\boldsymbol{X}$, which in the context of reaction kinetics is called *chemical master equation* (CME). Given some initial state $\boldsymbol{X}(0) = \boldsymbol{x}_0 \in \mathbb{N}_0^L$, let

$$p(\boldsymbol{x}, t) := \mathbb{P}(\boldsymbol{X}(t) = \boldsymbol{x} \mid \boldsymbol{X}(0) = \boldsymbol{x}_0) \tag{1.7}$$

denote the probability to find the process in state $\boldsymbol{x} = (x_1, \ldots, x_L)^\mathsf{T} \in \mathbb{N}_0^L$ at time t. In order to emphasize the dependence of this probability on the initial state $\boldsymbol{x}_0$, one can write $p(\boldsymbol{x}, t|x_0)$, but for the purpose of simplicity, we stick to the abbreviated notation. In case where the initial state is not deterministic but follows a distribution p_0, the notation $p(\boldsymbol{x}, t)$ serves as an abbreviation for $p(\boldsymbol{x}, t|\boldsymbol{X}(0) \sim p_0) := \mathbb{P}(\boldsymbol{X}(t) = \boldsymbol{x}|\boldsymbol{X}(0) \sim p_0)$.

Following the proceeding in [102, 127], we consider the infinitesimal time interval $[t, t+dt)$ for $dt > 0$ and the probability $p(\boldsymbol{x}, t+dt)$ to find the system in state $\boldsymbol{x}$ at time $t + dt$. In order to determine $p(\boldsymbol{x}, t + dt)$, we distinguish between three possible events that can trigger state $\boldsymbol{x}$ during $[t, t + dt)$:

(i) At time t the system was in state $\boldsymbol{x}$ and none of the reactions occurred within $[t, t + dt)$.

(ii) At time t the system was in state $\boldsymbol{x} - \boldsymbol{\nu}_k$ for some $k = 1, \ldots, K$ and exactly one reaction $\mathcal{R}_k$ occurred within $[t, t + dt)$.

(iii) More than one reaction fired within $[t, t+dt)$ yielding $\boldsymbol{x}$ at time $t+dt$.

Given the state $\boldsymbol{x}$, the waiting time for reaction $\mathcal{R}_k$ is assumed to be exponentially distributed with parameter $\alpha_k(\boldsymbol{x})$, such that the probability for reaction $\mathcal{R}_k$ *not* to occur in the considered infinitesimal time interval $[t, dt)$ is $\exp(-\alpha_k(\boldsymbol{x})dt)$, where $\exp(x) = e^x$ denotes the natural exponential function (see also Eq. (A.12) in Sect. A.1 of Appendix). Consequently, the probability for none of the reactions to take place within $[t, dt)$ conditioned on $\boldsymbol{X}(t) = \boldsymbol{x}$ is given by

$$\prod_{k=1}^{K} \exp(-\alpha_k(\boldsymbol{x})dt) = \prod_{k=1}^{K}(1 - \alpha_k(\boldsymbol{x})dt + o(dt)) = 1 - \sum_{k=1}^{K} \alpha_k(\boldsymbol{x})dt + o(dt)$$

(1.8)

because all reactions happen independently of each other.[3] The probability for the first event (i) is then given by the product of $p(\boldsymbol{x}, t)$ and (1.8). Similarly, the probability for the second event (ii) and a specific k is given by $p(\boldsymbol{x} - \boldsymbol{\nu}_k, t)$ times the probability that $\mathcal{R}_k$ occurs once conditioned on $\boldsymbol{X}(t) = \boldsymbol{x} - \boldsymbol{\nu}_k$. The latter is by definition of the propensity functions given by $\alpha_k(\boldsymbol{x} - \boldsymbol{\nu}_k)dt + o(dt)$. Finally, the probability for the third possible event

---

[3] The first equality results from writing the exponential function in form of a power series: $\exp(-\alpha_k(\boldsymbol{x})dt) = \sum_{i=0}^{\infty} \frac{(-\alpha_k(\boldsymbol{x})dt)^i}{i!} = 1 - \alpha_k(\boldsymbol{x})dt + \alpha_k(\boldsymbol{x})^2 dt^2 \mp \ldots = 1 - \alpha_k(\boldsymbol{x})dt + o(dt)$ for $dt \to 0$. The second equality follows from multiplying the factors and summarizing terms of order $o(dt)$.

(iii) is of order $o(dt)$ for $dt \to 0$ (cf. Assumption 1.4 and [99, 102]). By the law of total probability, we obtain

$$p(\boldsymbol{x}, t + dt) = p(\boldsymbol{x}, t)\left[1 - \sum_{k=1}^{K} \alpha_k(\boldsymbol{x})dt + o(dt)\right]$$

$$+ \sum_{k=1}^{K} p(\boldsymbol{x} - \boldsymbol{\nu}_k, t)\left[\alpha_k(\boldsymbol{x} - \boldsymbol{\nu}_k)dt + o(dt)\right] \qquad (1.9)$$

$$+ o(dt).$$

Subtracting $p(\boldsymbol{x}, t)$ from (1.9), dividing by dt, and taking the limit $dt \to 0$ (i.e., taking the derivative of $p(\boldsymbol{x}, t)$ with respect to t) give the *chemical master equation* (CME) (1.10).

The chemical master equation (CME):

$$\frac{dp(\boldsymbol{x}, t)}{dt} = \sum_{k=1}^{K}\left[\alpha_k(\boldsymbol{x} - \boldsymbol{\nu}_k)p(\boldsymbol{x} - \boldsymbol{\nu}_k, t) - \alpha_k(\boldsymbol{x})p(\boldsymbol{x}, t)\right] \qquad (1.10)$$

The negative terms in the right-hand side of (1.10) correspond to the "outflow" from the state $\boldsymbol{x}$ induced by the reactions (leading to a reduction of $p(\boldsymbol{x}, t)$), while the former positive terms in the sum correspond to the "inflow" induced by the reactions when the system comes from other states of the form $\boldsymbol{x} - \boldsymbol{\nu}_k$. We set

$$\alpha_k(\boldsymbol{x}) = 0 \quad \text{and} \quad p(\boldsymbol{x}, t) = 0 \quad \text{for } \boldsymbol{x} \notin \mathbb{N}_0^L$$

in order to exclude terms in the right-hand side of (1.10) where the argument $\boldsymbol{x} - \boldsymbol{\nu}_k$ contains negative entries.

The CME is actually not a single equation but a whole system of equations, containing one equation of type (1.10) for each state $\boldsymbol{x} \in \mathbb{X} \subset \mathbb{N}_0^L$. Letting

$$\ell_1^1 := \left\{v : \mathbb{X} \to \mathbb{R}_0^+ \,\Big|\, \sum_{\boldsymbol{x} \in \mathbb{X}} v(\boldsymbol{x}) = 1\right\}$$

be the set of non-negative functions on $\mathbb{X}$ that sum up to 1 (representing probability distributions on $\mathbb{X}$), we can define the operator

$$\mathcal{G} : \ell_1^1 \to \ell_1^1$$

by

$$(\mathcal{G}v)(\boldsymbol{x}) := \sum_{k=1}^{K} \left[\alpha_k(\boldsymbol{x} - \boldsymbol{\nu}_k)v(\boldsymbol{x} - \boldsymbol{\nu}_k) - \alpha_k(\boldsymbol{x})v(\boldsymbol{x}) \right]. \qquad (1.11)$$

For fixed t it holds $p(\cdot, t) \in \ell_1^1$, and with $p = p(\boldsymbol{x}, t)$, the CME then reads

$$\frac{dp}{dt} = \mathcal{G}p,$$

as an alternative formulation to the "state-wise" notation (1.10). The operator $\mathcal{G}$ is the infinitesimal generator of the process $\boldsymbol{X}$ (corresponding to the semigroup of forward transfer operators (see Sect. A.1 for details)).

One of the first formulations of the CME for the special case of auto-catalytic reactions was given by Delbrück in 1940 [47]. Further applications in the chemical context were given, e.g., by McQuarrie [178]. Gillespie then presented a rigorous physical and mathematical foundation of the CME in 1992 [102].

Existence of Solutions of the CME

The CME is a system of linear ordinary differential equations (ODEs) of large (possibly infinite) dimension given by the size $|\mathbb{X}|$ of the state space of the process $\boldsymbol{X} = (\boldsymbol{X}(t))_{t \geq 0}$. If the state space $\mathbb{X}$ is finite, the existence and uniqueness of solutions of the CME for given initial values $p_0(\boldsymbol{x}) = p(\boldsymbol{x}, t = 0)$ is guaranteed by standard results from the theory of ODEs, at least for the mass-action propensity functions discussed above. For infinite state spaces, on the other hand, more elaborated techniques are required to check the existence, uniqueness, and regularity of solutions to (1.10). Semigroup techniques can be applied [175], or the properties of so-called minimal non-negative solutions to (1.10) can be studied [94, 165, 196]. Here, we subsequently assume the following:

Assumption 1.5. For the considered propensity functions and the initial distribution $p_0 \in \ell_1^1$, the chemical master equation has a unique solution $p(\cdot, t) \in \ell_1^1$ for all $t \in [0, T]$ for some $T > 0$.

By the high dimensionality of the CME, finding its solution $p(\boldsymbol{x}, t)$ is not straightforward for most reaction networks. Different approaches for deriving the solution (or approximations of) will be presented in Sect. 1.3. Before that, we shortly examine the characteristic equations for the system's moments.

1.2.3 Moment Equations

Given the CME (1.10) one can directly derive evolution equations for the system's moments. Let

$$\mathbb{E}(\boldsymbol{X}(t)) := \sum_{\boldsymbol{x} \in \mathbb{X}} \boldsymbol{x} p(\boldsymbol{x}, t),$$

with $p(\boldsymbol{x}, t)$, defined in (1.7), be the *first-order moment* (or *mean* or *expectation*) of $\boldsymbol{X}(t)$. It is clear that $\mathbb{E}(\boldsymbol{X}(t)) \in (\mathbb{R}_+ \cup \{\infty\})^L$ exists for all t (i.e., $\boldsymbol{X}(t) \in \mathbb{X} \subset \mathbb{N}_0^L$ do not have any negative entries.[4] Remember that L is the number of different chemical species involved in the reaction network. Assuming that the expectation is actually finite for all t, we can multiply the CME (1.10) by $\boldsymbol{x}$ and sum over all $\boldsymbol{x}$ to get[5]

$$\frac{d}{dt}\mathbb{E}(\boldsymbol{X}(t)) = \sum_{k=1}^{K} \mathbb{E}(\alpha_k(\boldsymbol{X}(t)))\boldsymbol{\nu}_k. \qquad (1.12)$$

This is a system of equations of comparatively low dimension L. In contrast to the CME, however, it is in general nonlinear. Moreover, as soon as there is at least one bimolecular or higher-order reaction in the network, (1.12) is not a closed system of equations for the first-order moments but is coupled with higher-order moments.

In general, the time evolution equation for moments like

$$\mathbb{E}\Big(X_{l^1}(t) \cdot \ldots \cdot X_{l^d}(t)\Big) := \sum_{\boldsymbol{x}=(x_1,\ldots,x_L) \in \mathbb{X}} x_{l^1} \cdot \ldots \cdot x_{l^d} p(\boldsymbol{x}, t)$$

for $d \in \mathbb{N}$ and $\{l^1, \ldots, l^d\} \subset \{1, \ldots, L\}$ is given by

$$\frac{d}{dt}\mathbb{E}\Big(X_{l^1}(t) \cdot \ldots \cdot X_{l^d}(t)\Big) = \sum_{k=1}^{K} \mathbb{E}\Big(\alpha_k(\boldsymbol{X}(t))(X_{l^1}(t) + \nu_{l^1 k}) \cdot \ldots \cdot (X_{l^d}(t) + \nu_{l^d k})\Big)$$

$$- \sum_{k=1}^{K} \mathbb{E}\Big(\alpha_k(\boldsymbol{X}(t)) X_{l^1}(t) \cdot \ldots \cdot X_{l^d}(t)\Big),$$

---

[4]A random variable $\boldsymbol{X}(t) : (\Omega, \mathcal{E}, \mathbb{P}) \to (\mathbb{R}^L, \mathcal{B}(\mathbb{R}^L))$ is called *quasi-integrable* (with respect to $\mathbb{P}$) if at least one of the two expectations $\mathbb{E}(\boldsymbol{X}(t)^+)$ and $\mathbb{E}(\boldsymbol{X}(t)^-)$ is finite, where $(\boldsymbol{X}(t)^+)_l := \max\{X_l(t), 0\}$ and $(\boldsymbol{X}(t)^-)_l := -\min\{X_l(t), 0\}$. Here, it holds $\boldsymbol{X}(t)^- = \boldsymbol{0}$ for all t, such that $\mathbb{E}(\boldsymbol{X}(t)^-) = \boldsymbol{0}$ is finite for all t.
[5]$\sum_{\boldsymbol{x}} \alpha_k(\boldsymbol{x} - \boldsymbol{\nu}_k)p(\boldsymbol{x} - \boldsymbol{\nu}_k, t)\boldsymbol{x} - \sum_{\boldsymbol{x}} \alpha_k(\boldsymbol{x})p(\boldsymbol{x}, t)\boldsymbol{x} = \sum_{\boldsymbol{x}} \alpha_k(\boldsymbol{x})p(\boldsymbol{x}, t)(\boldsymbol{x} + \boldsymbol{\nu}_k) - \sum_{\boldsymbol{x}} \alpha_k(\boldsymbol{x})p(\boldsymbol{x}, t)\boldsymbol{x} = \sum_{\boldsymbol{x}} \alpha_k(\boldsymbol{x})p(\boldsymbol{x}, t)\boldsymbol{\nu}_k = \mathbb{E}(\alpha_k(\boldsymbol{X}(t)))\boldsymbol{\nu}_k.$

which for a second-order moment $\mathbb{E}(X_l^2(t))$ reduces to

$$\frac{d}{dt}\mathbb{E}\left(X_l^2(t)\right) = \sum_{k=1}^{K}\left[2\mathbb{E}(\alpha_k(\boldsymbol{X}(t))X_l(t))\nu_{lk} + \mathbb{E}(\alpha_k(\boldsymbol{X}(t)))\nu_{lk}^2\right].$$

We can observe that in case of linear reaction propensity functions α_k, the evolution equation for a moment of order m contains only moments of order up to m. This means that for linear reaction networks (containing solely reactions of order zero or one) and standard mass-action propensity functions α_k, all the moment equations up to a certain order form a self-contained finite system of linear ODEs, which can directly be solved by standard analytical or numerical methods. In [88], the evolution equations for the mean and variance of a general system of first-order reactions are derived explicitly (by means of the moment-generating function method). Given linear reaction propensity functions α_k, for example, the first-order moment equation (1.12) becomes

$$\frac{d}{dt}\mathbb{E}(\boldsymbol{X}(t)) = \sum_{k=1}^{K}\alpha_k\left(\mathbb{E}(\boldsymbol{X}(t))\right)\boldsymbol{\nu}_k,$$

which is an L-dimensional system of linear equations and can easily be written in matrix form and solved by matrix exponentiation.

In case of a nonlinear reaction network, however, also moments of order higher than the one under consideration appear on the right-hand side of the evolution equations. In this case, the set of coupled differential equations becomes infinite, and exact solution methods can in general not be applied. Instead, estimations of the moments by Monte Carlo simulations of the reaction jump process come into play (see Sect. 1.3).

Example 1.6 (Degradation). A very simple example of a reaction network is given by the process of degradation of a single species $\mathcal{S}_1$ under the reaction

$$\mathcal{R}_1 : \mathcal{S}_1 \xrightarrow{\gamma_1} \emptyset.$$

Here, the reaction jump process is of the form $\boldsymbol{X} = (X_1(t))_{t\geq 0}$, where $X_1(t) \in \mathbb{N}_0$ is the number of $\mathcal{S}_1$-particles at time t and the state-change vector is actually a scalar given by $\nu_{11} = -1$. Given the initial number of particles $X_1(0) = x_0 \in \mathbb{N}$, the path-wise representation of the process is

$$X_1(t) = x_0 - \mathcal{U}_1\left(\int_0^t \gamma_1 X_1(s)\,ds\right)$$

for a unit-rate Poisson process $\mathcal{U}_1$. This means that, given the number $X_1(t)$ of $\mathcal{S}_1$-particles at time t, the waiting time for the next reaction event (which reduces the population size by one) follows an exponential distribution with mean $\gamma_1 X_1(t)$, where $\gamma_1 > 0$ is the reaction rate constant. Consequently, with a decreasing population size, the average waiting time for the next reaction event increases.

Fixing the initial state x_0, the state space $\mathbb{X} = \{0, \ldots, x_0\}$ of the process is finite with $M = x_0 + 1$ elements, and the CME has the form

$$\frac{d}{dt}p(x,t) = \begin{cases} -\gamma_1 p(x,t) & \text{for } x = x_0, \\ \gamma_1 p(x+1,t) & \text{for } x = 0, \\ \gamma_1 p(x+1,t) - \gamma_1 p(x,t) & \text{otherwise.} \end{cases}$$

In order to obtain a matrix-vector notation of the CME, we can set $\boldsymbol{p}(t) := (p(0,t), p(1,t), \ldots, p(x_0,t)) \in \mathbb{R}_+^M$ and define a suitable generator matrix $G \in \mathbb{R}_+^{M,M}$ (with G^T being the matrix representation of the operator $\mathcal{G}$ defined in (1.11)) by

$$G := \begin{pmatrix} 0 & 0 & 0 & \cdots & 0 \\ \gamma_1 & -\gamma_1 & 0 & \cdots & 0 \\ & \ddots & \ddots & & \\ & & \ddots & \ddots & \\ 0 & \cdots & 0 & \gamma_1 & -\gamma_1 \end{pmatrix},$$

such that it holds $\boldsymbol{p}^\mathsf{T}(t) = \boldsymbol{p}(0)^\mathsf{T} \exp(Gt)$ as a compact reformulation of the CME (see also passage "Finite state space" in Sect. 1.3.1).

The evolution equation for the first-order moment is given by

$$\frac{d}{dt}\mathbb{E}(X_1(t)) = -\gamma_1 \mathbb{E}(X_1(t))$$

and has the solution $\mathbb{E}(X_1(t)) = e^{-\gamma_1 t}\mathbb{E}(X_1(0))$. Figure 1.3 shows two realizations of the degradation process as well as the time-dependent first moment $\mathbb{E}(X_1(t))$. $\diamond$

1.3 Computation of Expectations

Given the dynamics of the reaction jump process $(\boldsymbol{X}(t))_{t \geq 0}$ characterized by the path-wise representation (1.6) or the CME (1.10), we are interested in

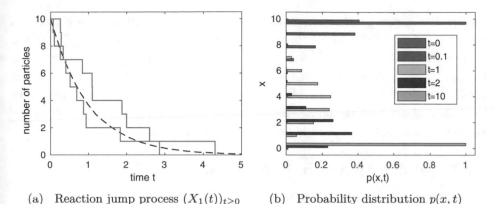

(a)   Reaction jump process $(X_1(t))_{t\geq0}$ (b)   Probability distribution $p(x, t)$

Figure 1.3. Degradation process. **(a)** Two individual realizations (blue and red) and time-dependent first-order moment $\mathbb{E}(X_1(t))$ (dashed line) of the reaction jump process $(X_1(t))_{t\geq0}$ containing the number of $\mathcal{S}_1$-molecules under degradation $\mathcal{R}_1 : \mathcal{S}_1 \xrightarrow{\gamma_1} \emptyset$ (see Example 1.6) with rate constant $\gamma_1 = 1$ and initial state $X_1(0) = 10$. **(b)** Corresponding probability distribution function $p(x, t) = \mathbb{P}(X_1(t) = x | X_1(0) = 10)$ (as solution of the related CME) for different time points $t \geq 0$

computing the expectation value of an observable of the reaction network. An *observable* is any function $F : \mathbb{X} \to \mathbb{R}$ for which the expectation value

$$\mathbb{E}(F, t) := \sum_{x \in \mathbb{X}} F(x)p(x, t) \tag{1.13}$$

exists for all $t \in [0, T]$. In case of a finite state space $\mathbb{X}$, this condition is trivially fulfilled for all functions $F : \mathbb{X} \to \mathbb{R}$. For the general infinite state space $\mathbb{X} = \mathbb{N}_0^L$, all functions $F \in \ell^\infty(\mathbb{X})$ that are uniformly bounded on $\mathbb{X}$ clearly belong to the class of observables. Note that by choosing $F = \mathbb{1}_x$ with

$$\mathbb{1}_x(y) := \begin{cases} 1 & \text{if } y = x, \\ 0 & \text{otherwise,} \end{cases}$$

we obtain $\mathbb{E}(F, t) = p(x, t)$. That is, finding the solution $p(x, t)$ of the CME is contained in the investigations as a special case. Likewise, setting $F = \text{id}$ (i.e., $F(x) = x$) gives $\mathbb{E}(F, t) = \mathbb{E}(X(t))$, such that the computation of the first-order moment of the system is also part of the analysis in this section.

In some cases it might be possible to derive a differential equation for $\mathbb{E}(F, t)$ as a function of time and to solve this equation in order to determine $\mathbb{E}(F, t)$. In Sect. 1.2.3 this has been done for $F = \text{id}$ and $\mathbb{E}(F, t) = \mathbb{E}(X(t))$. Yet we have seen that the resulting equations can only be directly solved for special types of reaction networks. More generally, one approach to compute $\mathbb{E}(F, t)$ from (1.13) is to find the probability function $p(\boldsymbol{x}, t)$ (or approximations of it) by directly solving the CME (1.10) as a set of linear ODEs. This approach is described in Sect. 1.3.1. Alternatively, one can find approximations of $\mathbb{E}(F, t)$ by Monte Carlo simulations[6] of the reaction jump process $\boldsymbol{X}$ based on its path-wise representation (1.6). The corresponding stochastic simulation methods are introduced in Sect. 1.3.2.

1.3.1 Direct Approach: Solving the CME

As noted before, from the mathematical point of view, the CME (1.10) is a system of linear ODEs, which suggests to apply standard analytical (or numerical) methods for solving it directly. However, despite its simple structure, the CME contains one equation for every possible population state $\boldsymbol{x} \in \mathbb{X} \subset \mathbb{N}_0^L$, such that the system of equations is in general infinite. Moreover, its dimension increases exponentially with the number L of involved species. It is therefore not surprising that an exact analytic solution of the CME exists again only for certain classes of systems and some special reaction dynamics. We provide a short overview of these cases here; for a more detailed review, see Section 3.4 in [209].

Linear Reaction Networks

Special results for solving the CME exist for linear reaction networks (containing no bimolecular or higher-order reaction, i.e., $\sum_l s_{lk} \leq 1$) under the additional condition that $\sum_l s'_{lk} \leq 1$ holds for all reactions, where s_{lk}, s'_{lk} are the stoichiometric coefficients (cf. (1.1)). In [138], the explicit formula for the probability distribution $p(\boldsymbol{x}, t)$, as an exact solution of the CME, is expressed in terms of the convolution of multinomial and product Poisson distributions with time-dependent parameters. The parameters evolve according to the traditional reaction rate equations, which will be considered in Sect. 2.1. The first results in this regard were presented in [43, 91] for linear reaction networks fulfilling particular initial conditions.

The restriction $\sum_l s'_{lk} \leq 1$ means that there is not more than one product molecule for each reaction. This excludes linear reactions of autocatalytic

---

[6]The term *Monte Carlo method* is used for a wide class of algorithms that are based on the following idea: According to the law of large numbers, the running average of n repeated independent realizations of a random variable converges to the expectation value for $n \to \infty$. The term *Monte Carlo simulation* is used, when the random variable refers to the realization of a stochastic process in a certain time interval, i.e., for each realization of the random variable, an entire trajectory is computed.

type $\mathcal{S}_l \to \mathcal{S}_l + \mathcal{S}_{l'}$ or splitting/unbinding $\mathcal{S}_l \to \mathcal{S}_{l'} + \mathcal{S}_{l''}$ ($l \neq l', l''$) which are definitely relevant in biological application.

Other Special Systems

For nonlinear systems with infinite state space, no analytic solutions are known in general. Still the CME has been solved for some other special types of systems, e.g., systems with only one reaction (single reaction networks) [47, 136, 163] or enzyme kinetic systems (Michaelis-Menten kinetics) with only one enzyme [13]. Furthermore, there are systems where no time-dependent solution is known, but one can at least derive the exact equilibrium distribution $\pi(\boldsymbol{x})$ (cf. page 190 in Appendix) given by the solution of the steady-state equation

$$0 = \sum_{k=1}^{K} \left[\alpha_k(\boldsymbol{x} - \boldsymbol{\nu}_k)\pi(\boldsymbol{x} - \boldsymbol{\nu}_k) - \alpha_k(\boldsymbol{x})\pi(\boldsymbol{x}) \right] \qquad (1.14)$$

which results from setting $\frac{dp(\boldsymbol{x},t)}{dt} = 0$ in the CME. Such a steady-state solution has been derived, for example, in [112] for a gene regulatory system and in [218] for enzyme kinetics.

Finite State Space

There exist reaction networks where the number of particles is naturally bounded for all species such that the state space $\mathbb{X}$ of the reaction jump process $\boldsymbol{X}$ reduces to a finite set. An example is given by the degradation process of Example 1.6 when considering a fixed initial state, with the state space given by $\mathbb{X} = \{0, \ldots, x_0\}$ where $x_0 \in \mathbb{N}$ is the initial number of $\mathcal{S}_1$-particles. Also the network of binding and unbinding $\mathcal{S}_1 + \mathcal{S}_2 \leftrightarrow \mathcal{S}_3$ as considered in Example 1.3 has a natural upper bound for its population size when assuming the initial state to be bounded.

Given that the state space $\mathbb{X}$ is finite with $|\mathbb{X}| = M$, one can define the probability vector $\boldsymbol{p}(t) := (p(\boldsymbol{x}, t))_{\boldsymbol{x} \in \mathbb{X}} \in \mathbb{R}_+^M$ and rewrite the CME in matrix-vector notation as

$$\frac{d}{dt}\boldsymbol{p}^{\mathsf{T}}(t) = \boldsymbol{p}^{\mathsf{T}}(t)G, \qquad (1.15)$$

(cf. (A.9) in Appendix) where $G = (g(\boldsymbol{x}, \boldsymbol{y}))_{\boldsymbol{x},\boldsymbol{y} \in \mathbb{X}} \in \mathbb{R}^{M,M}$ is the generator matrix containing the transition rates between the states. Equivalently, we can write $\frac{d}{dt}\boldsymbol{p}(t) = G^{\mathsf{T}}\boldsymbol{p}(t)$, i.e., G^{T} is the matrix representation of the operator $\mathcal{G}$ defined in (1.11). For $\boldsymbol{x}, \boldsymbol{y} \in \mathbb{X}$ with $\boldsymbol{x} \neq \boldsymbol{y}$ the matrix elements are given by

$$g(\boldsymbol{x}, \boldsymbol{y}) := \begin{cases} \alpha_k(\boldsymbol{x}) & \text{if } \boldsymbol{y} = \boldsymbol{x} + \boldsymbol{\nu}_k, \\ 0 & \text{otherwise,} \end{cases}$$

and for the diagonal entries it holds $g(\boldsymbol{x}, \boldsymbol{x}) := -\sum_{\boldsymbol{y} \neq \boldsymbol{x}} g(\boldsymbol{x}, \boldsymbol{y})$.

Using Eq. (1.15), the solution can be computed by matrix exponentiation

$$\boldsymbol{p}^{\mathsf{T}}(t) = \boldsymbol{p}^{\mathsf{T}}(0) \exp(Gt).$$

However, the size M of the system is often very large, which renders the matrix exponentiation computationally expensive. For example, given a system of three species, each of which has a bounded population size of not more than 100 particles, there are in total $M = 10^3$ possible states and hence $M = 10^3$ equations in the CME. Actually, even the most efficient numerical methods for matrix exponentiation, such as those developed in [2, 183], fail to solve Eq. (1.15) for practically relevant settings. Therefore, Eq. (1.15) is rarely used in application [209].

Solving the CME for relatively large systems requires more sophisticated approaches, like the *finite state projection method* by Munsky and Khammash [186] which can also be used in case of an infinite state space (see the following paragraph).

State Space Truncation

Whenever none of the former settings applies, deriving an exact solution of the CME is in general impossible, and reasonable approximation methods are necessary. One approach, called *finite state projection*, aims at truncating the state space and finding approximate solutions by using matrix exponentiation on the resulting finite subspace. The basic idea is to choose the projection subspace such that it contains the most likely states while excluding states of less probability. In [94] the approximation error resulting from finite state projection is studied rigorously, and error estimates are given.

The *finite state projection algorithm* is an adaptive approach where only those probabilities are stored that are sufficiently different from zero [186, 220, 221]. Based on an error estimation, the projection space is iteratively increased until a targeted level of accuracy is reached. It has been shown that the approximation quality improves monotonically when expanding the state space. Different variants of this algorithm exist (see [52] for an overview). For example, there are more variable frameworks where the projection space may depend on time, which is reasonable because also the probability distribution of the process is time-dependent.

Problems arise when the truncated state space needs to be large to obtain an adequate accuracy because this again renders the matrix exponentiation intractable. Then, further model reduction methods can be applied in order to reduce the number of degrees of freedom and to turn the CME into a computationally tractable problem, e.g., tensor or product approximation methods [137, 142] and Galerkin projection methods [48, 139]. The latter are shortly summarized in the following (quite technical) paragraph, which is aimed at readers with profound mathematical background.

Galerkin Projection Methods

Galerkin-based approaches represent the solution $p(x, t)$ of the CME in terms of a finite dimensional ansatz space which may depend on time. In [50, 63], for example, an orthogonal basis of discrete Charlier polynomials is chosen and adapted over time. This method can break down in case of low regularity of the solution, which is often the case for the CME. In order to overcome this breakdown, adaptive Galerkin methods based on a local basis (instead of one global basis) were proposed (see [48] for local polynomial bases and [154] for local radial basis functions).

In [139] an approach to directly solve the CME by a dynamical low-rank approximation is presented, which also can be understood as a Galerkin ansatz. This approach substantially reduces the number of degrees of freedom. It is based on the Dirac-Frenkel-McLachlan variational principle, which originally has been used in quantum mechanics to reduce high-dimensional Schrödinger equations. Given the truncated state space

$$\left\{ x \in \mathbb{N}_0^L \,\middle|\, 0 \le x_l \le x_l^{\max} \text{ for } l = 1, \dots, L \right\}$$

for chosen maximum population sizes $x_l^{\max} \in \mathbb{N}$, $l = 1, \dots, L$, the Galerkin ansatz is of the form

$$\hat{p}(x, t) := \sum_j a_j(t) U_j(x, t), \quad U_j(x, t) := \prod_{l=1}^L u_{j_l}^{(l)}(t, x_l), \qquad (1.16)$$

with time-dependent basis functions $u_{j_l}^{(l)} : \{0, \dots, x_l^{\max}\} \times [0, \infty) \to \mathbb{R}$ and coefficients $a_j(t) \in \mathbb{R}$ and with the sum running over all multi-indices $j = (j_1, \dots, j_L)$ where $1 \le j_l \le J_l$ for a chosen rank vector $J = (J_1, \dots, J_L)$. In order to make the representation (1.16) unique, additional orthonormality constraints can be imposed for the basis functions $u_{j_l}^{(l)}$. The function $\hat{p}$ defined by (1.16) serves as an approximation of the exact solution p of the CME, $\hat{p}(\boldsymbol{x}, t) \approx p(\boldsymbol{x}, t)$.

In order to keep $\hat{p}(\boldsymbol{x}, t)$ on the ansatz manifold $\mathcal{M}$ of chosen basis functions when propagating it in time, the time-derivative $\frac{d}{dt}\hat{p}$ must be an element of tangent space $\mathcal{T}_{\hat{p}}\mathcal{M}$ of $\mathcal{M}$ at $\hat{p}$. This excludes to simply set $\frac{d}{dt}\hat{p} = \mathcal{G}\hat{p}$ because $\mathcal{G}\hat{p}$ is generally not contained in $\mathcal{T}_{\hat{p}}\mathcal{M}$. The Dirac-Frenkel-McLachlan variational principle consists of setting

$$\frac{d}{dt}\hat{p} = \underset{\delta\hat{p}\in\mathcal{T}_{\hat{p}}\mathcal{M}}{\arg\min}\|\delta\hat{p} - \mathcal{G}\hat{p}\|_2. \tag{1.17}$$

In [139] it is shown how to derive equations of motion for the coefficients a_j and the basis functions $u_{j_l}^{(l)}$ based on the variational principle (1.17). Moreover, it is proven that the error between the exact solution p of the CME and the dynamical low-rank approximation $\hat{p}$ of the truncated CME can be made arbitrarily small by expanding the state space and the dimension of the manifold $\mathcal{M}$.

In general, a large truncated state space is required when the population size varies within a broad range and can reach high levels. For such situations there exist other approximation methods which do not cut the state space but define approximative reaction dynamics within an extended continuous domain. Such approaches are described in Chap. 2.

1.3.2 Indirect Approach: Stochastic Simulation

As we have seen, finding the probabilities $p(\boldsymbol{x}, t)$ by directly solving the CME can be a very demanding task. Hence, alternative ways to estimate the expectation $\mathbb{E}(F, t) = \sum_{\boldsymbol{x}\in\mathbb{X}} F(\boldsymbol{x})p(\boldsymbol{x}, t)$ of an observable F are required. The most prominent option results from the fact that the CME propagates probability distributions according to the reaction jump process $\boldsymbol{X} = (\boldsymbol{X}(t))_{t\geq 0}$. That is, if the initial values $\boldsymbol{X}(0)$ of the reaction jump process $\boldsymbol{X}$ are distributed according to $p_0 = p(\cdot, 0)$, formally denoted by $\boldsymbol{X}(0) \sim p_0$, then the state $\boldsymbol{X}(t)$ is distributed according to $p(\cdot, t)$. As a direct consequence, we get

$$\mathbb{E}(F, t) = \mathbb{E}\big(F(\boldsymbol{X}(t)) \big| \boldsymbol{X}(0) \sim p_0\big), \tag{1.18}$$

where the right-hand side no longer contains $p(\cdot, t)$. This opens the door to Monte Carlo methods for the numerical approximation of $\mathbb{E}(F, t)$: Assume that we have an algorithmic scheme providing us with a set $\boldsymbol{x}_1, \boldsymbol{x}_2, \ldots, \boldsymbol{x}_n$

of states $x_i \in \mathbb{X}$ that is (asymptotically for $n \to \infty$) distributed according to p_0. For each t we can define the *running mean*

$$E_n(t) := \frac{1}{n} \sum_{i=1}^{n} F(X(t, x_i))$$

as an approximation of $\mathbb{E}(F, t)$, where $(X(t, x_i))_{t \geq 0}$ denotes a realization of the reaction jump process starting at time $t = 0$ in x_i. The convergence $E_n(t) \to \mathbb{E}(F, t)$ for $n \to \infty$ is guaranteed by the law of large numbers. The convergence rate is of order $n^{-1/2}$, which results from the central limit theorem [15].

This means that, given the ability to sample the initial distribution, we can approximate the evolution of expectation values via many realization of the reaction jump process X. The arising question is how to construct these realizations. As the process X defined by (1.6) is a Markov jump process on the state space $\mathbb{X} \subset \mathbb{N}_0^I$, it can be characterized in terms of jump times and transition probabilities in the following way (see Sect. A.1.3 for the general theoretical background). Let

$$\tau(t) := \inf\{s > 0 | X(t + s) \neq X(t)\}$$

denote the *residual lifetime* in state $X(t)$, i.e., the time that remains before the process performs a jump. The distribution of $\tau(t)$ is given by

$$\mathbb{P}(\tau(t) > s | X(t) = x) = \exp(-\lambda(x)s),$$

where

$$\lambda(x) := \sum_{k=1}^{K} \alpha_k(x)$$

is the *jump rate function* of X. These residual lifetimes recursively define the *jump times* T_j, $j = 0, 1, 2, \ldots$, of the process X by $T_0 := 0$ and

$$T_{j+1} := T_j + \tau(T_j).$$

The states of the process X at the jump times T_j define the *embedded Markov chain* $\hat{X} = (\hat{X}_j)_{j \in \mathbb{N}_0}$ by

$$\hat{X}_j := X(T_j), \quad j = 0, 1, 2, \ldots .$$

This embedded Markov chain $\hat{X}$ is a discrete-time Markov process on $\mathbb{X}$ with a stochastic transition matrix $\hat{P} = (\hat{p}(x, y))_{x, y \in \mathbb{X}}$, where

$$\hat{p}(x, y) := \mathbb{P}(X(t + \tau(t)) = y | X(t) = x)$$

is the probability for the process $\boldsymbol{X}$ to switch from $\boldsymbol{x} \in \mathbb{X}$ to $\boldsymbol{y} \in \mathbb{X}$ given that a jump occurs. In terms of the propensity functions α_k, these transition probabilities are given by

$$\hat{p}(\boldsymbol{x}, \boldsymbol{y}) = \begin{cases} \dfrac{\alpha_k(\boldsymbol{x})}{\lambda(\boldsymbol{x})} & \text{if } \boldsymbol{y} = \boldsymbol{x} + \boldsymbol{\nu}_k, \\ 0 & \text{otherwise.} \end{cases}$$

As the reaction jump process $\boldsymbol{X}$ is constant in between the jump times T_j, it inversely is fully characterized by the states of the embedded Markov chain $\hat{\boldsymbol{X}}$ and the sequence $(T_j)_{j \geq 0}$ of jump times. In order to construct a trajectory $(\boldsymbol{x}(t))_{t \geq 0}$ of the process $\boldsymbol{X}$, it is therefore sufficient to create the jump times and the states at these jump times (as a realization $(\hat{\boldsymbol{x}}_j)_{j \in \mathbb{N}_0}$ of $\hat{\boldsymbol{X}}$), which can be done in the following recursive manner. Let $\hat{\boldsymbol{x}}_0 \in \mathbb{X}$ be an initial state and set $T_0 = 0$. Given a jump time T_j and a state $\hat{\boldsymbol{x}}_j$ of the embedded Markov chain for $j \in \mathbb{N}_0$, set

$$T_{j+1} = T_j + \tau,$$

where τ is exponentially distributed with parameter $\lambda(\hat{\boldsymbol{x}}_j)$, and

$$\hat{\boldsymbol{x}}_{j+1} = \hat{\boldsymbol{x}}_j + \boldsymbol{\nu}_{k^*},$$

where the index $k^* \in \{1, \ldots, K\}$ is a statistically independent integer random variable with point probabilities $\alpha_k(\hat{\boldsymbol{x}}_j)/\lambda(\hat{\boldsymbol{x}}_j)$ for $k = 1, \ldots, K$. The realization $(\boldsymbol{x}(t))_{t \geq 0}$ of $\boldsymbol{X}$ is then defined as

$$\boldsymbol{x}(t) := \hat{\boldsymbol{x}}_j \quad \text{for } T_j \leq t < T_{j+1}.$$

This two-step recursive procedure has been introduced by Gillespie in 1976 [99] and is the basis for the *stochastic simulation algorithm* (SSA) as a numerical schemes to construct statistically exact realizations of the reaction jump process $\boldsymbol{X}$. Different variants of the SSA exist. One of them is *Gillespie's direct method* which makes use of the inversion generating method to obtain the random numbers τ and k^* in each iteration step [101]:[7] Generate two independent random numbers ψ_1, ψ_2 from the standard uniform distribution $U(0, 1)$ and set

$$\tau = \frac{1}{\lambda(\boldsymbol{x})} \ln\left(\frac{1}{\psi_1}\right), \tag{1.19}$$

---

[7]Gillespie's algorithm is essentially the same as the *kinetic Monte Carlo* method [240].

which gives an exponentially distributed waiting time $\tau \sim \exp(\lambda(\boldsymbol{x}))$, as well as

$$k^* = \text{smallest integer satisfying } \sum_{k=1}^{k^*} \alpha_k(\boldsymbol{x}) > \psi_2 \lambda(\boldsymbol{x}). \qquad (1.20)$$

In total, the standard SSA consists of the following steps:

Stochastic simulation algorithm (SSA):

1. Initialize time $t \leftarrow t_0$ and state $\boldsymbol{x} \leftarrow \boldsymbol{x}_0$ and choose a time horizon $T > t_0$.

2. Calculate $\alpha_k(\boldsymbol{x})$ for all $k = 1, \ldots, K$ and their sum $\lambda(\boldsymbol{x})$.

3. Generate τ and k^* according to (1.19) and (1.20), respectively.

4. Execute the next reaction by replacing $t \leftarrow t + \tau$ and $\boldsymbol{x} \leftarrow \boldsymbol{x} + \boldsymbol{\nu}_{k^*}$.

5. If $t \geq T$, end the simulation. Otherwise return to step 2.

Note that the generated time steps τ are exact and not a finite approximation to an infinitesimal dt as used for numerical ODE-solvers or simulations of stochastic differential equations. This means that the trajectories generated by the SSA are statistically exact realizations of the stochastic process $\boldsymbol{X}$. All trajectories plotted in Figs. 1.2, 1.3, and 1.4 of this chapter have been created by means of Gillespie's SSA.

Example 1.7 (Birth-Death Process). Expanding the degradation Example 1.6, we consider one species $\mathcal{S}_1$ undergoing two reactions

$$\mathcal{R}_1 : \mathcal{S}_1 \xrightarrow{\gamma_1} \emptyset, \quad \mathcal{R}_2 : \emptyset \xrightarrow{\gamma_2} \mathcal{S}_1.$$

This means that a new $\mathcal{S}_1$-particle can arise at any time by means of reaction $\mathcal{R}_2$ and the expected waiting time for such a particle formation is given by $1/\gamma_2$ time units. At the same time, every existing particle can disappear by reaction $\mathcal{R}_1$ independently of the other particles. Each particle has an expected lifetime of $1/\gamma_1$ time units. The first-order moment $\mathbb{E}(X_1(t))$ now evolves according to

$$\frac{d}{dt}\mathbb{E}(X_1(t)) = -\gamma_1\mathbb{E}(X_1(t)) + \gamma_2,$$

with the solution given by $\mathbb{E}(X_1(t)) = e^{-\gamma_1 t}\left(\frac{\gamma_2}{\gamma_1}(e^{\gamma_1 t} - 1) + \mathbb{E}(X_1(0))\right)$. In Fig. 1.4, a trajectory of $(X_1(t))_{t \geq 0}$ and the time-dependent mean $\mathbb{E}(X_1(t))$ are shown. In this example, the state space $\mathbb{X} = \mathbb{N}_0$ is not bounded. $\diamond$

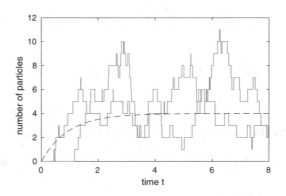

Figure 1.4. Birth-death process. Two individual realizations (blue and red, from Gillespie's SSA) and time-dependent first-order moment $\mathbb{E}(X_1(t))$ (dashed line) of the reaction jump process $(X_1(t))_{t\geq0}$ containing the number of $\mathcal{S}_1$-molecules for the birth-death process of Example 1.7 with the two reactions $\mathcal{R}_1 : \mathcal{S}_1 \xrightarrow{\gamma_1} \emptyset$ and $\mathcal{R}_2 : \emptyset \xrightarrow{\gamma_2} \mathcal{S}_1$, where $\gamma_1 = 1$, $\gamma_2 = 4$ and $X_1(0) = 0$

Modifications of the SSA

Several modifications of the SSA have been developed, each of them producing statistically exact dynamics. In the so-called first-reaction method, which was also proposed by Gillespie [99], random reaction times τ_k are drawn for *all* reactions in each iteration step (with τ_k following an exponential distribution with parameter $\alpha_k(\boldsymbol{x})$), executing then only the one which occurs *first*.

The first-reaction method uses K random numbers per iteration. For systems with many reactions, it is therefore less efficient than the standard SSA which only requires two random numbers per iteration.

In contrast, the *next-reaction method* proposed by Gibson and Bruck [98] can be significantly faster, especially if there are many reactions or many species involved in the system. The method reduces the quantity of required random numbers per iteration step to one by storing the next reaction times of all reactions in a priority cue. In return, it is more challenging to code than the previous simulation methods.

Further methods exist to improve the performance of exact stochastic simulation or to extend the applicability of the code, like the *modified next reaction method* [6] which allows extensions to time-dependent rate constants; the *modified direct method* [32] and the *sorting direct method* [177] which use a special indexing of the reactions according to their propensity functions to make the simulation more efficient; or the *logarithmic direct method* [168], to mention only a few of them.

Such improvements are useful, but they do not decrease the number of iteration steps necessary (on average) to reach a given time horizon. In case of rapidly occurring reaction events, the time increase τ in each iteration step is small, which induces a long runtime, even if the reaction events are determined efficiently in each step. This motivates to quit the exactness of the SSA in order to find faster approximate simulation strategies. One of them is the so-called τ-leaping method which is based on the idea of aggregating several reaction events per iteration step [105].

Approximation by τ-Leaping

Suppose that the state $\boldsymbol{X}(t)$ of the system at some time t is known to be $\boldsymbol{x} \in \mathbb{N}_0^L$. For some $\tau > 0$ (which now is not the jump time anymore), denote by $R_k(\boldsymbol{x}, \tau)$ the random number of $\mathcal{R}_k$-reactions occurring in the subsequent time interval $[t, t + \tau]$. It then holds

$$\boldsymbol{X}(t + \tau) = \boldsymbol{x} + \sum_{k=1}^{K} R_k(\boldsymbol{x}, \tau)\boldsymbol{\nu}_k, \tag{1.21}$$

as an analog of (1.4).

Given the state $\boldsymbol{X}(t) = \boldsymbol{x}$ at time t, let $\tau > 0$ be small enough such that during the time interval $[t, t + \tau)$ the propensity functions of all reactions are likely to stay roughly constant, that is, they fulfill the *leap condition*

$$\alpha_k(\boldsymbol{X}(s)) \approx \alpha_k(\boldsymbol{x}) \quad \forall s \in [t, t+\tau], \forall k = 1, \dots, K. \tag{1.22}$$

In this case it makes sense to consider the approximation

$$R_k(\boldsymbol{x}, \tau) = \mathcal{U}_k \left(\int_t^{t+\tau} \alpha_k(\boldsymbol{X}(s)) ds \right)$$
$$= \mathcal{U}_k(\alpha_k(\boldsymbol{x})\tau) + \mathcal{O}(\tau^2).$$

where the last equality results from the Taylor expansion $\int_t^{t+\tau} \alpha_k(\boldsymbol{X}(s)) ds = \alpha_k(\boldsymbol{X}(t))\tau + \mathcal{O}(\tau^2)$ of the integral, together with $\boldsymbol{X}(t) = \boldsymbol{x}$.[8] This means that $R_k(\boldsymbol{x}, \tau)$ can be approximated by a Poisson random variable with mean $\alpha_k(\boldsymbol{x})\tau$ and (1.21) can be replaced by

$$\boldsymbol{X}(t + \tau) \approx \boldsymbol{x} + \sum_{k=1}^{K} \mathcal{U}_k(\alpha_k(\boldsymbol{x})\tau)\boldsymbol{\nu}_k, \tag{1.23}$$

---

[8] The asymptotic notation $\mathcal{O}(\tau^2)$ represents some function decreasing not significantly slower than τ^2, such that $\limsup_{\tau \to 0} \left| \frac{\mathcal{O}(\tau^2)}{\tau^2} \right| < \infty$.

where $\mathcal{U}_k(\alpha_k(\boldsymbol{x})\tau)$ are statistically independent Poisson variables with mean (and variance) $\alpha_k(\boldsymbol{x})\tau$. Equation (1.23) is an approximation of (1.6) and obviously suggests to apply basic numerical schemes – like the explicit Euler method [26] – for approximative simulations of $\boldsymbol{X}$. The resulting *Euler-τ-leaping* algorithm works as follows.

Euler-τ-leaping algorithm:

1. Initialize time t_0 and state $\boldsymbol{x}_0$ and choose a time horizon $T > t_0$ and a lag time $\tau > 0$. Set $j = 0$.

2. Calculate $\alpha_k(\boldsymbol{x}_j)$ for all $k = 1, \ldots, K$.

3. Generate for each k a Poisson random number $\mathcal{U}_k(\alpha_k(\boldsymbol{x}_j)\tau)$.

4. Set $t_{j+1} = t_j + \tau$ and $\boldsymbol{x}_{j+1} = \boldsymbol{x}_j + \sum_{k=1}^{K} \mathcal{U}_k(\alpha_k(\boldsymbol{x})\tau)\boldsymbol{\nu}_k$.

5. Augment $j \rightarrow j + 1$ and end the simulation in case of $t_j \geq T$. Otherwise, return to step 2.

An example, where the τ-leaping algorithm is applied, can be found on page 30. Alternative calculation schemes have been derived, e.g., in [104] (midpoint τ-leaping) and [128] (second-order tau-leaping) to improve the algorithm's accuracy for a given τ. An error analysis and comparison of the different approaches is given in [8].

The accuracy of τ-leaping depends on how well the leap condition (1.22) is satisfied. In the uncommon trivial case where the propensity functions are all independent of the system's state $\boldsymbol{x}$, the leap condition is exactly satisfied for any value of τ, and the τ-leaping method turns out to be exact. In the more common situation where the propensities depend linearly or quadratically on the molecular population, τ-leaping is not exact because the propensities are affected by reactions that occur within the time interval $[t, t + \tau)$.

For a practical implementation of τ-leaping, one needs a rule to determine an adequate value of τ for a given reaction network. On the one hand, choosing the step size too small renders the algorithm inefficient. Especially if τ is smaller than the average time advance in the SSA, τ-leaping is less efficient than the exact simulation. Increasing τ, on the other hand, reduces the accuracy of the algorithm. As originally suggested in [104], it is reasonable to require the *expected relative change* in the propensity functions during the leap to be bounded by some specified control parameter. Concrete schemes to calculate the largest τ that satisfies this requirement can be found in [30, 107]. They are based on the constraint

$$|\alpha_k\left(\boldsymbol{x} + \boldsymbol{\nu}_{\text{total}}(\boldsymbol{x}, \tau)\right) - \alpha_k(\boldsymbol{x})| \leq \varepsilon\lambda(\boldsymbol{x}) \quad \forall k = 1, \ldots, K$$

as a specification of the leap condition (1.22), where ε is a predefined error control parameter ($0 < \varepsilon \ll 1$) and $\boldsymbol{\nu}_{\text{total}}(\boldsymbol{x}, \tau) := \sum_{k=1}^{K} R_k(\boldsymbol{x}, \tau)\boldsymbol{\nu}_k$.

Applied to so-called *stiff systems* which include multiple time scales, the fastest of which are stable, these schemes can result in unnecessarily small values for τ. For such systems, *implicit* τ-leaping methods have been derived which are more efficient than the standard explicit τ-leaping scheme [31, 169, 195].

The number $R_k(\boldsymbol{x}, \tau)$ of firings of each reaction $\mathcal{R}_k$ during the chosen time interval of length τ is approximated by a Poisson variable which can have arbitrarily large sample values. It is thereby possible that a reaction fires so many times within this time interval that more reactant molecules are consumed than are actually available and the population is driven to a value below zero. With increasing population size and small values of τ, this problem becomes more and more unlikely, although it will never be ruled out. Different approaches exist to overcome this problem.

Binomial τ-Leaping

One idea is to further approximate $R_k(\boldsymbol{x}, \tau)$ by a binomial random variable with the same mean $\alpha_k(\boldsymbol{x})\tau$, but with the number of trials (i.e., the maximum number of successes of the binomial random variable) bounded by the present number of reactant molecules (see Example 1.8). This method is called *binomial τ-leaping* [34, 166, 191, 227] and uses the recursion

$$\boldsymbol{X}(t + \tau) \approx \boldsymbol{x} + \sum_{k=1}^{K} B_k(v_k, \alpha_k(\boldsymbol{x})\tau/v_k)\boldsymbol{\nu}_k, \tag{1.24}$$

where generally $B(v, q)$ denotes a binomial distribution with parameters $v \in \mathbb{N}$ (number of trials) and $q \in [0, 1]$ (success probability) and v_k is the upper bound for the number of firings of reaction $\mathcal{R}_k$. For a first-order reaction of type $R_1 : S_1 \to S_2$, e.g., one can set $v_1 = x_1$, and for a second-order reaction of type $R_1 : S_1 + S_2 \to S_3$, a reasonable choice is given by $v_1 := \min\{x_1, x_2\}$ (see Example 1.8 for an application).[9]

There are two weak points of binomial τ-leaping: On the one hand, the bound for the number of firings during the leap is often *overly restrictive*

---

[9]More generally, set $v_k = \min\{\lfloor -x_l/\nu_{lk}\rfloor | l = 1, \ldots, L$ with $\nu_{lk} < 0\}$ for reactions with $\min\{\nu_{lk}|l = 1, \ldots, L\} < 0$.

because the number of reactants might simultaneously be increased by another reaction; on the other hand, if several reactions exist with a common consumed reactant, the individual limitations of reaction extents may *not be restrictive enough*, because a combination of the reactions may still render the population negative. Finding suitable restrictions which take all possible combinations of reactions during a leap into account is a challenging task.

Further methods to resolve the problem of negative populations include the *modified Poisson τ-leaping* algorithm proposed in [27] which combines Poisson approximations for "uncritical" reactions with exact stochastic simulation of the "critical" reactions or post-leap checks for updating the step size during simulation [7].

Other Approaches to Accelerate SSA

Another type of algorithm is the so-called *R-leaping* method [16], which is also illustrated in Example 1.8. Here, one defines a value $N_{\mathcal{R}} \in \mathbb{N}$ for the total number of reaction firings in each iteration step (across all reactions) and assumes that these reactions do not substantially alter the value of any reaction propensity function. The time span τ for these $N_{\mathcal{R}}$ reactions to take place is a random variable which follows a gamma distribution with parameters $(N_{\mathcal{R}}, 1/\lambda(\boldsymbol{x}))$

$$\tau \sim \Gamma(N_{\mathcal{R}}, 1/\lambda(\boldsymbol{x})). \tag{1.25}$$

Within each iteration step, there are N_k firings of reaction $\mathcal{R}_k$ with $\sum_{k=1}^{K} N_k = N_{\mathcal{R}}$. The numbers N_k follow correlated binomial distributions depending on $N_{\mathcal{R}}$ (see [16] for details). An algorithm which combines the advantages of τ-leaping and R-leaping is proposed in [171].

Example 1.8 (Epidemiology: SIR Model). The stochastic models for reaction kinetics not only apply in the biochemical context but can also be used to describe interaction kinetics between individuals like human beings or animals. One interesting field of application is given by epidemic processes describing the spreading of infectious diseases within a population. The most fundamental model is the so-called *SIR model*, where S stands for *susceptible*, I for *infectious*, and R for *recovered*. Each infectious individual can pass on the disease to a susceptible individual, which is described by the reaction

$$\mathcal{R}_1: \quad S + I \xrightarrow{\gamma_1} 2I,$$

and each infectious individual can recover by

$$\mathcal{R}_2: \quad I \xrightarrow{\gamma_2} R,$$

thereby becoming immune. Figure 1.5 contains simulations of the associated reaction jump process using Gillespie's SSA (i.e., the exact simulation), standard τ-leaping, binomial τ-leaping, and finally R-leaping. The standard τ-leaping (with a fixed value for τ) and the R-leaping method cause problems when the number of susceptible or infectious individuals approaches zero because both methods have a very high probability to render the population negative. In order to circumvent these problems, we implemented additional constraints regarding the number of reactions per iteration step (namely, upper bounds depending on the actual population size). $\diamond$

Similar to τ-leaping R-leaping can induce negative numbers of particles when some species appear in low copy numbers. However, the problem of negative particle numbers and inaccurate or inefficient results naturally vanishes when the population of all reactants is large. In this case, given that each reaction changes the population only by a few molecules, the effect of an individual reaction is proportionally small and many reaction events will be necessary to noticeably influence the propensity functions. In comparison to a system with a small population, reaction events occur more often but are individually less significant, such that the aggregation of reaction events by τ-leaping (or R-leaping) becomes reasonable. In this sense, τ-leaping and related approximate numerical methods build the natural connection to the approximation schemes for settings of large populations that are considered in Chap. 2.

1.4 Stochastic Versus Deterministic Modeling

So far, we discussed stochastic models for the reactions of a volume of well-mixed molecules and arrived at the reaction jump process (RJP) model and its distribution-based form, the chemical master equation (CME). The vast majority of books on systems biology or cellular dynamics, on the other hand, do *not* start with these kinds of models but first introduce *deterministic* dynamical models based on systems of ordinary differential equations in form of the so-called reaction rate equations (RREs). Solving the RRE is much easier, i.e., it produces significantly less computational effort and thus is feasible for systems with very many species and reactions. Moreover, many cellular processes can be described successfully by means of the RRE. So why did we start with RJP-based models, or, in other words, what are the advantages of RJP-based descriptions over simpler RRE-based ones?

In order to give an answer to this question, we consider a simple case of an epigenetic switching process. More precisely, we examine a genetic

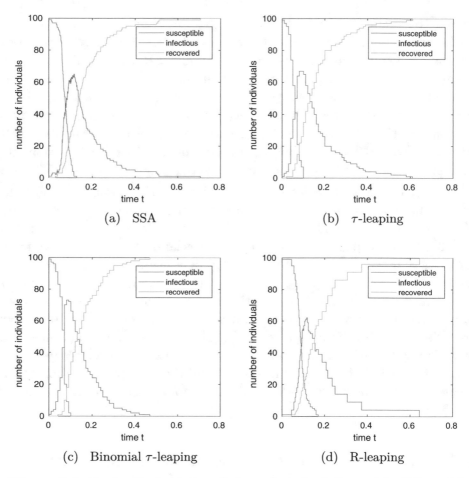

Figure 1.5. Stochastic simulation. Independent simulations of the SIR model given in Example 1.8 using (**a**) Gillespie's SSA, (**b**) standard Euler-τ-leaping based on (1.23), (**c**) binomial τ-leaping based on (1.24), (**d**) R-leaping. Initial state: $x_0 = (99, 1, 0)^{\mathsf{T}}$. Rate constants: $\gamma_1 = 1$, $\gamma_2 = 10$. For (**b**) and (**c**) additional constraints have been used to avoid negative populations. In (**a**) and (**d**) the time increments are random variables given by (1.19) and (1.25), respectively, while they are fixed to $\tau = 0.01$ for (**b**) and (**c**). The total number of reactions occurring in one iteration step is always 1 in (**a**), while it is a random number in (**b**) and (**c**), and it is fixed to $N_{\mathcal{R}} = 5$ in (**d**)

toggle switch model from [93], also analyzed in [182, 200]. This switch comprises two genes G_1 and G_2, and two types of protein P_1 and P_2. The expression of each of the two respective genes results in the production of the respective type of protein; gene G_1 produces protein P_1 and gene G_2 protein P_2. However, the two genes repress each others' expression, i.e., each protein can bind particular DNA sites upstream of the gene which codes for the other protein, thereby repressing its transcription.

The reaction model developed in [93] has the form of a birth-death process for the two proteins. Let $X_1(t)$ and $X_2(t)$ denote the number of proteins of type 1 and 2 at time t, respectively, and consider four reactions

$$\mathcal{R}_1 : \emptyset \longrightarrow P_1, \quad \mathcal{R}_2 : P_1 \longrightarrow \emptyset,$$

$$\mathcal{R}_3 : \emptyset \longrightarrow P_2, \quad \mathcal{R}_4 : P_2 \longrightarrow \emptyset,$$

with (non-mass-action-type) propensity functions given by

$$\alpha_1(\boldsymbol{x}) = \frac{a_1}{1 + (x_2/K_2)^{n_1}}, \quad \alpha_2(\boldsymbol{x}) = \gamma_1 x_1,$$

$$\alpha_3(\boldsymbol{x}) = \frac{a_2}{1 + (x_1/K_1)^{n_2}}, \quad \alpha_4(\boldsymbol{x}) = \gamma_2 x_2$$

for $\boldsymbol{x} = (x_1, x_2)^{\mathsf{T}}$ with x_1 and x_2 referring to the number of proteins of type P_1 and P_2, respectively. The corresponding state-change vectors are $\boldsymbol{\nu}_1 = (1,0)^{\mathsf{T}}$, $\boldsymbol{\nu}_2 = (-1,0)^{\mathsf{T}}$, $\boldsymbol{\nu}_3 = (0,1)^{\mathsf{T}}$, and $\boldsymbol{\nu}_4 = (0,-1)^{\mathsf{T}}$. Herein we use the following parameter values: $a_1 = 155$, $a_2 = 30$, $n_1 = 3$, $n_2 = 1$, $K_1 = K_2 = 1$, $\gamma_1 = \gamma_2 = 1$, in consistency with [182, 200]. A more detailed model (with more molecular species and mass-action-type propensity functions) will be studied in Sect. 2.3.

Stochastic RJP Model We start with the respective RJP/CME model in the domain $(x_1, x_2)^{\mathsf{T}} \in \mathbb{D} = \mathbb{N}_0^2$. If we investigate the process numerically by means of the SSA, we observe the following: The process has two metastable areas $\mathbb{D}_1 \subset \mathbb{D}$ and $\mathbb{D}_2 \subset \mathbb{D}$ located close to the states $(155,0)^{\mathsf{T}}$ and $(0,30)^{\mathsf{T}}$, respectively. Starting in $\boldsymbol{x}_0 = (155,0)^{\mathsf{T}}$ (or more generally in $\mathbb{D}_1$), the process remains for a random time period within $\mathbb{D}_1$ before it exits from the vicinity of $\mathbb{D}_1$ and makes the transition toward the second metastable area $\mathbb{D}_2$ (see Fig. 1.7). Once arrived there, it stays there for an even longer period of time (on average), before it eventually returns to $\mathbb{D}_1$. Actually, transitions between the two areas $\mathbb{D}_1$ and $\mathbb{D}_2$ are rare, especially the one from $\mathbb{D}_2$ to $\mathbb{D}_1$. As we can see in Fig. 1.6, the process has an equilibrium distribution that is concentrated around the two metastable areas $\mathbb{D}_1$ and $\mathbb{D}_2$ and is much higher in area $\mathbb{D}_2$ than in area $\mathbb{D}_1$. The metastable areas thus are the reason for the characteristic *bistable* dynamical behavior of the switch. In [182], this bistable dynamical behavior is investigated in more detail.

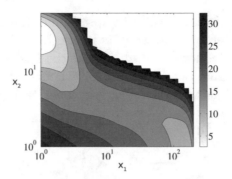

Figure 1.6. Equilibrium distribution for toggle switch model. Equilibrium distribution $\pi(\boldsymbol{x})$ (see also (1.14)) of the RJP model for the toggle switch with $a_1 = 155$, $a_2 = 30$, $n_1 = 3$, $n_2 = 1$, $K_1 = K_2 = 1$, $\gamma_1 = \gamma_2 = 1$. For ease of presentation, the quantity $F = -\log \pi$ is plotted against the particle numbers x_1 and x_2. F exhibits minima where π has maxima. The area in the top right corner is left white for display reasons; the values of π are exponentially small there

Deterministic RRE Model As we will see later in Sect. 2.1.1, the RRE description of the same reaction model leads to the following coupled system of ordinary differential equations depending on the four propensity functions α_k, $k = 1, \ldots, 4$ defined above

$$\frac{d}{dt}\boldsymbol{C}(t) = \sum_{k=1}^{4} \alpha_k(\boldsymbol{C}(t))\,\boldsymbol{\nu}_k,$$

where the integer protein particle numbers X_i are now replaced by real-valued protein concentrations $C_i \in \mathbb{R}$, $i = 1, 2$, $\boldsymbol{C} = (C_1, C_2)^{\mathsf{T}}$. This means that

$$\frac{d}{dt}C_1(t) = \frac{a_1}{1 + (C_2(t)/K_2)^{n_1}} - \gamma_1 C_1(t),$$

$$\frac{d}{dt}C_2(t) = \frac{a_2}{1 + (C_1(t)/K_1)^{n_2}} - \gamma_2 C_2(t)$$

(see also [200]). The structure of this dynamical model is much easier than that of the RJP model: Its state space is just two dimensional, and we do not have any stochastic fluctuations. When analyzing the RRE model, we first observe that it exhibits two essential stable fixed points at $(C_1, C_2) \approx (154.9, 0.2)$ and $(C_1, C_2) \approx (0.005, 29.8)$ that are located inside the two areas $\mathbb{D}_1$ and $\mathbb{D}_2$, respectively. That is, the deterministic dynamics, when started in the vicinity of these fixed points, will be asymptotically attracted there and never leave this metastable area. Thus, the RRE description of the epigenetic switch does not show bistable behavior, that is, the most characteristic dynamical feature of the genetic switch is lost! In contrast, the bistability is inherent in the reaction jump process and reproduced by the SSA simulations.

Stochastic Versus Deterministic Descriptions Obviously, for such metastable dynamics, the stochastic fluctuations of the RJP model are essential for reproducing the dynamical characteristics of the system which are completely lost for the deterministic RRE description. Can we build a dynamical model that is as low dimensional as the RRE description but still allows us to reproduce the essential bistability of the toggle switch? The answer is yes, in this case. The so-called chemical Langevin equation (CLE), which will be introduced in detail in Sect. 2.1.2, reads as follows:

$$dC(t) = \sum_{k=1}^{4} \alpha_k(C(t)) \, \nu_k dt + \sum_{k=1}^{4} \sqrt{\alpha_k(C(t))} \, \nu_k dB_k(t),$$

where the B_k denote standard Brownian motion processes. The CLE exhibits switching between the metastable areas (see Fig. 1.7). That is, the CLE correctly reproduces the bistable dynamical behavior while still being low dimensional. Yet, it has other disadvantages: Euler-Maruyama simulations [150] of the CLE may render the population negative because most of the time one of the two state components is close to zero.

This leads us to the questions that we want to study next: What is the relation between the RJP and RRE models, i.e., when may the deterministic RRE model be a good approximation of the stochastic dynamics given by the RJP model? And how does the CLE model fit in? Can we also understand how the CLE model can be derived from the RJP model and when we are allowed to use it without risking negative populations to occur?

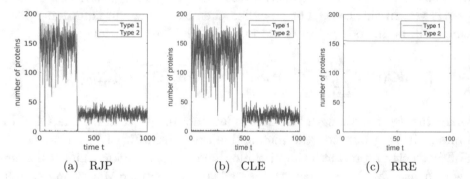

(a) RJP   (b) CLE   (c) RRE

Figure 1.7. Realizations for toggle switch model. Independent simulations of (**a**) the RJP and (**b**) the CLE (setting negative values to zero in each iteration step), both starting in $(155, 0)$. (**c**) Deterministic dynamics of the RRE given the same initial state $(155, 0)$. The deterministic model is unable to capture the characteristic behavior of the system, which is the rare switching between the two metastable domains

Chapter 2

Population Scaling

The reaction jump process $X = (X(t))_{t \geq 0}$ introduced in Chap. 1 is counting the number of particles of each chemical species involved in the well-mixed reactive system under consideration. Jumps of the process within its discrete state space are induced by chemical reactions. For a large molecular population, these discrete jumps become proportionally small. In this case, an appropriate rescaling of the dynamics can lead to effective approximative models, which are subject of the present chapter.

In order to illustrate the concept of small and large molecular populations, let us imagine the (simplistic) situation of a single DNA molecule in a sea of a large number of proteins—all of a single species. If the population of proteins is large, then the close proximity of the DNA molecule (with small volume V_0) will contain one DNA molecule ($X_1^{V_0} = 1$) with many proteins, say $X_2^{V_0}$ proteins. If we enlarge the focus and look at a larger volume V_1, then we will still count *one* copy of DNA ($X_1^{V_1} = 1$) but many more proteins $X_2^{V_1} \gg X_2^{V_0}$. Thus the number of DNA molecules is not increasing with the volume V but the number of proteins is. Or, in other words, the population of DNA does *not* scale with V while the large protein population does. In order to distinguish between the two cases, one has to consider the particle numbers for different volumes, that is, we have to consider a volume-dependent sequence of reactive particle systems of the same type.

S. Winkelmann, C. Schütte, *Stochastic Dynamics in Computational Biology*, Frontiers in Applied Dynamical Systems: Reviews and Tutorials 8, https://doi.org/10.1007/978-3-030-62387-6_2

Let us again consider well-mixed systems. That is, for each volume V the system is assumed to behave well-mixed in space (see Assumption 1.1), such that the system's state may be characterized by the number of particles of each species and the dynamics are given by a reaction jump process as described in Chap. 1. In our first step, we assume that the population size of each involved species scales with V, such that large values of V imply high numbers of particles for each species. Under this condition, the components of the reaction jump process can be equally scaled, and we will see that the dynamics become well approximated by continuous processes which are defined by ordinary differential equations (ODEs) or stochastic differential equations (SDEs)—this is a classical result proven by Kurtz [156–158, 160] and subject of Sect. 2.1. If, instead, the scaling of the particle's abundance varies for different species, a uniform scaling of the dynamics is inappropriate, and multiscale approaches are demanded, which are introduced in Sect. 2.2. An illustrative application is given in Sect. 2.3. The chapter is completed by a consideration of time-dependent particle abundance in Sect. 2.4.

2.1 Uniform Scaling: From Stochastic Jumps to Deterministic Dynamics

As in Chap. 1, let the system consist of L chemical species $\mathcal{S}_1, \ldots, \mathcal{S}_L$ undergoing K reactions $\mathcal{R}_1, \ldots, \mathcal{R}_K$. We consider the dynamics depending on the volume $V > 0$ as a free parameter and assume that the particle numbers of all species scale with the volume. The state of the system at time t is denoted by $\boldsymbol{X}^V(t) = (X_1^V(t), \ldots, X_L^V(t))^\mathsf{T}$ to indicate the dependency of the volume, where the component $X_l^V(t)$ refers to the number of particles of species $\mathcal{S}_l$ at time t given the volume V. Here (in Sect. 2.1) the state-change vectors $\boldsymbol{\nu}_k$ are assumed to be independent of V. In Sect. 2.2 also state-change vectors with entries that scale with the volume will be considered.

2.1.1 The Reaction Rate Equation

The first models for the dynamics of chemical reaction networks were formulated in terms of particle concentrations (also called *number densities* in physics and chemistry) instead of discrete particle numbers. In the 1860s, Guldberg and Waage developed the classical law of mass action for deterministic dynamics, stating that the rate of a reaction is proportional to the product of the concentrations of reactant molecules [115, 241, 242]. In the following, we will see how these classical deterministic reaction kinetics can be deduced from the stochastic model introduced in Chap. 1.

Scaling of Reaction Propensities

In case of volume exclusion, also the reaction propensities have to scale with the volume. We denote them by α_k^V in order to indicate their dependence on V. For the standard mass-action kinetics, the volume-scaled propensities are given by

$$
\alpha_k^V(\boldsymbol{x}) = \begin{cases} \gamma_k V \prod_{l=1}^{L} \frac{1}{V^{s_{lk}}} \binom{x_l}{s_{lk}} & \text{if } x_l \geq s_{lk} \text{ for all } l = 1, \ldots, L, \\ 0 & \text{otherwise,} \end{cases} \tag{2.1}
$$

where $\boldsymbol{x} \in \mathbb{N}_0^L$ contains the number of present particles for each species. Again, the variables s_{lk} for $l = 1, \ldots, L$ denote the stoichiometric coefficients of reaction $\mathcal{R}_k$ (see (1.1)). For example, we obtain $\alpha_k^V(\boldsymbol{x}) = V\gamma_k$ for a reaction of order zero, while for a first-order reaction of species $\mathcal{S}_l$ the equality $\alpha_k^V(\boldsymbol{x}) = \gamma_k x_l$ holds. For a second-order reaction of two different species $\mathcal{S}_l$ and $\mathcal{S}_{l'}$, the rescaled propensity is given by $\alpha_k^V(\boldsymbol{x}) = \frac{\gamma_k}{V} x_l x_{l'}$, which reflects the fact that inside a larger volume it takes more time for two reactant molecules to find each other. In order to extend the domain of α_k^V to $\mathbb{R}_+^L$, we can rewrite the first line of (2.1), replacing the binomial coefficient, as

$$
\alpha_k^V(\boldsymbol{x}) = \frac{\gamma_k}{V^{|a_k|-1}} \frac{1}{\prod_{l'=1}^{L}(s_{l'k}!)} \prod_{l=1}^{L} \prod_{i=0}^{s_{lk}-1} (x_l - i) \tag{2.2}
$$

for $\boldsymbol{x} \in \mathbb{R}_+^L$ with $x_l \geq s_{lk}$ for all l, where $|a_k| := \sum_{l=1}^{L} s_{lk}$ is the molecularity of the reaction. Again, we set $\alpha_k^V(\boldsymbol{x}) = 0$ for $\boldsymbol{x} \in \mathbb{R}_+^L$ with $x_l < s_{lk}$ for any l. These reaction propensity functions α_k^V will be called *microscopic propensities*.

As in the volume-independent setting of Chap. 1, the dimension of the reaction propensity has to be the inverse of time for each V. This means that the dimension of the rate constant γ_k has to depend on the order of the reaction. Actually, for a reaction of order zero, the dimension unit of γ_k is given by $[(Vt)^{-1}]$ where $[t]$ is the unit of time and $[V]$ is the unit of the volume. For a first-order reaction, the dimension unit of γ_k is $[t^{-1}]$, and for a second-order reaction, the dimension unit of γ_k is $[Vt^{-1}]$. In general, the unit of the rate constant for a reaction of order $|a_k| = n$ is $[V^{n-1}t^{-1}]$, such that the unit of the propensity in (2.2) reduces to $[\alpha_k^V(\boldsymbol{x})] = \frac{[\gamma_k]}{[V^{|a_k|-1}]} = [t^{-1}]$.

The Rescaled Process

For a given value V of the volume, we now consider the scaled process $\boldsymbol{C}^V = (\boldsymbol{C}^V(t))_{t \geq 0}$ of concentrations defined as

$$\boldsymbol{C}^V(t) := \frac{1}{V}\boldsymbol{X}^V(t)$$

with values $\boldsymbol{c} = (c_1, \ldots, c_L)^{\mathsf{T}} = \frac{1}{V}\boldsymbol{x} \in \mathbb{R}_+^L$, where $C_l^V(t) = c_l$ refers to the concentration of species $\mathcal{S}_l$ in numbers per unit volume.

$$\tilde{\alpha}_k^V(\boldsymbol{c}) := \alpha_k^V(\boldsymbol{c}V)$$

be the propensity function in terms of concentrations, which will be called *macroscopic propensities*. For propensities of mass-action type, it then holds

$$\frac{1}{V}\tilde{\alpha}_k^V(\boldsymbol{c}) = \frac{\gamma_k}{V^{|a_k|}}\frac{1}{\prod_{l'=1}^L (s_{l'}!)}\prod_{l=1}^L \prod_{i=0}^{s_{lk}-1}(c_l V - i)$$

$$= \frac{\gamma_k}{\prod_{l'=1}^L (s_{l'}!)}\prod_{l=1}^L \prod_{i=0}^{s_{lk}-1}\left(c_l - \frac{i}{V}\right)$$

$$= \tilde{\gamma}_k \prod_{l=1}^L c_l^{s_{lk}}\left(1 + \mathcal{O}(V^{-1})\right), \tag{2.3}$$

where

$$\tilde{\gamma}_k := \frac{\gamma_k}{\prod_{l=1}^L (s_{lk}!)} \tag{2.4}$$

are the rate constants of classical deterministic mass action [238].[1] The dimension of the classical rate constant $\tilde{\gamma}_k$ is again $[V^{n-1}t^{-1}]$ where $n = |a_k|$ is the molecularity of the reaction. The last line of (2.3) motivates the following assumption:

Assumption 2.1 (Classical Scaling of Propensity Functions). For the volume-dependent propensity functions $\tilde{\alpha}_k^V$, there exist suitable (locally Lipschitz continuous) limits $\tilde{\alpha}_k$ with

$$\frac{1}{V}\tilde{\alpha}_k^V \xrightarrow{V \to \infty} \tilde{\alpha}_k$$

uniformly on compact subsets of $\mathbb{R}_+^L$ for all $k = 1, \ldots, K$.

---

[1]In general, we use the notation $f(V) = \mathcal{O}(g(V))$ for $V \to \infty$ to indicate that it holds $\limsup_{V \to \infty} |f(V)/g(V)| < \infty$.

By (2.3) this assumption is fulfilled for mass-action kinetics, with the limit given by the classical mass-action propensity functions of deterministic reaction kinetics having the form

$$\tilde{\alpha}_k(\boldsymbol{c}) = \tilde{\gamma}_k \prod_{l=1}^{L} c_l^{s_{lk}}.$$

For example, the propensity function of a first-order reaction of species $\mathcal{S}_l$ fulfills

$$\frac{1}{V}\tilde{\alpha}_k^V(\boldsymbol{c}) = \frac{1}{V}\alpha_k^V(V\boldsymbol{c}) \stackrel{(2.2)}{=} \frac{\tilde{\gamma}_k}{V}Vc_l = \tilde{\gamma}_k c_l \stackrel{(2.4)}{=} \tilde{\gamma}_k c_l,$$

while for a second-order reaction of type $2\mathcal{S}_l \to \dots$ it holds

$$\frac{1}{V}\tilde{\alpha}_k^V(\boldsymbol{c}) = \frac{1}{V}\alpha_k^V(V\boldsymbol{c}) \stackrel{(2.2)}{=} \frac{\tilde{\gamma}_k}{2V^2}Vc_l(Vc_l - 1) \stackrel{(2.4)}{=} \tilde{\gamma}_k c_l^2 - \frac{\tilde{\gamma}_k}{V}c_l \xrightarrow{V\to\infty} \tilde{\gamma}_k c_l^2,$$

and accordingly for other reactions. Table 2.1 gives an overview of these deterministic mass-action propensity functions for elementary reactions up to order two. It also contains the volume-dependent propensities $\alpha_k^V(\boldsymbol{x})$ for $\boldsymbol{x} \in \mathbb{N}_0^L$ in terms of the rate constants $\tilde{\gamma}_k$, based on the general formula

$$\alpha_k^V(\boldsymbol{x}) = \tilde{\gamma}_k V \prod_{l=1}^{L} \frac{x_l!}{(x_l - s_{lk})! V^{s_{lk}}}.$$

Details about the relation between stochastic and deterministic reaction networks and their propensity functions can be found in [99, 100, 102, 105]. In particular, it is worth mentioning the following insight taken from [105]: "The propensity functions are grounded in molecular physics, and the formulas of classical chemical kinetics are approximate consequences of the formulas of stochastic chemical kinetics, not the other way around."

Table 2.1. Mass-action propensity functions of elementary chemical reactions in terms of particle concentrations

Order	Reaction	Microscopic propensity	Macroscopic propensity
0th	$\emptyset \xrightarrow{\tilde{\gamma}_0} \dots$	$\alpha_0^V(\boldsymbol{x}) = V\tilde{\gamma}_0$	$\tilde{\alpha}_0(\boldsymbol{c}) = \tilde{\gamma}_0$
1st	$\mathcal{S}_l \xrightarrow{\tilde{\gamma}_1} \dots$	$\alpha_1^V(\boldsymbol{x}) = \tilde{\gamma}_1 x_l$	$\tilde{\alpha}_1(\boldsymbol{c}) = \tilde{\gamma}_1 c_l$
2nd	$\mathcal{S}_l + \mathcal{S}_{l'} \xrightarrow{\tilde{\gamma}_2} \dots$	$\alpha_2^V(\boldsymbol{x}) = \frac{\tilde{\gamma}_2}{V} x_l x_{l'}$	$\tilde{\alpha}_2(\boldsymbol{c}) = \tilde{\gamma}_2 c_l c_{l'}$
2nd	$2\mathcal{S}_l \xrightarrow{\tilde{\gamma}_{2'}} \dots$	$\alpha_{2'}^V(\boldsymbol{x}) = \frac{\tilde{\gamma}_{2'}}{V} x_l(x_l - 1)$	$\tilde{\alpha}_{2'}(\boldsymbol{c}) = \tilde{\gamma}_{2'} c_l^2$

The Limit Process

As a direct analog to the path-wise representation (1.6) of the volume-independent reaction jump process $\boldsymbol{X}$ considered in Chap. 1, the time-change representation of the volume-dependent reaction jump process $\boldsymbol{X}^V = (\boldsymbol{X}^V(t))_{t\geq 0}$ is given by

$$\boldsymbol{X}^V(t) = \boldsymbol{X}^V(0) + \sum_{k=1}^K \mathcal{U}_k \left(\int_0^t \alpha_k^V \left(\boldsymbol{X}^V(s) \right) ds \right) \boldsymbol{\nu}_k, \qquad (2.5)$$

where $\mathcal{U}_k$ denotes independent unit-rate Poisson processes. When dividing (2.5) by V, we obtain the associated concentration version

$$\boldsymbol{C}^V(t) = \boldsymbol{C}^V(0) + \sum_{k=1}^K \frac{1}{V}\mathcal{U}_k \left(\int_0^t \alpha_k^V \left(V\boldsymbol{C}^V(s) \right) ds \right) \boldsymbol{\nu}_k$$

$$= \boldsymbol{C}^V(0) + \sum_{k=1}^K \frac{1}{V}\mathcal{U}_k \left(V \int_0^t \frac{1}{V}\tilde{\alpha}_k^V \left(\boldsymbol{C}^V(s) \right) ds \right) \boldsymbol{\nu}_k. \qquad (2.6)$$

The process $\boldsymbol{C}^V = (\boldsymbol{C}^V(t))_{t\geq 0}$ is still a stochastic process with jumps occurring at random time points which are determined by the Poisson processes $\mathcal{U}_k$. By the scaling, however, the size of jumps decreases with increasing V, while their frequency of occurrence increases. Let $\mathcal{U}_k^V(t) := \frac{1}{V}\mathcal{U}_k(Vt)$, $k = 1, \ldots, K$, denote the rescaled Poisson processes. In the large volume limit, these rescaled Poisson processes may be approximated by their first-order moments as it holds

$$\lim_{V \to \infty} \sup_{s \leq t} \left| \mathcal{U}_k^V(s) - s \right| = 0 \quad \text{a.s.}$$

for each k and $t \geq 0$, a fact which follows from the law of large numbers (see [160]). We obtain the following convergence result for the process $\boldsymbol{C}^V$. A formal derivation is given in Sect. A.3.3 (Theorem A.14).

Large volume limit/convergence: Let $C^V(0) \to C(0)$ for $V \to \infty$. Then, it holds $\lim_{V \to \infty} C^V = C$ in the sense of

$$\lim_{V \to \infty} \sup_{t \leq T} \left| C^V(t) - C(t) \right| = 0 \quad \text{a.s.}$$

for every $T > 0$, where the process $C = (C(t))_{t \geq 0}$ is given by

$$C(t) = C(0) + \sum_{k=1}^{K} \int_0^t \tilde{\alpha}_k(C(s)) ds \, \nu_k \qquad (2.7)$$

with the limit propensities $\tilde{\alpha}_k$ given in Assumption 2.1.

As for the order of convergence, it has been shown that

$$C^V(t) = C(t) + \mathcal{O}\left(V^{-1/2}\right) \qquad (2.8)$$

uniformly in t (cf. [70, 160]).

Remark 2.2. In probability theory, the $\mathcal{O}$-notation refers to convergence in probability. More precisely, Eq. (2.8) means the following: For each $\varepsilon > 0$ there exist a finite $m > 0$ and a finite $V^* > 0$ such that $\mathbb{P}(|C^V(t) - C(t)| > mV^{-1/2}) < \varepsilon$ for all $V > V^*$.

Equation (2.7) is called *reaction rate equation*. It is a deterministic equation of motion, which often appears in form of the following ordinary differential equation.

The reaction rate equation (RRE)

$$\frac{dC(t)}{dt} = \sum_{k=1}^{K} \tilde{\alpha}_k(C(t)) \nu_k \qquad (2.9)$$

Example 1.7 (continued). Consider again a single species S_1 undergoing a birth-death process triggered by the two reactions

$$\mathcal{R}_1 : S_1 \xrightarrow{\gamma_1} \emptyset, \quad \mathcal{R}_2 : \emptyset \xrightarrow{\gamma_2} S_1.$$

In this case, the reaction jump process X^V has dimension one and we directly write $X^V = X_1^V$. According to Table 2.1, the volume-dependent propensities of these two reactions have the form $\alpha_1^V(x_1) = \gamma_1 x_1$ and $\alpha_2^V(x_1) = V\gamma_2$. Figure 2.1 shows the trajectories of the rescaled jump processes $C_1^V = X_1^V/V$ for different values of V ($V = 10, 10^2, 10^3$), where $X_1^V(t)$ refers to the number of $\mathcal{S}_1$-particles at time t. For comparison also the limit process $C_1 = (C_1(t))_{t\geq 0}$ is plotted. It describes the concentration of $\mathcal{S}_1$-particles evolving according to the RRE (2.9) which here becomes

$$\frac{dC_1(t)}{dt} = -\gamma_1 C_1(t) + \gamma_2. \tag{2.10}$$

One can see the decrease of the jump size in C_1^V for increasing V, as well as the convergence of the process C_1^V to the deterministic limit C_1. $\diamond$

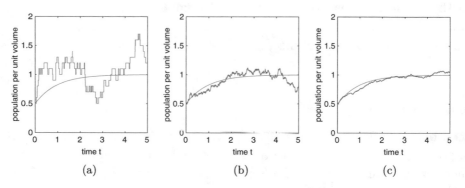

Figure 2.1. Rescaled birth-death process. Gillespie simulations of the rescaled jump processes $C_1^V = X_1^V/V$ (blue trajectories) for the birth-death process of Example 1.7 in comparison to the deterministic limit process C_1 (red lines) given by the RRE (2.10) for different values of V. The rate constants are set to $\gamma_1 = \gamma_2 = 1$, and the initial state is $C_1^V(0) = C_1(0) = 0.5$. **(a)** $V = 10$. **(b)** $V = 10^2$. **(c)** $V = 10^3$

Approximation of First-Order Moments for Finite V

Until now, we have considered realizations of the reaction jump process (RJP) and trajectories of the reaction rate equation (RRE), that is, we have compared path-wise objects parametrized by time. Based on its stochastic nature, every realization of the RJP will look somewhat different, but in the limit $V \to \infty$ almost every realization will become indistinguishable from the (unique) deterministic RRE trajectory $(\boldsymbol{C}(t))_{t\geq 0}$ with the speed of convergence scaling as $\mathcal{O}(V^{-1/2})$. Next, let us consider the expectation $\mathbb{E}\left(\boldsymbol{C}^V(t)\right)$ of the rescaled RJP $\boldsymbol{C}^V(t) = \boldsymbol{X}^V(t)/V$. This expectation is a unique object, which—for large volume $V > 0$—should be close to the value $C(t)$. More precisely, under the condition $C(0) = X^V(0)/V$, one finds that

$$\mathbb{E}\left(C^V(t)\right) = C(t) + \mathcal{O}\left(V^{-1}\right) \tag{2.11}$$

for all $t \geq 0$ (cf. [156–158]). That is, we get a higher order of convergence of the expectation compared to path-wise convergence (see also (2.21) and its derivation in Appendix A.5.1).

In case of a linear reaction network, we even obtain equality, i.e., it holds $\mathbb{E}\left(X^V(t)\right) = VC(t)$ for all t independently of V. This is due to the fact that, by linearity of the propensity functions α_k, the ODE (1.12) for the first-order moment can be written as

$$\frac{d}{dt}\mathbb{E}\left(X^V(t)\right) = \sum_{k=1}^{K} \alpha_k^V\left(\mathbb{E}\left(X^V(t)\right)\right)\nu_k. \tag{2.12}$$

Using in addition the equality $\alpha_k^V(x) = V\tilde{\alpha}_k(x/V)$, Eq. (2.12) is the same ODE as the RRE (2.9) after multiplication by V. Therefore, given consistent initial states $VC(0) = \mathbb{E}(X^V(0))$, the solutions agree for all times for the special setting of solely linear reaction propensities.

More generally, for a nonlinear reaction network (containing reactions of order two or more), the approximation error decreases according to (2.11), which is demonstrated in the following Example:

Example 1.3 (continued).   Consider again the nonlinear reaction network of binding $S_1 + S_2 \xrightarrow{\gamma_1} S_3$ and unbinding $S_3 \xrightarrow{\gamma_2} S_1 + S_2$ of the three species S_1, S_2, S_3 (see page 5). The volume-scaled propensities are

$$\alpha_1^V(x) = \frac{\gamma_1}{V}x_1 x_2, \quad \alpha_2^V(x) = \gamma_2 x_3,$$

and in terms of concentrations—noting that $\tilde{\gamma}_k = \gamma_k$ for the considered reactions—we obtain

$$\tilde{\alpha}_1(c) = \gamma_1 c_1 c_2, \quad \tilde{\alpha}_2(c) = \gamma_2 c_3$$

(see Table 2.1). Figure 2.2 shows the time-dependent expectation $\mathbb{E}\left(X_3^V(t)\right)$ of the number of S_3-particles rescaled by V for $V = 1, 2, 10, 100$ in comparison to $C_3(t)$. The expectation $\mathbb{E}\left(X_3^V\right)$ is a component of the vector $\mathbb{E}\left(X^V\right)$ which evolves according to (1.12). For large values of V, we observe $\mathbb{E}\left(X_3^V(t)\right)/V \approx C_3(t)$ for all t, in consistency with (2.11). $\diamond$

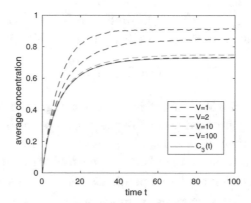

Figure 2.2. Rescaled system of binding and unbinding: Approximation by deterministic dynamics. Dashed lines: rescaled components $\mathbb{E}\left(X_3^V(t)\right)/V$ of the solution $\mathbb{E}(\boldsymbol{X}^V)$ of (2.12) for the example of binding and unbinding and different values of V. Solid line: component C_3 of the solution $\boldsymbol{C}$ of the RRE (2.9). The expectation $\mathbb{E}(\boldsymbol{X}^V)$ has been estimated by Monte Carlo simulations of the reaction jump processes $\boldsymbol{X}^V$. Initial states: $\boldsymbol{X}^V(0) = (V, V, 0)^{\mathsf{T}}$ and $\boldsymbol{C}(0) = (1, 1, 0)^{\mathsf{T}}$

2.1.2   The Chemical Langevin Equation

In order to capture not only the first-order moment of $\boldsymbol{X}^V$ but also its variance, we use again the equation

$$\boldsymbol{C}^V(t) = \boldsymbol{C}^V(0) + \sum_{k=1}^{K} \frac{1}{V}\mathcal{U}_k\left(V\int_0^t \frac{1}{V}\tilde{\alpha}_k^V\left(\boldsymbol{C}^V(s)\right)ds\right)\boldsymbol{\nu}_k,$$

which is exactly fulfilled by the rescaled RJP $\boldsymbol{C}^V = \frac{1}{V}\boldsymbol{X}^V$ (see Eq. (2.6)), but this time we consider the second-order approximation for the Poisson processes $\mathcal{U}_k$ given by

$$\frac{1}{V}\mathcal{U}_k(Vt) = t + \frac{1}{\sqrt{V}}W_k(t) + \mathcal{O}\left(\frac{\log(V)}{V}\right) \qquad (2.13)$$

for standard Brownian motion processes (Wiener processes) $W_k = (W_k(t))_{t\geq 0}$ [9, 160]. Thus, a higher-order approximation (including now terms of order $V^{-1/2}$) of the process $\boldsymbol{C}^V = \frac{1}{V}\boldsymbol{X}^V$ is given by $\boldsymbol{C}^V(t) \approx \tilde{\boldsymbol{C}}^V(t)$

with

$$\tilde{C}^{V}(t) = \tilde{C}^{V}(0) + \sum_{k=1}^{K} \int_{0}^{t} \tilde{\alpha}_{k}\left(\tilde{C}^{V}(s)\right) ds\, \boldsymbol{\nu}_{k}$$

$$+ \sum_{k=1}^{K} \frac{1}{\sqrt{V}} W_{k}\left(\int_{0}^{t} \tilde{\alpha}_{k}\left(\tilde{C}^{V}(s)\right) ds\right) \boldsymbol{\nu}_{k},$$

where W_{k}, $k = 1, \ldots, K$, are independent standard Brownian motions in $\mathbb{R}$. Under some additional assumptions on the compatibility of the processes C^{V} and W_{k} (see [9, 70]), there exist standard Brownian motions B_{k} with

$$\tilde{C}^{V}(t) = \tilde{C}^{V}(0) + \sum_{k=1}^{K} \int_{0}^{t} \tilde{\alpha}_{k}\left(\tilde{C}^{V}(s)\right) ds\, \boldsymbol{\nu}_{k}$$

$$+ \sum_{k=1}^{K} \frac{1}{\sqrt{V}} \int_{0}^{t} \sqrt{\tilde{\alpha}_{k}\left(\tilde{C}^{V}(s)\right)} dB_{k}(s)\, \boldsymbol{\nu}_{k}, \tag{2.14}$$

where the last line now contains a sum of Itô integrals (see Sect. A.2 in the Appendix for some basic explanations). Replacing $\tilde{\alpha}_{k}(\boldsymbol{c})$ by $\frac{1}{V}\alpha_{k}^{V}(V\boldsymbol{c})$ and setting $\tilde{\boldsymbol{X}}^{V}(t) := V\tilde{\boldsymbol{C}}^{V}(t)$, a scalar transformation back to numbers rather than concentrations (i.e., multiplying (2.14) by V) gives

$$\tilde{\boldsymbol{X}}^{V}(t) = \tilde{\boldsymbol{X}}^{V}(0) + \sum_{k=1}^{K} \int_{0}^{t} \alpha_{k}^{V}\left(\tilde{\boldsymbol{X}}^{V}(s)\right) ds\, \boldsymbol{\nu}_{k}$$

$$+ \sum_{k=1}^{K} \int_{0}^{t} \sqrt{\alpha_{k}^{V}\left(\tilde{\boldsymbol{X}}^{V}(s)\right)} dB_{k}(s)\, \boldsymbol{\nu}_{k}. \tag{2.15}$$

A switch from integral notation to differential notation leads to the Itô stochastic differential equation (2.16), which is called *chemical Langevin equation*.[2]

The chemical Langevin equation (CLE)

$$d\tilde{\boldsymbol{X}}^{V}(t) = \sum_{k=1}^{K} \alpha_{k}^{V}\left(\tilde{\boldsymbol{X}}^{V}(t)\right) \boldsymbol{\nu}_{k}\, dt + \sum_{k=1}^{K} \sqrt{\alpha_{k}^{V}\left(\tilde{\boldsymbol{X}}^{V}(t)\right)} \boldsymbol{\nu}_{k}\, dB_{k}(t) \quad (2.16)$$

---

[2]In the literature also the integral version (2.15) is called *chemical Langevin equation*.

Stochastic processes given by an SDE like (2.16) will be called *diffusion processes* in the following (see also Definition A.7 in Appendix A.2), and approximations of the reaction jump process by such SDE-processes are termed *diffusion approximations*.

The CLE (2.16) is the counterpart of the RRE (2.9) which defines the deterministic dynamics of C. Compared to the RRE (see the asymptotic equation (2.8)), the CLE delivers a higher order of approximation of the reaction jump process X^V, as it holds

$$C^V(t) = \tilde{C}^V(t) + \mathcal{O}\left(\frac{\log(V)}{V}\right),$$

which follows from (2.13) [70, 159]. Taking the limit $V \to \infty$ in Eq. (2.14) directly leads back to the integral notation (2.7) of the deterministic RRE. On the scale of particle numbers, the CLE appears on the mesoscopic level, while the RRE can be seen as the macroscopic modeling approach.

Example 1.3 (continued). For the system of binding and unbinding, we have seen that in case of $V = 100$ the rescaled mean $\mathbb{E}(X^V)/V$ of the reaction jump process X^V is well approximated by the solution C of the RRE (2.9) (cf. Figure 2.2). Now we compare the reaction jump process to the process $\tilde{X}^V$ defined by the CLE (2.16). In Fig. 2.3 independent trajectories of the two processes are depicted together with their time-dependent mean and standard deviation. For the chosen volume $V = 100$, the difference in the shape of the trajectories is still visible: While X^V is piecewise constant and switches at random time points between discrete states, the diffusion process $\tilde{X}^V$ is continuous in space and time. On the other hand, V is large enough for the first- and second-order moments to be indistinguishable (see the solid and dashed lines in Fig. 2.3).   ◇

Remark 2.3. Note that by its white noise components dB_k the CLE can possibly drive the population to unphysical negativity. Several attempts exist to handle this problem. In [208], e.g., a *complex CLE* is presented which extends the domain of the CLE to complex space and still admits a physical interpretation. Alternatively, there exist adaptive methods which switch back to a discrete description whenever the population size reaches a low level. We will come back to such adaptive methods in Sect. 2.4.

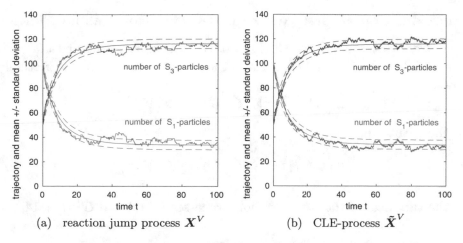

(a)   reaction jump process $\boldsymbol{X}^V$ (b)   CLE-process $\tilde{\boldsymbol{X}}^V$

Figure 2.3. Rescaled system of binding and unbinding: Approximation by continuous stochastic dynamics. (**a**) Gillespie simulation of the reaction jump process $\boldsymbol{X}^V$ and (**b**) Euler-Maruyama simulation of the process $\tilde{\boldsymbol{X}}^V$ defined by the CLE (2.16) for the system of binding $\mathcal{S}_1 + \mathcal{S}_2 \to \mathcal{S}_3$ and unbinding $\mathcal{S}_3 \to \mathcal{S}_1 + \mathcal{S}_2$ and $V = 100$. Initial state: 100 $\mathcal{S}_1$-particles, 100 $\mathcal{S}_2$-particles, 50 $\mathcal{S}_3$-particles. The number of $\mathcal{S}_2$-particles is not displayed because it agrees with the number of $\mathcal{S}_1$-particles for all times. The additional solid and dashed lines indicate the corresponding mean and standard deviations depending on time. Reaction rates constants: $\gamma_1 = 10^{-1}$, $\gamma_2 = 10^{-2}$

Alternative Notation

With the drift vector field $\mu^V : \mathbb{R}^L \to \mathbb{R}^L$, $\mu^V = (\mu_l^V)_{l=1,\ldots,L}$, defined as

$$\mu^V(\boldsymbol{x}) := \sum_{k=1}^{K} \alpha_k^V(\boldsymbol{x}) \boldsymbol{\nu}_k \qquad (2.17)$$

and the function $\sigma^V : \mathbb{R}^L \to \mathbb{R}^{L,K}$, $\sigma^V = (\sigma_{lk}^V)_{l=1,\ldots,L,k=1,\ldots,K}$, defined as

$$\sigma_{lk}^V(\boldsymbol{x}) := \sqrt{\alpha_k^V(\boldsymbol{x})} \nu_{lk} \qquad (2.18)$$

giving the strength of noise, the CLE (2.16) reads

$$d\tilde{\boldsymbol{X}}^V(t) = \mu^V\left(\tilde{\boldsymbol{X}}^V(t)\right) dt + \sigma^V\left(\tilde{\boldsymbol{X}}^V(t)\right) d\boldsymbol{W}(t)$$

for a K-dimensional Wiener process $(\boldsymbol{W}(t))_{t \geq 0}$.

Approximation of First- and Second-Order Moments for Finite V

In contrast to the RRE, the CLE not only approximates the first-oder moments of the RJP (see (2.11)) but also the second-order moments. Actually, it holds

$$\mathbb{E}\left(C^V(t)\right) = \mathbb{E}\left(\tilde{C}^V(t)\right) + \mathcal{O}\left(V^{-2}\right)$$

for the means and

$$\mathbb{V}\left(C^V(t)\right) = \mathbb{V}\left(\tilde{C}^V(t)\right) + \mathcal{O}\left(V^{-2}\right)$$

for the variances of the concentration processes $C^V(t)$ and $\tilde{C}^V(t)$ [114].

Comparison of RJP, CLE, and RRE

Table 2.2 gives an overview of the modeling approaches described so far. For well-mixed systems, the reaction jump process (RJP) model is the *root model*, i.e., it gives the most precise description of the reactive dynamics of the system under consideration. In particular, it allows to describe effects resulting from individual discrete events and from stochastic fluctuations like bursting or the switching behavior illustrated in Sect. 1.4 for the case of the genetic toggle switch. Its main disadvantage lies in the huge computational effort needed for numerical simulation of the RJP for medium and large populations that induce very small time steps and for cases with many different reactions. Furthermore, the inherent randomness often requires to generate a large number of independent realizations in order to get reasonable approximations of expectation values (sampling problem).

In comparison to the RJP model, the chemical Langevin equation (CLE) is approximative only, i.e., in general it requires medium population levels of all species involved for acceptable accuracy. The CLE no longer incorporates discrete events. Still, some aspects of the stochastic fluctuations like switching can be reproduced, but with no guarantees that switching rates are accurate. Regarding numerical simulation, the CLE has similar disadvantages as the RJP: small temporal step sizes for large population levels, as well as the sampling problem. In addition, simulation of the CLE may produce unphysical solutions such as negative concentrations.

The reaction rate equation (RRE) also is approximative only. In general, it requires quite large population levels of all species involved. The RRE is a deterministic model, i.e., it exhibits neither discrete events nor stochastic fluctuations, and therefore any stochastically induced switching behavior is lost completely. Numerical simulation of the RRE, on the other hand, is very efficient: There are well-described and elaborated numerical algorithms that allow for large time steps including adaptive accuracy control (cf. [49] for this and many other numerical tools available), as well as for very large numbers of species and very long simulation time spans. Furthermore, its deterministic nature implies that one trajectory is all what is needed (no sampling problem).

Table 2.2. Comparison of RJP, CLE, and RRE

	RJP	CLE	RRE
Type	Stochastic, continuous in time, discrete in space	Stochastic, continuous in time and space	Deterministic, continuous in time and space
Properties	Precise (root model), captures discrete events, fluctuations and switching behavior	Approximative, includes fluctuations and switching, negative concentrations possible	Approximative, fluctuations and stochastically induced switching are lost
Numerical effort	Large computational effort for large populations, sampling problem	Medium computational effort for large populations, sampling problem	Numerically very efficient, no sampling problem

2.1.3 Chemical Fokker-Planck and Liouville Equations

Diffusion approximations of Markov jump processes have been investigated in the 1970s by Kurtz [157, 160]. Gillespie also gave a derivation of the CLE based on approximations of Poisson variables [103]. However, there exist alternative derivations of the CLE that go back to Kramers and Moyal in the 1940s and consider a Taylor series expansion of the related CME instead of the Poisson characterization (2.5) (cf. [153, 185]). Given such a *Kramers-Moyal expansion* of the CME (1.10) with the variable $\boldsymbol{x}$ considered as continuous and $p(\boldsymbol{x}, t)$ being replaced by $\rho^V(\boldsymbol{x}, t)$ (see Appendix A.5.1 for details), a truncation after the second-order term gives the partial differential equation

$$\frac{\partial}{\partial t}\rho^V(\boldsymbol{x}, t) = -\sum_{l=1}^{L}\frac{\partial}{\partial x_l}\left[\mu_l^V(\boldsymbol{x})\rho^V(\boldsymbol{x}, t)\right] + \sum_{l,l'=1}^{L}\frac{\partial^2}{\partial x_l\partial x_{l'}}\left[\Sigma_{ll'}^V(\boldsymbol{x})\rho^V(\boldsymbol{x}, t)\right]$$

$$(2.19)$$

which is called *chemical Fokker-Planck equation (CFPE)*. Here, $\mu^V(\boldsymbol{x}) = (\mu_l^V(\boldsymbol{x}))_{l=1,\dots,L} \in \mathbb{R}^L$ is the drift vector defined in (2.17) and $\boldsymbol{\Sigma}^V(\boldsymbol{x}) = (\Sigma_{l,l'}^V(\boldsymbol{x}))_{l,l'=1,\dots,L} \in \mathbb{R}^{L,L}$ is the diffusion matrix defined as $\boldsymbol{\Sigma}^V := \frac{1}{2}\sigma^V(\sigma^V)^{\mathsf{T}}$ for σ^V given in (2.18), i.e., it holds

$$\Sigma_{l,l'}^V(\boldsymbol{x}) := \frac{1}{2}\sum_{k=1}^K \sigma_{lk}^V(\boldsymbol{x})\sigma_{l'k}^V(\boldsymbol{x})$$

$$= \frac{1}{2}\sum_{k=1}^K \alpha_k^V(\boldsymbol{x})\nu_{lk}\nu_{l'k}.$$

For each time t, the function $\rho^V(\boldsymbol{x}, t)$ is a probability density function depending on the continuous state variable $\boldsymbol{x} \in \mathbb{R}^L$ (in contrast to $p(\cdot, t)$ which is defined on the discrete state space $\mathbb{N}_0^L$). The CFPE is equivalent to the CLE in the sense that the probability density function of the process $\tilde{\boldsymbol{X}}^V$ defined by the CLE (2.16) exactly obeys the CFPE (2.19), i.e., it holds

$$\mathbb{P}\left(\tilde{\boldsymbol{X}}^V(t) \in B \,\middle|\, \tilde{\boldsymbol{X}}^V(0) \sim \rho_0^V\right) = \int_B \rho^V(\boldsymbol{x}, t)\, d\boldsymbol{x}$$

for ρ^V given by (2.19), the initial distribution $\rho_0^V := \rho^V(\cdot, 0)$ and $B \subset \mathbb{R}^L$ (more precisely $B \in \mathcal{B}$ where $\mathcal{B}$ denotes the Borel σ-algebra on $\mathbb{R}^L$).

From the CFPE (2.19) one can derive ordinary differential equations for the moments of the process $\tilde{\boldsymbol{X}}^V$ described by the CLE. By multiplying Eq. (2.19) by $\boldsymbol{x}$ and integrating over the state space $\mathbb{X}$, for example, one obtains an equation for the first-order moment $\mathbb{E}(\tilde{\boldsymbol{X}}^V)$. In comparison with the corresponding equations for the reaction jump process $\boldsymbol{X}^V$, which were given in Sect. 1.2.3, it turns out that for moments of up to order two, these equations agree exactly. However, since the equations are in general coupled to higher-order moments for which the derived equations do not agree, the moments of the CLE process and the reaction jump process are generally not consistent. Only for linear reaction networks the moment equations will always decouple, such that the moments up to order two of both processes agree exactly [209]. For nonlinear reaction networks, a high-level approximation of mean and variance is again induced by large values of V.

The analog of the CFPE (2.19) for deterministic dynamics is given by the (generalized) *Liouville equation*

$$\frac{\partial}{\partial t}\rho(\boldsymbol{c}, t) = -\sum_{l=1}^L \frac{\partial}{\partial c_l}\left[\tilde{\mu}_l(\boldsymbol{c})\rho(\boldsymbol{c}, t)\right] = -\mathrm{div}_{\boldsymbol{c}}\left[\sum_{k=1}^K \boldsymbol{\nu}_k\tilde{\alpha}_k(\boldsymbol{c})\rho(\boldsymbol{c}, t)\right], \quad (2.20)$$

where $\rho(\boldsymbol{c},t)$ is the density in terms of the concentration vector $\boldsymbol{c} = (c_1,\ldots,c_L)$ and the drift vector $\tilde{\boldsymbol{\mu}} = (\tilde{\mu}_l)_{l=1,\ldots,L}$ is defined as

$$\tilde{\boldsymbol{\mu}}(\boldsymbol{c}) := \sum_{k=1}^{K} \tilde{\alpha}_k(\boldsymbol{c})\boldsymbol{\nu}_k.$$

Here, $\boldsymbol{c}$ refers to possible values of the process $\boldsymbol{C}$ governed by the RRE (2.9), i.e., it holds

$$\mathbb{P}\left(\boldsymbol{C}(t) \in B \,|\, \boldsymbol{C}(0) \sim \rho_0\right) = \int_B \rho(\boldsymbol{c},t)\,d\boldsymbol{c}$$

for the deterministic process $\boldsymbol{C}$ defined by the RRE (2.9) and $\rho_0 = \rho(\cdot,0)$.

Remark 2.4. In a recent article [174], the authors give a mathematically rigorous proof for the convergence of the solution of the CME to the solution of the Liouville equation (2.20) in the population limit using the energy dissipation principle in combination with so-called Γ-limits, a novel technique that allows to show convergence of processes on discrete state spaces to continuous ones.

The RRE (2.9) and the CLE (2.16) are *path-wise formulations* of the dynamics, just like the reaction jump process given in (1.6). In contrast, the CFPE (2.19) and the Liouville equation (2.20) describe the *propagation of distributions* in time, just as the CME (1.10) does for the reaction jump process. See Fig. 2.4 for an overview.

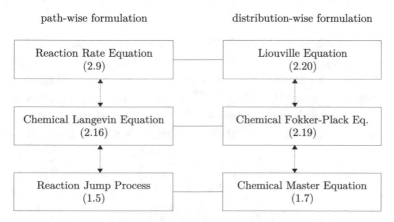

Figure 2.4. Overview of modeling approaches for well-mixed reaction kinetics. For each level, there is a path-wise formulation of the dynamics (boxes on the left) and an equation for the propagation of distributions (boxes on the right)

Evolution of Expectation Values

In Appendix A.5.1 it is demonstrated how the Liouville equation can be derived from the chemical master equation in the limit $V \to \infty$, provided that the propensities scale with the volume such that Assumption 2.1 is fulfilled. In particular, it is shown that the solution $\rho = \rho(\boldsymbol{c}, t)$ of the Liouville equation results from transport of the initial distribution $\rho_0 = \rho(\cdot, 0)$ along the trajectories of the RRE system. If we denote the flow operator associated with RRE by Φ^t such that the solution $\boldsymbol{C}$ of the RRE (2.9) for initial conditions $\boldsymbol{C}(0) = \boldsymbol{c}_0$ can be written

$$\boldsymbol{C}(t) = \Phi^t \boldsymbol{c}_0,$$

then the evolution of expectation values with respect to the solution $\rho = \rho(\boldsymbol{c}, t)$ of the Liouville equation satisfies

$$\mathbb{E}^{\text{Li}}(f, t) := \int f(\boldsymbol{c}) \rho(\boldsymbol{c}, t) \, d\boldsymbol{c} = \int f(\Phi^t \boldsymbol{c}) \rho_0(\boldsymbol{c}) \, d\boldsymbol{c} = \mathbb{E}^{\text{Li}}(f \circ \Phi^t, 0)$$

for a $\boldsymbol{c}$-dependent observable $f : \mathbb{R}^L \to \mathbb{R}$, where the superscript Li refers to "Liouville."

A similar expression is valid for the evolution of expectation values of $\boldsymbol{x}$-dependent observable $F : \mathbb{X} \to \mathbb{R}$ with respect to $\rho^V = \rho^V(\boldsymbol{x}, t)$ as solution of the chemical Fokker-Planck equation (2.19)

$$\mathbb{E}^{\text{FP}}(F, t) := \int F(\boldsymbol{x}) \rho^V(\boldsymbol{x}, t) \, d\boldsymbol{x} = \mathbb{E}\left(F(\tilde{\boldsymbol{X}}^V(t)) \Big| \tilde{\boldsymbol{X}}^V(0) \sim \rho_0^V \right),$$

where the index FP stands for "Fokker-Planck," $\tilde{\boldsymbol{X}}^V$ is the solution process of the CLE (2.16), and ρ_0^V denotes its initial distribution. A comparison of these identities with the related identity (1.18) for the evolution of expectation values with respect to the CME exhibits striking similarities. Consequently, we can again use these identities to approximate the expectation values by running averages in the framework of a Monte Carlo scheme

$$\mathbb{E}^{\text{Li}}(f, t) \approx \frac{1}{n} \sum_{i=1}^{n} f\left(\Phi^t \boldsymbol{c}_i\right),$$

$$\mathbb{E}^{\text{FP}}(F, t) \approx \frac{1}{n} \sum_{i=1}^{n} F\left(\tilde{\boldsymbol{X}}^V(t, \boldsymbol{x}_i)\right),$$

where the sequence $\boldsymbol{c}_1, \ldots, \boldsymbol{c}_n, \ldots$ or $\boldsymbol{x}_1, \ldots, \boldsymbol{x}_n, \ldots$ is (asymptotically) distributed according to ρ_0 or ρ_0^V, respectively, and $(\tilde{\boldsymbol{X}}^V(t, \boldsymbol{x}_i))_{t \geq 0}$ denotes a realization of the CLE process starting at time $t = 0$ in state $\boldsymbol{x}_i$. The convergence rate is again of order $n^{-1/2}$, just as in the CME case (see Sect. 1.3.2).

In the case where the initial probability distribution $P_0 = P_0(\boldsymbol{x})$ of the reaction jump process $\boldsymbol{X}$ scales with the volume V such that the limit distribution ρ_0 for $V \to \infty$ is concentrated at a single point $\boldsymbol{c}_0$, i.e., $\rho_0(\boldsymbol{c}) = \delta(\boldsymbol{c} - \boldsymbol{c}_0)$ formally, then we can show an even stronger result (see Appendix A.5.1 for details)

$$\mathbb{E}(F, t) = F(V\Phi^t \boldsymbol{c}_0) + \mathcal{O}\left(\frac{1}{V}\right), \tag{2.21}$$

where $\mathbb{E}(F, t)$ denotes the expectation value of the reaction jump process itself!

System Size Expansion

The CLE (2.16) defines a stochastic process which approximates the reaction jump process $\boldsymbol{X}^V$. Likewise, the RRE (2.7) defines a deterministic process as the large volume limit of the dynamics. For situations where only the distribution of the reaction jump process is of interest (while individual trajectories of the processes are not required), an alternative approach is to directly approximate the solution $p(\boldsymbol{x}, t)$ of the CME (1.10) by means of a *system size expansion* [226]. The expansion is based on the representation (van Kampen's ansatz)

$$\frac{X_l^V(t)}{V} = C_l(t) + V^{-1/2}\epsilon_l(t), \quad l = 1, \ldots, L, \tag{2.22}$$

where C_l are the components of the solution $\boldsymbol{C} = (C_l)_{l=1,\ldots,L}$ to the RRE (2.7) and $\epsilon_l = (\epsilon_l(t))_{t \geq 0}$, $l = 1, \ldots, L$, are stochastic processes in $\mathbb{R}$ whose statistical properties determine the differences between the deterministic process $\boldsymbol{C}$ and the reaction jump process $\boldsymbol{X}^V$. The term "system size" refers to the volume V. By system size expansion, one obtains a relation between $p(\boldsymbol{x}, t)$ and the distribution $\Pi(\boldsymbol{\epsilon}, t)$ (probability density) of $\boldsymbol{\epsilon} = (\epsilon_l)_{l=1,\ldots,L} \in \mathbb{R}^L$, as well as a Fokker-Planck equation for the process $\boldsymbol{\epsilon} = (\boldsymbol{\epsilon}(t))_{t \geq 0}$ of the form

$$\partial_t \Pi(\boldsymbol{\epsilon}, t) = \left[\mathcal{L}^{(0)} + V^{-1/2}\mathcal{L}^{(1)} + V^{-1}\mathcal{L}^{(2)} + \mathcal{O}(V^{-3/2})\right] \Pi(\boldsymbol{\epsilon}, t) \tag{2.23}$$

for certain differential operators $\mathcal{L}^{(0)}$, $\mathcal{L}^{(1)}$, $\mathcal{L}^{(2)}$ (acting on Π as a function of $\boldsymbol{\epsilon}$). A truncation of the expansion (2.23) in zeroth order leads to the special case of *linear noise approximation* (LNA) given by the Fokker-Planck equation

$$\partial_t \tilde{\Pi}(\boldsymbol{\epsilon}, t) = \mathcal{L}^{(0)} \tilde{\Pi}(\boldsymbol{\epsilon}, t), \tag{2.24}$$

where $\mathcal{L}^{(0)} = -\partial_\epsilon J^{(0)} \epsilon + \frac{1}{2} \partial_\epsilon^2 J^{(1)}$ is a Fokker-Planck operator[3] with linear coefficients $J^{(0)}, J^{(1)} \in \mathbb{R}^{L,L}$ that depend on the solution of the reaction rate equation and on the propensity functions.[4] For (2.24) the analytical solution is known (given by multivariate normal distributions) (see the work by van Kampen [233, 234]). This solution can be combined with the solution C of the RRE to obtain an approximation of $p(\boldsymbol{x}, t)$ (as exact solution of the CME) based on (2.22) for macroscopic/large volumes. For mesoscopic volumes, a system size expansion including terms of order $V^{-1/2}$ delivers a higher-order correction (cf. [111]).

A more detailed review of system size expansion approach can be found in [209]. A hybrid approach combining the CME and the LNA for systems with multiple scales has been developed in [33].

Remark 2.5 (Numerical Effort). In comparison to Monte Carlo simulations of the reaction jump process (or the CLE as its approximation), the system size expansion can be a much more efficient way to determine the distribution or first few moments of the process. The linear noise approximation actually results in a finite set of ordinary differential equations which can be solved with less numerical effort than doing stochastic simulations of the RJP or the CLE.

Path-Wise Simulation and Relation to τ-Leaping

Just as the CME, the CFPE and the Liouville equation cannot be solved analytically for most systems. Instead, the densities ρ^V and ρ have to be estimated by Monte Carlo simulations of the underlying processes given by the CLE and RRE, respectively. As for the solution ρ of the Liouville equation, only the initial distribution must be sampled. The deterministic trajectories can then be approximated by standard numerical ODE solvers (e.g., Runge-Kutta methods [26]). A suitable method for numerical simulations of the CLE process is the Euler-Maruyama algorithm [150].[5] Given a discrete time step $\tau > 0$, an approximative realization of the CLE process is recursively determined by

$$\tilde{\boldsymbol{X}}^V(t+\tau) = \tilde{\boldsymbol{X}}^V(t) + \sum_{k=1}^{K} \xi_k \left(\tilde{\alpha}_k^V(\tilde{\boldsymbol{X}}^V(t))\tau, \tilde{\alpha}_k^V(\tilde{\boldsymbol{X}}^V(t))\tau \right) \boldsymbol{\nu}_k, \qquad (2.25)$$

---

[3]Shorthand for: $\mathcal{L}^{(0)} = -\sum_{l=1}^{L} \frac{\partial}{\partial \epsilon_l}(J^{(0)}\epsilon)_l + \frac{1}{2}\sum_{l,l'=1}^{L} J_{ll'}^{(1)} \frac{\partial}{\partial \epsilon_l} \frac{\partial}{\partial \epsilon_{l'}}$.

[4]$J^{(0)}$ is the Jacobian of the reaction rate equation.

[5]Note that there exist more efficient numerical SDE solvers than Euler-Maruyama (see [126] for an introduction). Anyway, the Euler-Maruyama recursion is well suited to demonstrate the parallelism to the τ-leaping method.

where $\xi_k(\mu_k, \sigma_k^2) \sim \mathcal{N}(\mu_k, \sigma_k^2)$ denote independent normal random variables with mean μ_k and variance σ_k^2.[6] This is an approximation of the corresponding τ-leaping formula

$$\boldsymbol{X}^V(t+\tau) = \boldsymbol{X}^V(t) + \sum_{k=1}^{K} \mathcal{U}_k \left(\alpha_k^V \left(\boldsymbol{X}^V(t) \right) \tau \right) \boldsymbol{\nu}_k \qquad (2.26)$$

from Sect. 1.3.2 (see (1.23)). By the central limit theorem, the approximation of the Poisson variables $\mathcal{U}_k \left(\alpha_k^V \left(\boldsymbol{X}^V(t) \right) \tau \right)$ by the normal random variables ξ_k becomes plausible whenever the volume V is large enough for all reactions to fire several times within the time interval $[t, t + \tau]$, i.e.

$$\alpha_k^V \left(\boldsymbol{X}^V(t) \right) \tau \gg 1 \text{ for all } k = 1, \dots, K. \qquad (2.27)$$

In this sense, τ-leaping can be interpreted as the bridge to the approximative continuous dynamics given by the CLE and the RRE [99, 100, 102, 105, 106] (see Fig. 2.5 for an overview of the path-wise algorithms).

In contrast to the SSA simulations of the CME, the numerical effort of CLE simulations is independent of the number of particles and their reactivity. For highly reactive systems, the CLE simulations can thus be much more efficient than the CME simulations. On the other hand, CLE simulations typically demand much more computational effort than numerical simulations based on the RRE model. For the RRE, very efficient numerical algorithms, e.g., based on linear-implicit Runge-Kutta schemes [49], including adaptive accuracy and step size control, have been developed that in many cases allow for very large time steps that often are many orders of magnitude larger than the ones permitted in CLE-based simulations.

2.2 Hybrid Models for Multiple Population Scales

Approximations by the CLE or the RRE require the whole reactive system to scale with the volume. In many applications, however, there exist multiple population scales, such that only part of the system may be approximated by continuous dynamics, while some discrete stochasticity of other components needs to be maintained because it is of crucial relevance for the characteristics of the system. In this case, the different modeling approaches can be combined with each other, leading to effective hybrid models for multiscale systems.

---

[6]In (2.25) it holds $\mu_k = \sigma_k^2$ for all k.

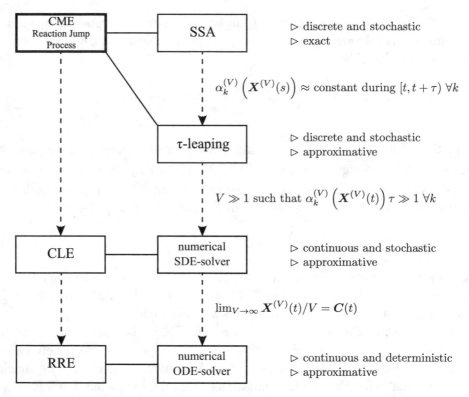

Figure 2.5. Numerical schemes for different volumes. Given that the propensities scale with the volume such that Assumption 2.1 is fulfilled, the reaction jump process (1.6) can be approximated by the CLE process (2.15) for large volume systems. In the thermodynamic limit, noise vanishes and the system can be approximated by the deterministic process defined by the RRE (2.9). For each level of modeling, there exist path-wise algorithms to create trajectories of the corresponding process

In Sects. 2.2.1–2.2.3 we introduce the path-wise formulation of the hybrid process, which gives a nice characterization by a single combined equation. This is the basis for numerical simulations of the hybrid process, i.e., indirect solutions of the hybrid system. In Sect. 2.2.4 a hybrid master equation (a CME coupled to an ODE) is formulated, which may be directly solvable by numerical integration.

2.2.1   Piecewise-Deterministic Reaction Processes

The following investigations are based on [85, 250, 251]. As before, let $X_l^V(t) \in \mathbb{N}_0$ denote the number of particles of species $\mathcal{S}_l$ at time $t \geq 0$ given the system's volume $V > 0$. Each reaction $\mathcal{R}_k$ is specified by a volume-dependent state-change vector $\boldsymbol{\nu}_k^V$ and a propensity function $\alpha_k^V(\boldsymbol{x})$ which defines the probability per unit of time for the reaction to occur given that the system is in state $\boldsymbol{x} \in \mathbb{N}_0^L$. In contrast to the derivation of the RRE (2.9), we now assume that only part of the molecular population scales with the system's volume, while some of the species appear in low copy numbers independent of V. Without loss of generality let $\mathcal{S}_1, \ldots, \mathcal{S}_{l'}$ for some $l' < L$ be these low-abundant species. For partial approximation of the dynamics, only the components $X_{l'+1}^V, \ldots, X_L^V$ referring to the large copy number species are scaled by $1/V$, while the others are left unscaled, i.e., we consider the process $\boldsymbol{\Theta}^V = (\boldsymbol{\Theta}^V(t))_{t \geq 0}$ with $\boldsymbol{\Theta}^V(t) = \left(\Theta_1^V(t), \ldots, \Theta_L(t)\right)^{\mathsf{T}}$ defined as

$$\Theta_l^V(t) := \begin{cases} X_l^V(t) & \text{for } l = 1, \ldots, l', \\ \frac{1}{V} X_l^V(t) & \text{for } l = l'+1, \ldots, L. \end{cases} \tag{2.28}$$

Analogously to the time-change representation (1.6) of the unscaled process, the dynamics of the partially scaled process $\boldsymbol{\Theta}^V$ are characterized by

$$\boldsymbol{\Theta}^V(t) = \boldsymbol{\Theta}^V(0) + \sum_{k=1}^K \mathcal{U}_k \left(\int_0^t \tilde{\alpha}_k^V \left(\boldsymbol{\Theta}^V(s) \right) ds \right) \tilde{\boldsymbol{\nu}}_k^V \tag{2.29}$$

for partially scaled state-change vectors $\tilde{\boldsymbol{\nu}}_k^V$ defined as

$$\tilde{\nu}_{lk}^V := \begin{cases} \nu_{lk}^V & \text{for } l = 1, \ldots, l', \\ \frac{1}{V} \nu_{lk}^V & \text{for } l = l'+1, \ldots, L, \end{cases}$$

and adapted reaction propensity functions $\tilde{\alpha}_k^V$ given by

$$\tilde{\alpha}_k^V(\boldsymbol{\theta}) := \alpha_k^V(\boldsymbol{x}) \tag{2.30}$$

for $\boldsymbol{\theta} = (\theta_l)_{l=1,\ldots,L}$, where $\theta_l = x_l$ for $l = 1, \ldots, l'$ and $\theta_l = x_l/V$ for $l = l'+1, \ldots, L$. As before, the $\mathcal{U}_k$ denotes independent, unit-rate Poisson processes. In the following, we again assume that α_k^V is also defined for non-integer $\boldsymbol{x} \in \mathbb{R}_+^L$.

We consider the situation where the state-change vectors $\tilde{\boldsymbol{\nu}}_k^V$ can be separated into two classes: those which converge to non-zero limit vectors $\tilde{\boldsymbol{\nu}}_k$ for $V \to \infty$, belonging to reactions which in the large volume limit still induce jumps in the dynamics, and those where $V\tilde{\boldsymbol{\nu}}_k^V$ converges for $V \to \infty$, referring to reactions which in the large volume limit induce a continuous flow. We define $\mathbb{K} := \{1, \ldots, K\}$ and

$$\mathbb{K}_0 := \{ k \in \mathbb{K} | \tilde{\boldsymbol{\nu}}_k^V \xrightarrow{V \to \infty} \tilde{\boldsymbol{\nu}}_k \text{ for some } \tilde{\boldsymbol{\nu}}_k \neq 0 \},$$

$$\mathbb{K}_1 := \{ k \in \mathbb{K} | V \cdot \tilde{\boldsymbol{\nu}}_k^V \xrightarrow{V \to \infty} \tilde{\boldsymbol{\nu}}_k \text{ for some } \tilde{\boldsymbol{\nu}}_k \neq 0 \}, \tag{2.31}$$

and we assume the equality

$$\mathbb{K}_1 = \mathbb{K} \setminus \mathbb{K}_0,$$

i.e., all reactions of the network belong to either $\mathbb{K}_0$ or $\mathbb{K}_1$. Parallel to Assumption 2.1, we formulate the following condition for the propensity functions.

Assumption 2.6 (Adapted Scaling of Propensity Functions). For the volume-dependent propensity functions α_k^V there exist suitable (locally Lipschitz continuous) limits $\tilde{\alpha}_k$ with

$$\tilde{\alpha}_k^V \xrightarrow{V \to \infty} \tilde{\alpha}_k$$

uniformly on compact subsets of $\mathbb{R}_+^L$ for $k \in \mathbb{K}_0$ and

$$\frac{1}{V} \tilde{\alpha}_k^V \xrightarrow{V \to \infty} \tilde{\alpha}_k$$

uniformly on compact subsets of $\mathbb{R}_+^L$ for $k \in \mathbb{K}_1$.

Given this separation, we can split the sum in (2.29) into two sums and write

$$\boldsymbol{\Theta}^V(t) = \boldsymbol{\Theta}^V(0) + \sum_{k \in \mathbb{K}_0} \mathcal{U}_k \left(\int_0^t \tilde{\alpha}_k^V \left(\boldsymbol{\Theta}^V(s) \right) ds \right) \tilde{\boldsymbol{\nu}}_k^V$$

$$+ \sum_{k \in \mathbb{K}_1} \frac{1}{V} \mathcal{U}_k \left(V \int_0^t \frac{1}{V} \tilde{\alpha}_k^V \left(\boldsymbol{\Theta}^V(s) \right) ds \right) V \tilde{\boldsymbol{\nu}}_k^V. \tag{2.32}$$

In comparison to (2.6), where the overall scaling of the reaction jump process led to the factor V^{-1} in front of the Poisson processes $\mathcal{U}_k$, this factor is here contained in the partially rescaled state-change vectors $\tilde{\boldsymbol{\nu}}_k$. Applying the convergence results of Sect. 2.1 to the second sum (2.32) and using the assumptions above indicate that, as $V \to \infty$, the process $\boldsymbol{\Theta}^V$ converges to a process $\boldsymbol{\Theta} = (\boldsymbol{\Theta}(t))_{t \geq 0}$ fulfilling the following combined equation.

Piecewise-deterministic reaction process (PDRP)

$$\boldsymbol{\Theta}(t) = \boldsymbol{\Theta}(0) + \sum_{k \in \mathbb{K}_0} \mathcal{U}_k \left(\int_0^t \tilde{\alpha}_k(\boldsymbol{\Theta}(s))\, ds \right) \tilde{\boldsymbol{\nu}}_k + \sum_{k \in \mathbb{K}_1} \int_0^t \tilde{\alpha}_k(\boldsymbol{\Theta}(s)) \tilde{\boldsymbol{\nu}}_k\, ds$$

(2.33)

Before specifying the convergence properties, we describe and illustrate the dynamics defined by this combined Eq. (2.33). The process $\boldsymbol{\Theta}$ exhibits a *piecewise deterministic* behavior: After starting at time $t = t_0 = 0$ in $\boldsymbol{\Theta}(0)$, it follows a deterministic motion given by

$$\frac{d\boldsymbol{\Theta}(t)}{dt} = \mu(\boldsymbol{\Theta}(t)), \tag{2.34}$$

with the vector field $\mu : \mathbb{R}^L \to \mathbb{R}^L$ given by

$$\mu(\boldsymbol{\theta}) := \sum_{k \in \mathbb{K}_1} \tilde{\alpha}_k(\boldsymbol{\theta}) \tilde{\boldsymbol{\nu}}_k, \tag{2.35}$$

until t reaches the first time t_1 of a jump induced by the Poisson processes $\mathcal{U}_k$, $k \in \mathbb{K}_0$. The waiting time $\tau_1 = t_1 - t_0$ for this first jump is distributed according to

$$\mathbb{P}(\tau_1 > t) = \exp\left(-\int_0^t \tilde{\lambda}(\boldsymbol{\Theta}(s))\, ds \right), \quad s \geq 0, \tag{2.36}$$

where

$$\tilde{\lambda}(\boldsymbol{\theta}) := \sum_{k \in \mathbb{K}_0} \tilde{\alpha}_k(\boldsymbol{\theta})$$

is the overall jump propensity. The jump itself is of the form

$$\boldsymbol{\Theta}(t_1) \to \boldsymbol{\Theta}(t_1) + \tilde{\boldsymbol{\nu}}_k$$

for some $k \in \mathbb{K}_0$ and refers to the occurrence of a discrete reaction. Given that the discrete reaction happens at time $t_1 > 0$, the conditional probability that this reaction will be $\mathcal{R}_k$ is given by

$$\frac{\tilde{\alpha}_k(\boldsymbol{\Theta}(t_1))}{\tilde{\lambda}(\boldsymbol{\Theta}(t_1))}$$

for every $k \in \mathbb{K}_0$. After the jump, the deterministic evolution restarts with a new initial state $\boldsymbol{\Theta}(t_1)$. Let the resulting sequence of jump times be denoted by $(t_j)_{j=1,2,\ldots}$, where $t_j < t_{j+1}$.

By construction it holds

$$\nu_{lk}^V = \tilde{\nu}_{lk} = 0 \quad \forall k \in \mathbb{K}_1, \; l = 1, \ldots, l',$$

such that the deterministic flow is constrained to the components Θ_l, $l = l' + 1, \ldots, L$, of high-abundant species. In fact, we can decompose Θ according to

$$\Theta(t) = (\boldsymbol{Y}(t), \boldsymbol{C}(t))$$

into a discrete component $\boldsymbol{Y}(t) := (\Theta_1(t), \ldots, \Theta_{l'}(t)) \in \mathbb{N}^{l'}$ and a continuous component $\boldsymbol{C}(t) := (\Theta_{l'+1}(t), \ldots, \Theta_L(t)) \in \mathbb{R}^{L-l'}$ and write $\mu(\boldsymbol{\theta}) = (0, \mu_{\boldsymbol{y}}(\boldsymbol{c}))$ for $\boldsymbol{\theta} = (\boldsymbol{y}, \boldsymbol{c}) \in \mathbb{N}^{l'} \times \mathbb{R}^{L-l'}$ and a suitable vector field $\mu_{\boldsymbol{y}} : \mathbb{R}^{L-l'} \to \mathbb{R}^{L-l'}$. In between two consecutive jumps, the discrete component $\boldsymbol{Y} = (\boldsymbol{Y}(t))_{t \geq 0}$ is constant, while the continuous flow of $\boldsymbol{C} = (\boldsymbol{C}(t))_{t \geq 0}$ is given by the ODE

$$\frac{d}{dt}\boldsymbol{C}(t) = \mu_{\boldsymbol{y}}(\boldsymbol{C}(t)), \quad t_j \leq t < t_{j+1}, \tag{2.37}$$

which depends on the actual state $\boldsymbol{Y}(t_j) = \boldsymbol{y}$ of the discrete component.

The stochastic jumps, on the other hand, can affect both the low-abundant and the high-abundant species. More precisely, we define $\mathbb{K}_c \subset \mathbb{K} = \{1, \ldots, K\}$ to be the subset of those indices belonging to reactions which do *not affect* the low copy number species, i.e.

$$\mathbb{K}_c := \{k \in \mathbb{K} \mid \nu_{lk} = 0 \; \forall l = 1, \ldots, l'\}, \tag{2.38}$$

and set $\mathbb{K}_d := \mathbb{K} \setminus \mathbb{K}_c$. The reactions $\mathcal{R}_k$, $k \in \mathbb{K}_c$ are referred to as the "continuous" reactions, while $\mathbb{K}_d$ is the index set of the "discrete" reactions. It then holds $\mathbb{K}_d \subset \mathbb{K}_0$, while reactions in $\mathbb{K}_c$ can belong to both classes $\mathbb{K}_0$ and $\mathbb{K}_1$ depending on whether the net change ν_{lk}^V of a continuous species $\mathcal{S}_l$, $l \in \{l'+1, \ldots, L\}$ scales with V or not. This means that the "continuous" process $\boldsymbol{C}$ actually does not need to have globally time-continuous trajectories, but its continuous flow can be interrupted by discrete jump events at the times t_j. The term "continuous" rather refers to the continuous state space of $\boldsymbol{C}$, as opposed to the discrete state space of the (piecewise constant) jump process $\boldsymbol{Y}$.

The following example has no biochemical background but is a purely artificial system that aims to illustrate the different types of reactions and their relevance for the multiscale reaction kinetics.

Example 2.7 (Artificial Network). We consider an artificial reaction network consisting of two species $\mathcal{S}_1$ and $\mathcal{S}_2$, where $\mathcal{S}_1$ appears in low copy numbers while the abundance of $\mathcal{S}_2$ is proportional to the volume $V > 0$. The process is given by $\boldsymbol{X}^V = (\boldsymbol{X}^V(t))_{t \geq 0}$ with $\boldsymbol{X}^V(t) = (X_1^V(t), X_2^V(t))^{\mathsf{T}} \in \mathbb{N}_0^2$. Here, $X_1^V(t)$ refers to the (volume-independent) number of $\mathcal{S}_1$-particles at time t and $X_2^V(t)$ refers to the (volume-dependent) number of $\mathcal{S}_2$-particles at time t. The population is affected by six reactions

$$
\begin{aligned}
\mathcal{R}_1 : &\quad \mathcal{S}_1 \xrightarrow{\gamma_1} \mathcal{S}_2, \\
\mathcal{R}_2 : &\quad \mathcal{S}_1 \xrightarrow{\gamma_2} \mathcal{S}_1 + \mathcal{S}_2, \\
\mathcal{R}_3 : &\quad \mathcal{S}_1 \xrightarrow{\gamma_3} n\mathcal{S}_2, \\
\mathcal{R}_4 : &\quad \mathcal{S}_1 \xrightarrow{\gamma_4} \mathcal{S}_1 + n\mathcal{S}_2, \\
\mathcal{R}_5 : &\quad \mathcal{S}_2 \xrightarrow{\gamma_5} \emptyset, \\
\mathcal{R}_6 : &\quad \emptyset \xrightarrow{\gamma_6} \mathcal{S}_1,
\end{aligned}
\tag{2.39}
$$

where γ_k $(k = 1, \ldots, 6)$ are the rate constants and $n \in \mathbb{N}$ scales with the volume, i.e., it holds $\lim_{V \to \infty} n/V = c_0$ for some $c_0 > 0$. Each of the reactions $\mathcal{R}_1, \ldots, \mathcal{R}_4$ induces an increase in the $\mathcal{S}_2$-population, but with different effects on the state of the system: While $\mathcal{R}_1$ and $\mathcal{R}_3$ eliminate the $\mathcal{S}_1$-particle in order to transform it into one or several $\mathcal{S}_2$-particles, reactions $\mathcal{R}_2$ and $\mathcal{R}_4$ do not affect the $\mathcal{S}_1$-population. For the reactions $\mathcal{R}_3$ and $\mathcal{R}_4$, the increase in the number of $\mathcal{S}_2$-particles induced by a single reaction is large (namely, of order V), while it is small for $\mathcal{R}_1$ and $\mathcal{R}_2$. This means that a high abundance of $\mathcal{S}_2$-particles can result from rarely/occasionally appearing reactions $\mathcal{R}_3$ and $\mathcal{R}_4$ or from frequently appearing reactions $\mathcal{R}_1$ and $\mathcal{R}_2$ (frequency of order V). A scaling of the $\mathcal{R}_1$-propensity with V, however, would lead to a disproportionately fast extinction of the low-abundant $\mathcal{S}_1$-particles and is thereby inappropriate. Thus, the following scaling of the propensity functions is reasonable

$$
\alpha_1^V(\boldsymbol{x}) = \gamma_1 x_1, \quad \alpha_2^V(\boldsymbol{x}) = V \gamma_2 x_1, \quad \alpha_3^V(\boldsymbol{x}) = \gamma_3 x_1,
$$

$$
\alpha_4^V(\boldsymbol{x}) = \gamma_4 x_1, \quad \alpha_5^V(\boldsymbol{x}) = \gamma_5 x_2, \quad \alpha_6^V(\boldsymbol{x}) = \gamma_6,
$$

where $\boldsymbol{x} = (x_1, x_2)^{\mathsf{T}} \in \mathbb{N}_0^2$ denotes the state of the system. The state-change vectors are given by

$$
\boldsymbol{\nu}_1^V = \begin{pmatrix} -1 \\ 1 \end{pmatrix}, \quad \boldsymbol{\nu}_2^V = \begin{pmatrix} 0 \\ 1 \end{pmatrix}, \quad \boldsymbol{\nu}_3^V = \begin{pmatrix} -1 \\ n \end{pmatrix},
$$

$$\boldsymbol{\nu}_4^V = \begin{pmatrix} 0 \\ n \end{pmatrix}, \quad \boldsymbol{\nu}_5^V = \begin{pmatrix} 0 \\ -1 \end{pmatrix}, \quad \boldsymbol{\nu}_6^V = \begin{pmatrix} 1 \\ 0 \end{pmatrix},$$

such that $\mathbb{K}_c = \{2, 4, 5\}$ and $\mathbb{K}_d = \{1, 3, 6\}$. We consider the partially scaled process $\boldsymbol{\Theta}^V = \left(\Theta_1^V, \Theta_2^V \right)$ defined as

$$\Theta_1^V (t) := X_1^V (t), \quad \Theta_2^V (t) := \frac{1}{V} X_2^V (t)$$

(see (2.28)). The partially scaled vectors $\tilde{\boldsymbol{\nu}}_k^V$ are given by

$$\tilde{\boldsymbol{\nu}}_1^V = \begin{pmatrix} -1 \\ 1/V \end{pmatrix}, \quad \tilde{\boldsymbol{\nu}}_2^V = \begin{pmatrix} 0 \\ 1/V \end{pmatrix}, \quad \tilde{\boldsymbol{\nu}}_3^V = \begin{pmatrix} -1 \\ n/V \end{pmatrix},$$

$$\tilde{\boldsymbol{\nu}}_4^V = \begin{pmatrix} 0 \\ n/V \end{pmatrix}, \quad \tilde{\boldsymbol{\nu}}_5^V = \begin{pmatrix} 0 \\ -1/V \end{pmatrix}, \quad \tilde{\boldsymbol{\nu}}_6^V = \begin{pmatrix} 1 \\ 0 \end{pmatrix},$$

and it holds

$$\tilde{\boldsymbol{\nu}}_1^V \xrightarrow{V \to \infty} \begin{pmatrix} -1 \\ 0 \end{pmatrix} =: \tilde{\boldsymbol{\nu}}_1, \quad \tilde{\boldsymbol{\nu}}_3^V \xrightarrow{V \to \infty} \begin{pmatrix} -1 \\ c_0 \end{pmatrix} =: \tilde{\boldsymbol{\nu}}_3,$$

$$\tilde{\boldsymbol{\nu}}_4^V \xrightarrow{V \to \infty} \begin{pmatrix} 0 \\ c_0 \end{pmatrix} =: \tilde{\boldsymbol{\nu}}_4, \quad \tilde{\boldsymbol{\nu}}_6^V \xrightarrow{V \to \infty} \begin{pmatrix} 1 \\ 0 \end{pmatrix} =: \tilde{\boldsymbol{\nu}}_4,$$

while

$$V \cdot \tilde{\boldsymbol{\nu}}_2^V \xrightarrow{V \to \infty} \begin{pmatrix} 0 \\ 1 \end{pmatrix} =: \tilde{\boldsymbol{\nu}}_2, \quad V \cdot \tilde{\boldsymbol{\nu}}_5^V \xrightarrow{V \to \infty} \begin{pmatrix} 0 \\ -1 \end{pmatrix} =: \tilde{\boldsymbol{\nu}}_5,$$

which gives $\mathbb{K}_0 = \{1, 3, 4, 6\}$ and $\mathbb{K}_1 = \{2, 5\}$. The adapted reaction propensities in terms of values $\boldsymbol{\theta} = (\theta_1, \theta_2)$ of the process $\boldsymbol{\Theta}^V$ are

$$\tilde{\alpha}_1^V (\boldsymbol{\theta}) = \gamma_1 \theta_1, \quad \tilde{\alpha}_2^V (\boldsymbol{\theta}) = V \gamma_2 \theta_1, \quad \tilde{\alpha}_3^V (\boldsymbol{\theta}) = \gamma_3 \theta_1,$$

$$\tilde{\alpha}_4^V (\boldsymbol{\theta}) = \gamma_4 \theta_1, \quad \tilde{\alpha}_5^V (\boldsymbol{\theta}) = V \gamma_5 \theta_2, \quad \tilde{\alpha}_6^V (\boldsymbol{\theta}) = \gamma_6$$

(see (2.30)). Taking the limit $V \to \infty$, we observe that Assumption 2.6 is fulfilled with

$$\tilde{\alpha}_1(\boldsymbol{\theta}) = \gamma_1 \theta_1, \quad \tilde{\alpha}_2(\boldsymbol{\theta}) = \gamma_2 \theta_1, \quad \tilde{\alpha}_3(\boldsymbol{\theta}) = \gamma_3 \theta_1,$$

$$\tilde{\alpha}_4(\boldsymbol{\theta}) = \gamma_4 \theta_1, \quad \tilde{\alpha}_5(\boldsymbol{\theta}) = \gamma_5 \theta_2, \quad \tilde{\alpha}_6(\boldsymbol{\theta}) = \gamma_6.$$

We obtain $\lim_{V\to\infty} \Theta^V \to \Theta$ in distribution (see paragraph "Convergence" on page 65 and Sect. A.4.2) where $\Theta = (\Theta(t))_{t\geq0}$ follows the combined equation (2.33) with the above-stated sets $\mathbb{K}_0$ and $\mathbb{K}_1$. That is, in the large volume limit, all four reactions $\mathcal{R}_1, \mathcal{R}_3, \mathcal{R}_4, \mathcal{R}_6$ cause jumps in the population, while $\mathcal{R}_2$ and $\mathcal{R}_5$ induce a continuous flow in the concentration $(C(t))_{t\geq0} = (\Theta_2(t))_{t\geq0}$ of $\mathcal{S}_2$-particles between the jumps. Reaction $\mathcal{R}_1$ (just as reaction $\mathcal{R}_6$) induces a jump in the discrete variable $(Y(t))_{t\geq0} = (\Theta_1(t))_{t\geq0}$ of $\mathcal{S}_1$-abundance, though it has no impact on the continuous variable $(C(t))_{t\geq0}$ in the large volume limit. In contrast, each occurrence of $\mathcal{R}_4$ instantaneously changes the concentration $C(t)$ by a significant amount without affecting the $\mathcal{S}_1$-population. Finally, reaction $\mathcal{R}_3$ induces jumps in both components of the process. Figure 2.6 shows trajectories of the resulting jump process and its approximation by the combined system (2.33). For the purpose of comparability, we plot the abundance $(V \cdot C(t))_{t\geq0}$ instead of the concentration $(C(t))_{t\geq0}$.

Note that in this example there is no feedback of the high-abundant species $\mathcal{S}_2$ onto the low-abundant $\mathcal{S}_1$. In Sect. 2.3, a multiscale reaction network with feedback will be investigated. $\diamond$

Remark 2.8. Reactions like $\mathcal{R}_3$ and $\mathcal{R}_4$ in Example 2.7 are often considered as "unphysical" because many particles are produced simultaneously by one firing of the reaction. In reality, such reactions actually consist of several reactions, each of which changes the number of particles only by a small amount. For example, reaction $\mathcal{R}_3$ will decompose into many split-up reactions of the $\mathcal{S}_1$-particle until it is decomposed into $\mathcal{S}_2$-particles, and $\mathcal{R}_4$ can be seen as a sequence of n reactions of type $\mathcal{R}_2$.

Such a piecewise-deterministic reaction process (PDRP) given by (2.33) belongs to the class of *piecewise-deterministic Markov processes (PDMP)* which can be defined in a more general setting (see Sect. A.4.1 in the Appendix).

Convergence

It has been shown in [40] that in the partial thermodynamic limit (i.e., when the number of high-abundant species tends to infinity), the partially scaled reaction jump process (2.28) converges in distribution to the PDRP given by (2.33) (see Theorem A.16 for details). In [85], even path-wise convergence (with respect to the Skorokhod topology, see Sect. A.3.2) of the partially scaled processes is proven, which furthermore leads to point-wise convergence, i.e., under appropriate conditions it holds that $\Theta^V(t) \xrightarrow{V\to\infty}$

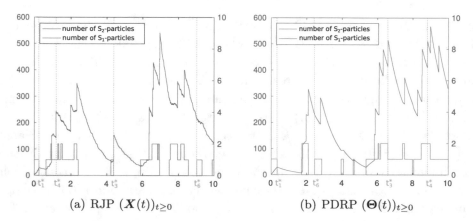

(a) RJP $(\boldsymbol{X}(t))_{t\geq 0}$ (b) PDRP $(\boldsymbol{\Theta}(t))_{t\geq 0}$

Figure 2.6. Multiscale reaction kinetics. Independent simulations of the artificial network (2.39) of Example 2.7 for rate constants $\gamma_k = 1$, $k = 1, \dots, 6$. (**a**) Gillespie simulation of the reaction jump process $(\boldsymbol{X}(t))_{t\geq 0}$ for $V = 100$. (**b**) Stochastic simulation (see page 68) of the PDRP $(\boldsymbol{\Theta}(t))_{t\geq 0}$ given by the combined Eq. (2.33), with the blue curve referring to $(V\Theta_2(t))_{t\geq 0}$. Initial state $X_1^V(0) = \Theta_1(0) = 1$ (number of $\mathcal{S}_1$-particles), $X_2^V(0) = \Theta_2(0) = 0$ (number/concentration of $\mathcal{S}_2$-particles). The marked times t_k^* identify some exemplary jumps in the population ($\mathcal{S}_1$-population and/or $\mathcal{S}_2$-population) induced by the reactions $\mathcal{R}_k$, $k \in \{1, 3, 4, 6\}$. In the partial large volume limit of the PDRP (see (**b**)), both $\mathcal{R}_3$ and $\mathcal{R}_4$ still cause jumps in the $\mathcal{S}_2$-abundance, while $\mathcal{R}_1$ and $\mathcal{R}_6$ lose their instantaneous impact onto the $\mathcal{S}_2$-population and only affect the trend in its deterministic evolution by modifying the number of $\mathcal{S}_1$-particles. This is due to the different scaling of the state-change vectors $\boldsymbol{\nu}_k$. Although the realizations (**a**) and (**b**) are different from each other, they still own the same characteristics

$\Theta(t)$ almost surely for each $t \geq 0$, i.e.

$$\mathbb{P}\left(\boldsymbol{\Theta}^V(t) \xrightarrow{V\to\infty} \boldsymbol{\Theta}(t)\right) = 1 \quad \forall t \geq 0.$$

An error bound and the rate of convergence have been derived in [140] in terms of marginal distributions of the discrete species and conditional expectations of the continuous species.

Numerical Simulation of Piecewise-Deterministic Reaction Processes

The jumps in the hybrid model of a piecewise-deterministic reaction process cannot be simulated by the standard SSA (Sect. 1.3.2) because the propensities of the discrete reactions continuously change in time. Simulation methods for reaction networks with propensities that depend *explicitly* on time have been derived in [6]. The propensity functions are integrated over time until a target value is reached and the corresponding reaction occurs. In our setting, however, the propensities are not explicit functions of time, but they depend on the temporal evolution of the continuous component $(C(t))_{t \geq 0}$ of the process, which is implicitly defined by the ODE (2.37). In [3] such hybrid processes are simulated by simultaneously integrating two ODEs, one for the deterministic flow of the continuous component and another for the propensities. This approach will be described here. The derivation of the algorithm can be summarized as follows (see also [180]).

Given that $\Theta(t) = \theta$ at some time t, the probability for the next jump to occur within the time interval $[t, t + \tau)$ is given by

$$1 - \exp\left(- \int_t^{t+\tau} \tilde{\lambda}(\Theta(s)) \, ds \right)$$

where $\tilde{\lambda}(\theta) := \sum_{k \in \mathbb{K}_0} \tilde{\alpha}_k(\theta)$ (see (2.36)). That is, the waiting time for the next discrete reaction follows an exponential distribution with time-dependent rates. Just as in Sect. 1.3.2, where the standard SSA was derived, we can apply the standard inversion generating method [101] to generate a random waiting time from this distribution: Draw a random number r_1 from $U(0, 1)$ and solve

$$\ln\left(\frac{1}{r_1} \right) = \int_t^{t+\tau} \tilde{\lambda}(\Theta(s)) \, ds$$

in order to obtain τ. For the corresponding algorithmic implementation, define the integrated rate function

$$g(\tau | t) := \int_t^{t+\tau} \tilde{\lambda}(\Theta(s)) \, ds$$

for $\tau \geq 0$, such that

$$\frac{d}{d\tau} g(\tau | t) = \tilde{\lambda}(\Theta(t + \tau)),$$

and let $\mathbb{1}_{\mathbb{K}_0}$ be the indicator function of the set $\mathbb{K}_0$, i.e.

$$\mathbb{1}_{\mathbb{K}_0}(k) = \begin{cases} 1 & \text{if } k \in \mathbb{K}_0, \\ 0 & \text{otherwise.} \end{cases}$$

The resulting numerical workflow is the following.

Stochastic simulation for PDRPs:

1. Initialize time $t \leftarrow t_0$ and state $\mathbf{\Theta}(t_0) \leftarrow \mathbf{\Theta}_0$ and choose a time horizon $T > t_0$.

2. Generate a random number r_1 from $U(0,1)$ and set $g(0|t) = 0$.

3. Solve the coupled ODE system starting at $\tau = 0$

$$\frac{d}{d\tau}\mathbf{\Theta}(t + \tau) = \sum_{k \in \mathbb{K}_1} \tilde{\alpha}_k(\mathbf{\Theta}(t + \tau))\tilde{\boldsymbol{\nu}}_k, \quad \frac{d}{d\tau}g(\tau|t) = \tilde{\lambda}(\mathbf{\Theta}(t + \tau))$$

   until the first time $\tau = \tau^*$ where $g(\tau^*|t) = \ln\left(\frac{1}{r_1}\right)$.

4. Generate another random number r_2 from $U(0,1)$ and choose k^* to be the smallest integer satisfying

$$\sum_{k=1}^{k^*} \mathbb{1}_{\mathbb{K}_0}(k)\tilde{\alpha}_k(\mathbf{\Theta}(t + \tau^*)) > r_2 \sum_{k=1}^{K} \mathbb{1}_{\mathbb{K}_0}(k)\tilde{\alpha}_k(\mathbf{\Theta}(t + \tau^*)).$$

5. Execute the next discrete reaction by replacing

$$\mathbf{\Theta}(t + \tau^*) \leftarrow \mathbf{\Theta}(t + \tau^*) + \tilde{\boldsymbol{\nu}}_{k^*}$$

   and update the time $t \leftarrow t + \tau^*$.

6. If $t \geq T$, end the simulation. Otherwise return to 2.

Regarding Step 3 in the simulation scheme for piecewise-deterministic reaction processes, we have to note that the coupled ODE system will usually be too complex to be solved analytically, such that numerical integration methods are required. One approach is to use temporal discretization with an adaptive time stepping scheme, e.g., based on linear-implicit Runge-Kutta schemes [49], including adaptive accuracy control. Using these schemes, the typical time step $\Delta t > 0$ will be significantly larger than the expected random jump time for the continuous reactions $\mathcal{R}_k$, $k \in \mathbb{K}_1$, thus making the simulation much more efficient than a Gillespie simulation of the entire reaction network (including all reactions $\mathcal{R}_k$, $k \in \mathbb{K}$).

Applying, for example, the explicit Euler scheme [26] leads to the following implementation of Step 3: Given $\mathbf{\Theta}(t)$ and $g(0|t) = 0$, repeat for

$n = 0, \ldots, n^*$

$$\Theta(t + (n+1)\Delta t) = \Theta(t + n\Delta t) + \Delta t \sum_{k \in \mathbb{K}_1} \tilde{\alpha}_k(\Theta(t + n\Delta t))\tilde{\nu}_k,$$

$$g((n+1)\Delta t | t) = g(n\Delta t | t) + \Delta t \tilde{\lambda}(\Theta(t + n\Delta t)), \qquad (2.40)$$

where n^* refers to the smallest integer with $g((n^* + 1)\Delta t | t) \geq \ln\left(\frac{1}{r_1}\right)$. Calculate $\delta t > 0$ such that $g(n^*\Delta t + \delta t | t) = \ln\left(\frac{1}{r_1}\right)$ by setting

$$\delta t = \frac{\ln\left(\frac{1}{r_1}\right) - g(n^*\Delta t | t)}{\tilde{\lambda}(\Theta(t + n^*\Delta t))}.$$

Set $\tau^* = n^*\Delta t + \delta t$ and

$$\Theta(t + \tau^*) = \Theta(t + n^*\Delta t) + \delta t \sum_{k \in \mathbb{K}_1} \tilde{\alpha}_k(\Theta(t + n^*\Delta t))\tilde{\nu}_k. \qquad (2.41)$$

Then continue with Step 4.

Remark 2.9. Due to the stiffness of the given coupled ODE system, the standard Euler method (being an explicit scheme for numerical integration) will require too small step sizes because of the well-known step size restrictions for explicit integration schemes. Therefore, implicit Runge-Kutta methods of higher order [49] will be more suitable and much more efficient for simulations of PDRPs compared to the scheme presented above.

Apart from the numerical integration error, which arises from Step 3 in the presented stochastic simulation algorithm, this type of numerical scheme for PDRPs is exact. However, it can become computationally very expensive if the calculation of the integrated propensities is difficult. An alternative is rejection-based algorithms as proposed in [225, 239]. One of them is the so-called <u>Ex</u>trande method (which stands for <u>ex</u>tra <u>r</u>eaction <u>a</u>lgorithm for <u>n</u>etworks in <u>d</u>ynamic <u>e</u>nvironments) which introduces an extra "virtual" reaction R_{K+1} into the system. This virtual reaction does not affect the system's state of molecule numbers, and its firing represents the rejection. The propensity of the virtual reaction changes with time in a way to keep the total propensity (here the sum $\tilde{\lambda}$) constant. The resulting augmented system can then be simulated by the standard SSA (cf. [239] for details and [60] for practical use in applications).

2.2.2   Piecewise Chemical Langevin Equation

In between two consecutive jump times, the dynamics of the piecewise-deterministic reaction process $\boldsymbol{\Theta}$ are given by the RRE

$$\frac{d\boldsymbol{\Theta}(t)}{dt} = \sum_{k \in \mathbb{K}_1} \tilde{\alpha}_k(\boldsymbol{\Theta}(t))\tilde{\boldsymbol{\nu}}_k$$

(cf. (2.34)). In order to enhance the approximation quality of the hybrid process and to keep some stochasticity also in the evolution of the high-abundant species, a nearby approach is to replace this deterministic flow by a randomized flow in form of a chemical Langevin equation (2.16).

As before, we assume a two-scale separation of the species' abundance and consider the partially scaled process $(\boldsymbol{\Theta}^V(t))_{t\geq 0}$ defined in (2.28). The reactions are again separated according to their type of convergence for $V \to \infty$ (see Definition (2.31)) into discrete reactions $\{\mathcal{R}_k | k \in \mathbb{K}_0\}$ and continuous reactions $\{\mathcal{R}_k | k \in \mathbb{K}_1\}$, where $\mathbb{K} = \mathbb{K}_0 \cup \mathbb{K}_1$. As a direct analog to the time-change representation (2.33) of the related piecewise-deterministic reaction process, we define the *piecewise chemical Langevin equation* (PCLE) (2.42) as follows.

Piecewise chemical Langevin equation (PCLE)

$$\boldsymbol{\Theta}(t) = \boldsymbol{\Theta}(0) + \sum_{k \in \mathbb{K}_0} \mathcal{U}_k \left(\int_0^t \tilde{\alpha}_k(\boldsymbol{\Theta}(s))\,ds \right) \tilde{\boldsymbol{\nu}}_k$$
$$+ \sum_{k \in \mathbb{K}_1} \int_0^t \tilde{\alpha}_k(\boldsymbol{\Theta}(s))\tilde{\boldsymbol{\nu}}_k\,ds + \sum_{k \in \mathbb{K}_1} \int_0^t \sqrt{\tilde{\alpha}_k(\boldsymbol{\Theta}(s))}\tilde{\boldsymbol{\nu}}_k\,dW_k(s)$$

(2.42)

The process defined by the PCLE (2.42) is a special type of *hybrid diffusion process* (see Appendix A.4.3 and [41]). In between two random jump times t_j and t_{j+1}, which are determined by the Poisson processes $\mathcal{U}_k$, the dynamics of $\boldsymbol{\Theta}$ now follow an Itô diffusion process given by the last two summands of (2.42). These two summands are just a variation of the CLE (2.14) in terms of concentrations. In analogy to the piecewise-deterministic setting, the randomized flow is restricted to the continuous component of the process. That is, for $t \in [t_j, t_{j+1})$ the discrete component $\boldsymbol{Y}$ of the process $\boldsymbol{\Theta} = (\boldsymbol{Y}(t), \boldsymbol{C}(t))_{t\geq 0}$ is constant, while $\boldsymbol{C} = (\boldsymbol{C}(t))_{t\geq 0}$ follows the SDE

$$d\boldsymbol{C}(t) = \mu_{\boldsymbol{y}}(\boldsymbol{C}(t))\,dt + \sigma_{\boldsymbol{y}}(\boldsymbol{C}(t))\,d\boldsymbol{W}(t), \quad t_j \leq t < t_{j+1},$$

for suitable drift and noise terms $\mu_y : \mathbb{R}^{L-l'} \to \mathbb{R}^{L-l'}$ and $\sigma_y : \mathbb{R}^{L-l'} \to \mathbb{R}^{L-l',m}$, respectively, and an m-dimensional Wiener process $(\boldsymbol{W}(t))_{t \geq 0}$, where we reordered the reactions if required such that $\mathbb{K}_1 = \{1, \ldots, m\}$ for some $m \leq K$. Just as in the ODE (2.37), the drift vector $\mu_y(\boldsymbol{c}) \in \mathbb{R}^{L-l'}$ is implicitly defined by $\mu(\boldsymbol{\theta}) = (0, \mu_y(\boldsymbol{c}))$ for $\boldsymbol{\theta} = (\boldsymbol{y}, \boldsymbol{c})$ with the vector field μ given in (2.35). Similarly, we can consider the diffusion matrix $\sigma(\boldsymbol{\theta}) \in \mathbb{R}^{L,m}$ given by

$$\sigma_{lk}(\boldsymbol{\theta}) := \sqrt{\tilde{\alpha}_k(\boldsymbol{\theta})}\tilde{\nu}_{lk}, \quad l = 1, \ldots, L, \ k = 1, \ldots, m, \qquad (2.43)$$

and set

$$\sigma(\boldsymbol{\theta}) = \begin{pmatrix} \mathbf{0} \\ \sigma_y(\boldsymbol{c}) \end{pmatrix}$$

to implicitly define the diffusion matrix $\sigma_y(\boldsymbol{c}) \in \mathbb{R}^{L-l',m}$, where $\mathbf{0} \in \mathbb{R}^{l',m}$ is a matrix with zero entries.

Numerical Simulation of Hybrid Diffusion Processes

The algorithm for PDRPs given on page 68 can directly be transferred to hybrid diffusion processes defined by the PCLE (2.42) by replacing the ODE $\frac{d}{d\tau}\Theta(t+\tau) = \sum_{k \in \mathbb{K}_1} \tilde{\alpha}(\Theta(t+\tau))\tilde{\nu}_k$ given in Step 3 by the SDE

$$d\Theta(t+\tau) = \sum_{k \in \mathbb{K}_1} \tilde{\alpha}_k(\Theta(t+\tau))\tilde{\nu}_k \, d\tau + \sum_{k \in \mathbb{K}_1} \sqrt{\tilde{\alpha}_k(\Theta(t+\tau))}\tilde{\nu}_k \, dW_k(\tau), \quad (2.44)$$

rendering the following algorithmic scheme

Stochastic simulation for the PCLE:

1. Initialize time $t \leftarrow t_0$ and state $\boldsymbol{\Theta}(t_0) \leftarrow \boldsymbol{\Theta}_0$ and choose a time horizon $T > t_0$.

2. Generate a random number r_1 from $U(0,1)$ and set $g(0|t) = 0$.

3. Solve the coupled SDE-ODE system starting at $\tau = 0$

$$d\boldsymbol{\Theta}(t+\tau) = \sum_{k \in \mathbb{K}_1} \tilde{\alpha}(\boldsymbol{\Theta}(t+\tau))\tilde{\boldsymbol{\nu}}_k \, d\tau + \sum_{k \in \mathbb{K}_1} \sqrt{\tilde{\alpha}_k(\boldsymbol{\Theta}(t+\tau))}\tilde{\boldsymbol{\nu}}_k \, dW_k(\tau),$$

$$\frac{d}{d\tau}g(\tau|t) = \tilde{\lambda}(\boldsymbol{\Theta}(t+\tau))$$

   until the first time $\tau = \tau^*$ where $g(\tau^*|t) = \ln\left(\frac{1}{r_1}\right)$.

4. Generate another random number r_2 from $U(0,1)$ and choose k^* to be the smallest integer satisfying

$$\sum_{k=1}^{k^*} \mathbb{1}_{\mathbb{K}_0}(k)\tilde{\alpha}_k(\boldsymbol{\Theta}(t+\tau^*)) > r_2 \sum_{k=1}^{K} \mathbb{1}_{\mathbb{K}_0}(k)\tilde{\alpha}_k(\boldsymbol{\Theta}(t+\tau^*)).$$

5. Execute the next discrete reaction by replacing

$$\boldsymbol{\Theta}(t+\tau^*) \leftarrow \boldsymbol{\Theta}(t+\tau^*) + \tilde{\boldsymbol{\nu}}_{k^*}$$

   and update the time $t \leftarrow t + \tau^*$.

6. If $t \geq T$, end the simulation. Otherwise return to 2.

Similar to the piecewise-deterministic setting of the previous section, the third step of the scheme requires some numerical approximation. Using the standard explicit Euler-Maruyama scheme for the SDE (2.44) gives

$$\boldsymbol{\Theta}(t+(n+1)\Delta t) = \boldsymbol{\Theta}(t+n\Delta t) + \Delta t \sum_{k \in \mathbb{K}_1} \tilde{\alpha}_k(\boldsymbol{\Theta}(t+n\Delta t))\tilde{\boldsymbol{\nu}}_k$$

$$+ \sqrt{\Delta t} \sum_{k \in \mathbb{K}_1} \sqrt{\tilde{\alpha}_k(\boldsymbol{\Theta}(t+n\Delta t))}\tilde{\boldsymbol{\nu}}_k \, \xi_k,$$

where $\xi_k \sim \mathcal{N}(0,1)$ for $k \in \mathbb{K}_1$ are mutually independent standard Gaussian random variables. Again, Δt is an appropriate time-step size and $n =$

$0, 1, 2, \ldots$ is increasing as long as $g(t + n\Delta t|t) < \ln\left(\frac{1}{r_1}\right)$, where $g(t + n\Delta t|t)$ is recursively calculated as in (2.40). Finally, one has to determine $\Theta(t + \tau^*)$ according to (2.41).

Such a combination of an Euler-Maruyama discretization and Gillespie's algorithm has also been used in [53] in the context of agent-based interaction dynamics. In [59] different variants of the hybrid simulation algorithm are presented. Just as in the setting of piecewise-deterministic dynamics, also rejection-based algorithms can be applied to alternatively simulate a PCLE without numerical integration of the propensity functions [225, 239].

Example 2.10. As a reduced version of the artificial Example 2.7, let the low-abundant species $\mathcal{S}_1$ and the high-abundant species $\mathcal{S}_2$ undergo the reactions

$$\mathcal{R}_1 : \quad \mathcal{S}_1 \xrightarrow{\gamma_1} \mathcal{S}_1 + \mathcal{S}_2, \qquad \mathcal{R}_2 : \quad \mathcal{S}_2 \xrightarrow{\gamma_2} \emptyset,$$
$$\mathcal{R}_3 : \quad \mathcal{S}_1 \xrightarrow{\gamma_3} \emptyset, \qquad \mathcal{R}_4 : \quad \emptyset \xrightarrow{\gamma_4} \mathcal{S}_1,$$

with the partially volume-scaled propensities $\alpha_1^V(\boldsymbol{x}) = V\gamma_1 x_1$, $\alpha_2^V(\boldsymbol{x}) = \gamma_2 x_2$, $\alpha_3^V(\boldsymbol{x}) = \gamma_3 x_1$, $\alpha_4^V(\boldsymbol{x}) = \gamma_4$. Figure 2.7 shows trajectories for the pure reaction jump process $\boldsymbol{X}^V = (X_1^V, X_2^V)$ and the two hybrid processes $\boldsymbol{\Theta} = (Y, C)$ given by the PDRP and the PCLE. For the purpose of comparability, the simulations are not fully independent, but the same trajectory of the process component $X_1^V = Y$ referring to the number of $\mathcal{S}_1$-particles is used for all three models. As this process component is actually independent of V, we here denote it by X_1. The trajectory $(X_1(t))_{t \geq 0}$ (the red curve in Fig. 2.7) is produced in advance (uniquely for all three approaches) by a Gillespie simulation of the reduced network consisting of the two reactions $\mathcal{R}_3, \mathcal{R}_4$ only—an approach which is feasible for the given reaction network because the jump dynamics of the $\mathcal{S}_1$-population do not depend on state of the $\mathcal{S}_2$-population.

Based on the $\mathcal{S}_1$-trajectory $(X_1(t))_{t \geq 0}$, the trajectories for the $\mathcal{S}_2$-population are generated independently for each of the three models, using only the two reactions $\mathcal{R}_1$ and $\mathcal{R}_2$. For the pure RJP, this means a Gillespie simulation of X_2^V involving only the two reactions $\mathcal{R}_1, \mathcal{R}_2$, with the propensity for $\mathcal{R}_1$ at time t given by $V\gamma_1 X_1(t)$. For the PDRP the continuous component is constructed by an Euler simulation of the ODE

$$dC(t) = V\gamma_1 X_1(t)dt - \gamma_2 C(t)dt, \tag{2.45}$$

and for the PCLE an Euler-Maruyama simulation of the SDE

$$dC(t) = V\gamma_1 X_1(t)dt - \gamma_2 C(t)dt + \sqrt{V\gamma_1 X_1(t) - \gamma_2 C(t)}dW(t) \tag{2.46}$$

is used, choosing $V = 50$ for each model. In Fig. 2.7 the rescaled trajectories $VC(t)$ are plotted. We can see that the PCLE process more closely resembles the trajectory of the RJP by producing noise also for the concentration of the high-abundant species S_2. In contrast, the PDRP gives rise to purely deterministic dynamics in between the random jump times of the trajectory $(X_1(t))_{t\geq0}$.   ◇

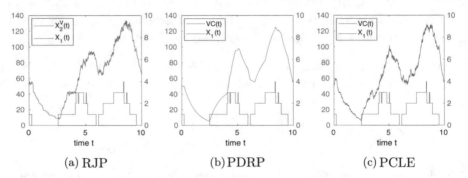

(a) RJP   (b) PDRP   (c) PCLE

Figure 2.7. Hybrid modeling approaches. Simulations of (**a**) the reaction jump process, (**b**) the piecewise-deterministic reaction process, and (c) the hybrid diffusion process defined by the PCLE for Example 2.10, using the same trajectory $(X_1^V(t))_{t\geq0} = (Y(t))_{t\geq0}$ for the number of S_1-particles in order to facilitate comparison. Red: number $X_1^V(t) = Y(t)$ of S_1-particles, blue: number of S_2-particles given by $X_2^V(t)$ or $VC(t)$ with C defined by (2.45) or (2.46), respectively $V = 50$

2.2.3   Three-Level Hybrid Process

The two hybrid modeling approaches considered hitherto (PDRP and PCLE) both assume a two-scale separation of the population and distinguish between low-abundant and high-abundant species. A nearby idea is to combine the approaches in order to get a joint model containing all three description levels: discrete jump processes for *low* copy number species, chemical Langevin dynamics containing stochastic fluctuation for *medium* copy number species, and deterministic reaction kinetics for *high* copy number species. The underlying idea is that the medium-abundant species possess a volume scaling different from the one of the high copy number species. For example, given that the high copy number species scale linear in V (as assumed in in the preceding sections), the medium copy number species could scale as $\sqrt{V}$. In this case, the adapted definition of the partially scaled process Θ^V is given by

$$\Theta_l^V(t) := \begin{cases} X_l^V(t) & \text{for } l = 1, \ldots, l', \\ \frac{1}{\sqrt{V}} X_l^V(t) & \text{for } l = l' + 1, \ldots, l'', \\ \frac{1}{V} X_l^V(t) & \text{for } l = l'' + 1, \ldots, L, \end{cases}$$

where $\mathcal{S}_1, \ldots, \mathcal{S}_{l'}$ are the low-abundant species, $\mathcal{S}_{l'+1}, \ldots, \mathcal{S}_{l''}$ are the medium-abundant species, and $\mathcal{S}_{l''+1}, \ldots, \mathcal{S}_L$ are the high-abundant species for some $1 < l' < l'' < L$. Likewise, the partially scaled state-change vectors read

$$\tilde{\nu}_{lk}^V := \begin{cases} \nu_{lk} & \text{for } l = 1, \ldots, l', \\ \frac{1}{\sqrt{V}} \nu_{lk} & \text{for } l = l' + 1, \ldots, l'', \\ \frac{1}{V} \nu_{lk} & \text{for } l = l'' + 1, \ldots, L. \end{cases}$$

Now we suppose that the state-change vectors $\tilde{\boldsymbol{\nu}}_k^V$ can be separated into *three* classes by the disjoint union

$$\mathbb{K} = \mathbb{K}_0 \cup \mathbb{K}_1 \cup \mathbb{K}_2,$$

where

$$\mathbb{K}_0 := \{ k \in \mathbb{K} | \tilde{\boldsymbol{\nu}}_k^V \xrightarrow{V \to \infty} \tilde{\boldsymbol{\nu}}_k \text{ for some } \tilde{\boldsymbol{\nu}}_k \neq 0 \},$$

$$\mathbb{K}_1 := \{ k \in \mathbb{K} | \sqrt{V} \cdot \tilde{\boldsymbol{\nu}}_k^V \xrightarrow{V \to \infty} \tilde{\boldsymbol{\nu}}_k \text{ for some } \tilde{\boldsymbol{\nu}}_k \neq 0 \}$$

and

$$\mathbb{K}_2 := \{ k \in \mathbb{K} | V \cdot \tilde{\boldsymbol{\nu}}_k^V \xrightarrow{V \to \infty} \tilde{\boldsymbol{\nu}}_k \text{ for some } \tilde{\boldsymbol{\nu}}_k \neq 0 \}.$$

In analogy to the two-scale setting (cf. Assumption 2.6), the accordingly rescaled propensity functions $\tilde{\alpha}_k^V$ are supposed to converge to locally Lipschitz continuous limit propensity functions $\tilde{\alpha}_k$.

This time, the sum in (2.29) is split into three components, and in the large volume limit we arrive at the joint Eq. (2.47) for the limit process $(\Theta(t))_{t \geq 0}$ consisting of a discrete and two continuous components. As in Sect. 2.2.2, $(W_k(t))_{t \geq 0}$ denotes independent Wiener processes in $\mathbb{R}$.

Three-level hybrid process

$$
\boldsymbol{\Theta}(t) = \boldsymbol{\Theta}(0) + \sum_{k \in \mathbb{K}_0} \mathcal{U}_k \left(\int_0^t \tilde{\alpha}_k(\boldsymbol{\Theta}(s)) \, ds \right) \tilde{\boldsymbol{\nu}}_k
$$

$$
+ \sum_{k \in \mathbb{K}_1} \int_0^t \tilde{\alpha}_k(\boldsymbol{\Theta}(s)) \tilde{\boldsymbol{\nu}}_k \, ds + \sum_{k \in \mathbb{K}_1} \int_0^t \sqrt{\tilde{\alpha}_k(\boldsymbol{\Theta}(s))} \tilde{\boldsymbol{\nu}}_k \, dW_k(s) \quad (2.47)
$$

$$
+ \sum_{k \in \mathbb{K}_2} \int_0^t \tilde{\alpha}_k(\boldsymbol{\Theta}(s)) \tilde{\boldsymbol{\nu}}_k \, ds
$$

Depending on the scaling of the state-change vectors, the dynamics given by (2.47) have different shape. Two fundamental cases are the following:

1. *Globally continuous dynamics of medium- and high-abundant species.* If the original state-change vectors $\boldsymbol{\nu}_k$ are all of order 1, then jumps in the limit process $(\boldsymbol{\Theta}(t))_{t \geq 0}$ only affect the low-abundant species and noise induced by the diffusion matrix function σ (cf. (2.43)) is restricted to the medium-abundant species.[7] The dynamics of the high copy number species are thus solely defined by the third line of (2.47), which is an ODE describing deterministic dynamics, but with coefficients that possibly depend not only on the discrete state of the low copy number species but also on the continuous stochastic flow of the medium copy number species. The components Θ_l, $l = l'+1, \ldots, L$, of the limit process $\boldsymbol{\Theta}$ are globally continuous.

2. *Jumps in medium- or high-abundant species.* Some entries ν_{lk} scale with V. More concisely, there are reactions $\mathcal{R}_k$ such that either ν_{lk} is of order $\sqrt{V}$ for some $l \in \{l' + 1, \ldots, L\}$ (i.e., for medium- or high-abundant species) or ν_{lk} is of order V for some $l \in \{l'' + 1, \ldots, L\}$ (i.e., for high-abundant species) or both.[8] In this case, jumps occur also in those components of $\boldsymbol{\Theta}$ that refer to medium- or high-abundant species, such that these components are only piecewise continuous.

The explicit form of the limit state-change vectors $\tilde{\boldsymbol{\nu}}_k$ and the corresponding propensity functions $\tilde{\alpha}_k$ depend on the concrete scaling of the different classes of species, which in turn depends on the application at hand.

---

[7] By definition it holds $\tilde{\nu}_{lk} = \lim_{V \to \infty} \sqrt{V} \frac{1}{V} \nu_{lk} = 0$ for $k \in \mathbb{K}_1$ and $l \in \{l' + 1, \ldots, L\}$.

[8] For a reaction $\mathcal{R}_k \in \mathbb{K}_c$, it is not possible that $\nu_{lk} \sim \sqrt{V}$ for some $l \in \{l'' + 1, \ldots, L\}$ while all other $\nu_{lk} \sim 1$ because this reaction would then belong to $\mathbb{K}_0$ *and* $\mathbb{K}_1$.

The joint Eq. (2.47) is quite general in the sense that its concrete shape and the resulting dynamics depend on the chosen scaling. In particular, it comprises the two hybrid models of PDRP and PCLE as special cases. The hybrid system defined by (2.47) combines the *discrete stochastic* dynamics of a Markov jump process with the *continuous stochastic* dynamics of a diffusion process and the *continuous deterministic* dynamics of an ODE process.

Remark 2.11 (Generalization). Besides this special scaling, we can more generally assume the existence of $d_2 > d_1 > 1$ such that the three classes of species are given by those with abundance of order V^{d_2}, those with abundance of order V^{d_1}, and those with abundance of order 1. The preceding definition of the partially scaled process can directly be transferred to this more general setting.

For stochastic simulations of the three-level hybrid process (2.47), the algorithms for PDRPs and PCLEs (see pages 68 and 72) can directly be combined with each other.

Example 2.12 (Three-Level Process). As an example for a reaction jump process with three population levels we consider the low-abundant species S_1 and S_2, the medium-abundant species S_3 and the high-abundant species S_4 undergoing the reactions

$$
\begin{array}{ll}
\mathcal{R}_1 : & S_1 \to S_2, \qquad\qquad \mathcal{R}_2 : \quad S_2 \to S_1, \\
\mathcal{R}_3 : & S_1 \to S_1 + S_3, \qquad \mathcal{R}_4 : \quad S_3 \to \emptyset, \\
\mathcal{R}_5 : & S_2 \to S_2 + S_4, \qquad \mathcal{R}_6 : \quad S_4 \to \emptyset,
\end{array}
$$

with the propensity α_3^V scaling with $\sqrt{V}$ ($\alpha_3^V(x) \sim \sqrt{V}$) and the propensity α_5^V scaling with V, while all others are independent of V. In this system, the total number $X_1(t) + X_2(t)$ of S_1- and S_2-particles is naturally kept constant for all times. The presence of S_1-particles induces a positive trend in the S_3-population, and the presence of S_2-particles induces a positive trend in the S_4-population (see Fig. 2.8). $\diamond$

2.2.4 Hybrid Master Equation

In the previous subsections, the hybrid dynamics for multiscale reactive systems have been derived and formulated in terms of the path-wise notation (2.5) of the reaction process. In Sect. 2.1.3 we have seen that model reductions for uniformly scaling reactive systems can also be derived based

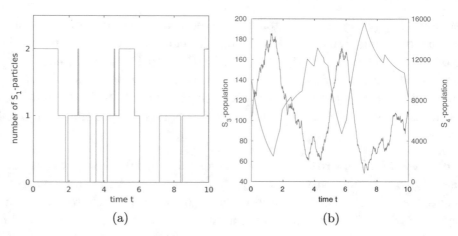

Figure 2.8. Three-level hybrid process. Hybrid simulation of the three-level process given by Eq. (2.47) for the reaction network of Example 2.12, keeping the discrete stochastic dynamics for the low-abundant S_1- and S_2-species, approximating the dynamics of the medium-abundant S_3-population by a diffusion process (CLE) and the dynamics of the high-abundant S_4-population by deterministic dynamics (RRE). Initial state: $\Theta_1(0) = 2$ (number of S_1-particles), $\Theta_2(0) = 0$ (number of S_2-particles), $\Theta_3(0) = 10^2$ (number of S_3-particles), and $\Theta_4(0) = 10^4$ (number of S_4-particles). The number of S_2-particles at time t is given by $\Theta_2(t) = 2-\Theta_1(t)$. (a) $(\Theta_1(t))_{t\geq 0}$. (b) $(\Theta_3(t))_{t\geq 0}$, $(\Theta_4(t))_{t\geq 0}$

on the related CME. In order to see how this works for systems with multiple population scales, let us leave the path-wise hybrid formulation behind and return to the evolution of probability distributions governed by the CME as our starting point. We aim to show how an approximative hybrid system can be derived from the CME directly.

The main idea when deriving reduced CMEs is to find a representation which couples the evolution of the distribution of the low-abundant species to the conditional expectations of the high-abundant species [125, 137, 181]. As in Sect. 2.2.1, let $S_1, \ldots, S_{l'}$ be the species arising in low copy numbers independent of V, while the other species $S_{l'+1}, \ldots, S_L$ have high abundance. Consider the related partition of the process $\boldsymbol{X}^V = (\boldsymbol{X}^V(t))_{t\geq 0}$ of the form

$$\boldsymbol{X}^V(t) = \begin{pmatrix} \boldsymbol{Y}^V(t) \\ \boldsymbol{Z}^V(t) \end{pmatrix}, \quad \boldsymbol{x} = \begin{pmatrix} \boldsymbol{y} \\ \boldsymbol{z} \end{pmatrix},$$

with $\boldsymbol{Y}^V(t) = \boldsymbol{y} \in \mathbb{N}_0^{l'}$ containing the number of entities of the low-abundant species and $\boldsymbol{Z}^V(t) = \boldsymbol{z} \in \mathbb{N}_0^{L-l'}$ containing the number of entities of the high-abundant species at time t.

As before, let $\mathbb{K}_c \subset \mathbb{K} = \{1, \ldots, K\}$ be the subset of those indices belonging to "continuous" reactions which do *not* affect the low copy number species, i.e.

$$\mathbb{K}_c := \{k \in \{1, \ldots, K\} \mid \nu_{lk} = 0 \; \forall l = 1, \ldots, l'\} \qquad (2.48)$$

as in (2.38). Correspondingly, let us again define by $\mathbb{K}_d := \mathbb{K} \setminus \mathbb{K}_c$ the index set of the "discrete" reactions.

In analogy to (1.7), let

$$p(\boldsymbol{x}, t) := \mathbb{P}(\boldsymbol{X}^V(t) = \boldsymbol{x} \mid \boldsymbol{X}^V(0) = \boldsymbol{x}_0)$$

be the probability to find the process $\boldsymbol{X}^V$ at time t in state $\boldsymbol{x} = (\boldsymbol{y}, \boldsymbol{z})^\mathsf{T}$ given some initial state $\boldsymbol{x}_0$.[9] First, consider the Bayesian decomposition

$$p(\boldsymbol{x}, t) = p(\boldsymbol{y}, \boldsymbol{z}, t) = p(\boldsymbol{z}, t \mid \boldsymbol{y}) p(\boldsymbol{y}, t),$$

where $p(\boldsymbol{y}, t) := \sum_{\boldsymbol{z}} p(\boldsymbol{y}, \boldsymbol{z}, t)$ is the marginal distribution and

$$p(\boldsymbol{z}, t \mid \boldsymbol{y}) := \mathbb{P}(\boldsymbol{Z}^V(t) = \boldsymbol{z} \mid \boldsymbol{Y}^V(t) = \boldsymbol{y}, \boldsymbol{X}^V(0) = \boldsymbol{x}_0)$$

is the conditional probability of finding the process $\boldsymbol{Z}^V$ in $\boldsymbol{z}$ at time t given that $\boldsymbol{Y}^V(t) = \boldsymbol{y}$. With $\boldsymbol{\nu}_k^{(y)} \in \mathbb{N}_0^{l'}$ and $\boldsymbol{\nu}_k^{(z)} \in \mathbb{N}_0^{L-l'}$ defined such that

$$\boldsymbol{\nu}_k = \begin{pmatrix} \boldsymbol{\nu}_k^{(y)} \\ \boldsymbol{\nu}_k^{(z)} \end{pmatrix},$$

the "full" CME (1.10) can be rewritten as

$$\begin{aligned}
\frac{d}{dt} p(\boldsymbol{y}, \boldsymbol{z}, t) = {} & \sum_{k=1}^{K} \alpha_k \left(\boldsymbol{y} - \boldsymbol{\nu}_k^{(y)}, \boldsymbol{z} - \boldsymbol{\nu}_k^{(z)} \right) p \left(\boldsymbol{y} - \boldsymbol{\nu}_k^{(y)}, \boldsymbol{z} - \boldsymbol{\nu}_k^{(z)}, t \right) \\
& - \sum_{k=1}^{K} \alpha_k \left(\boldsymbol{y}, \boldsymbol{z} \right) p(\boldsymbol{y}, \boldsymbol{z}, t).
\end{aligned} \qquad (2.49)$$

Summation with respect to $\boldsymbol{z}$ and incorporating (2.48) gives the reduced CME for the low-abundant species

$$\frac{d}{dt} p(\boldsymbol{y}, t) = \sum_{k \in \mathbb{K}_d} \left[\alpha_k^{(t)}(\boldsymbol{y} - \boldsymbol{\nu}_k^{(y)}) p \left(\boldsymbol{y} - \boldsymbol{\nu}_k^{(y)}, t \right) - \alpha_k^{(t)}(\boldsymbol{y}) p(\boldsymbol{y}, t) \right] \qquad (2.50)$$

---

[9]The probability $p(\boldsymbol{x}, t)$ actually depends on V and should be denoted by $p^V(\boldsymbol{x}, t)$. Anyway, we skip the upper index V for the purpose of readability, especially since we will introduce an alternative notation a few lines below (see (2.55).)

where

$$\alpha_k^{(t)}(\boldsymbol{y}) := \sum_{\boldsymbol{z}} \alpha_k(\boldsymbol{y}, \boldsymbol{z}) \, p(\boldsymbol{z}, t | \boldsymbol{y}). \qquad (2.51)$$

So far this does not involve any approximation, i.e., (2.50) is exact. However, in order to be able to evaluate $\alpha_k^{(t)}$ we need an approximation for the conditional distribution $p(\boldsymbol{z}, t | \boldsymbol{y})$. To this end, we first rewrite the full CME (2.49) in the following form

$$\frac{d}{dt} \Big[p(\boldsymbol{z}, t | \boldsymbol{y}) p(\boldsymbol{y}, t) \Big] \qquad (2.52)$$

$$= p(\boldsymbol{y}, t) \sum_{k \in \mathbb{K}_c} \Big[\alpha_k \Big(\boldsymbol{y}, \boldsymbol{z} - \boldsymbol{\nu}_k^{(z)} \Big) p \Big(\boldsymbol{z} - \boldsymbol{\nu}_k^{(z)}, t \Big| \boldsymbol{y} \Big)$$

$$- \alpha_k(\boldsymbol{y}, \boldsymbol{z}) \, p(\boldsymbol{z}, t | \boldsymbol{y}) \Big]$$

$$+ \sum_{k \in \mathbb{K}_d} \Big[\alpha_k \Big(\boldsymbol{y} - \boldsymbol{\nu}_k^{(y)}, \boldsymbol{z} - \boldsymbol{\nu}_k^{(z)} \Big) p \Big(\boldsymbol{z} - \boldsymbol{\nu}_k^{(z)}, t \Big| \boldsymbol{y} - \boldsymbol{\nu}_k^{(y)} \Big) p(\boldsymbol{y} - \boldsymbol{\nu}_k^{(y)}, t)$$

$$- \alpha_k(\boldsymbol{y}, \boldsymbol{z}) \, p(\boldsymbol{z}, t | \boldsymbol{y}) \, p(\boldsymbol{y}, t) \Big].$$

Next, we use the fact that the system scales differently for the low- and high-abundant species. That is, we switch to concentrations for the high-abundant species, setting

$$\boldsymbol{c} := \boldsymbol{z}/V.$$

Moreover, we introduce the smallness parameter $\varepsilon := 1/V$ that goes to zero for $V \to \infty$. In order to prepare the appropriate scaling of the propensities, we introduce the notation

$$\tilde{\alpha}_k^\varepsilon(\boldsymbol{y}, \boldsymbol{c}) := \alpha_k(\boldsymbol{y}, \boldsymbol{c}/\varepsilon)$$

and assume in analogy to Assumption 2.6 above—with $\mathbb{K}_d$ and $\mathbb{K}_c$ instead of $\mathbb{K}_0$ and $\mathbb{K}_1$—that

$$\tilde{\alpha}_k^\varepsilon(\boldsymbol{y}, \boldsymbol{c}) = \tilde{\alpha}_k(\boldsymbol{y}, \boldsymbol{c}), \qquad k \in \mathbb{K}_d, \qquad (2.53)$$

$$\varepsilon \tilde{\alpha}_k^\varepsilon(\boldsymbol{y}, \boldsymbol{c}) = \tilde{\alpha}_k(\boldsymbol{y}, \boldsymbol{c}), \qquad k \in \mathbb{K}_c, \qquad (2.54)$$

for suitable ε-independent propensities $\tilde{\alpha}_k$.[10] Furthermore, we define

$$p^\varepsilon(\boldsymbol{y}, \boldsymbol{c}, t) := p(\boldsymbol{y}, \boldsymbol{c}/\varepsilon, t) \qquad (2.55)$$

---

[10]In Assumption 2.6 we consider the limit of $\tilde{\alpha}_k^\varepsilon$ for $V \to \infty$, that is, $\varepsilon \to 0$. Assumptions (2.53) and (2.54) are just special cases of the same fundamental assumption.

and assume that it has a Bayesian decomposition given by

$$p^\varepsilon(\boldsymbol{y}, \boldsymbol{c}, t) = \rho^\varepsilon(\boldsymbol{c}, t|\boldsymbol{y})p^\varepsilon(\boldsymbol{y}, t),$$

where $p^\varepsilon(\boldsymbol{y}, t) := \int_{\mathbb{R}_+^{L-l'}} p^\varepsilon(\boldsymbol{y}, \boldsymbol{c}, t)\, d\boldsymbol{c}$. With this notation and the scaling of the propensities given in (2.53) and (2.54), the partitioned CME (2.52) takes the form

$$\frac{d}{dt}\left[\rho^\varepsilon(\boldsymbol{c}, t|\boldsymbol{y})p^\varepsilon(\boldsymbol{y}, t)\right] \tag{2.56}$$

$$= \frac{1}{\varepsilon}p^\varepsilon(\boldsymbol{y}, t) \sum_{k\in\mathbb{K}_c}\left[\tilde{\alpha}_k\left(\boldsymbol{y}, \boldsymbol{c} - \varepsilon\boldsymbol{\nu}_k^{(z)}\right)\rho^\varepsilon\left(\boldsymbol{c} - \varepsilon\boldsymbol{\nu}_k^{(z)}, t|\boldsymbol{y}\right)\right.$$

$$\left.-\tilde{\alpha}_k(\boldsymbol{y}, \boldsymbol{c})\rho^\varepsilon(\boldsymbol{c}, t|\boldsymbol{y})\right]$$

$$+ \sum_{k\in\mathbb{K}_d}\left[\tilde{\alpha}_k\left(\boldsymbol{y} - \boldsymbol{\nu}_k^{(y)}, \boldsymbol{c} - \varepsilon\boldsymbol{\nu}_k^{(z)}\right)\rho^\varepsilon\left(\boldsymbol{c} - \varepsilon\boldsymbol{\nu}_k^{(z)}, t|\boldsymbol{y} - \boldsymbol{\nu}_k^{(y)}\right)p^\varepsilon(\boldsymbol{y} - \boldsymbol{\nu}_k^{(y)}, t)\right.$$

$$\left.-\tilde{\alpha}_k(\boldsymbol{y}, \boldsymbol{c})\rho^\varepsilon(\boldsymbol{c}, t|\boldsymbol{y})\,p^\varepsilon(\boldsymbol{y}, t)\right].$$

Next, we proceed with a multiscale expansion of the distributions $\rho^\varepsilon(\boldsymbol{c}, t|\boldsymbol{y})$ and $p^\varepsilon(\boldsymbol{y}, t)$ in terms of the smallness parameter ε

$$\rho^\varepsilon(\boldsymbol{c}, t|\boldsymbol{y}) = \frac{1}{\sqrt{\varepsilon}}\exp\left(\frac{1}{\varepsilon}S^\varepsilon(\boldsymbol{c}, t|\boldsymbol{y})\right),$$

$$S^\varepsilon(\boldsymbol{c}, t|\boldsymbol{y}) = S_0(\boldsymbol{c}, t|\boldsymbol{y}) + \varepsilon S_1(\boldsymbol{c}, t|\boldsymbol{y}) + \mathcal{O}(\varepsilon^2),$$

$$p^\varepsilon(\boldsymbol{y}, t) = p^0(\boldsymbol{y}, t) + \varepsilon p^1(\boldsymbol{y}, t) + \mathcal{O}(\varepsilon^2),$$

where we assume that the conditional probability distribution in $\boldsymbol{c}$ for given $\boldsymbol{y}$ is ε-concentrated around the (unique) maximum of the (so-called) eikonal function S_0. This kind of asymptotic expansion is explained in more detail in [181] as well as in Appendix A.5.1, where we also outlined how to insert the expansion in the CME, utilize the scaling assumptions (2.54) and (2.53), and consider the terms of highest order to get an equation for S_0. Here, this procedure yields the following result: for every $\boldsymbol{y}$ and every t, the function $S_0(\cdot, t|\boldsymbol{y})$ has a unique maximum at the point $\Phi_{\boldsymbol{y}}^t \boldsymbol{c}_0$ which is transported in time along the trajectory of the associated reaction rate equation

$$\frac{d}{dt}\boldsymbol{C}(t) = \sum_{k\in\mathbb{K}_c}\boldsymbol{\nu}_k^{(z)}\tilde{\alpha}_k(\boldsymbol{y}, \boldsymbol{C}(t)). \tag{2.57}$$

Here, $\boldsymbol{c}_0$ denotes the unique location of the maximum of $S_0(\cdot, t = 0|\boldsymbol{y})$ and $\Phi_{\boldsymbol{y}}^t$ is the flow operator of (2.57).

This result also allows deriving the evolution equation for $p^0(\boldsymbol{y}, t)$: Integrating the ε-scaled CME (2.56) over $\boldsymbol{c}$, inserting the multiscale expansion for $p^\varepsilon(\boldsymbol{y}, t)$, and comparing the leading-order terms yields

$$\frac{d}{dt} p^0(\boldsymbol{y}, t)$$
$$= \sum_{k \in \mathbb{K}_d} \left[\tilde{\alpha}_k \left(\boldsymbol{y} - \boldsymbol{\nu}_k^{(y)}, \Phi^t_{\boldsymbol{y} - \boldsymbol{\nu}_k^{(y)}} \boldsymbol{c}_0 \right) p^0 \left(\boldsymbol{y} - \boldsymbol{\nu}_k^{(y)}, t \right) \right. \tag{2.58}$$
$$\left. - \tilde{\alpha}_k \left(\boldsymbol{y}, \Phi^t_{\boldsymbol{y}} \boldsymbol{c}_0 \right) p^0(\boldsymbol{y}, t) \right]$$

as a reduced CME for the low-abundant species, which approximates the reduced CME (2.50).

Further detailed analysis shows that the solution of (2.57) is only valid on the time scale $\mathcal{O}(\varepsilon)$, while the discrete-stochastic dynamics given by (2.58) are valid on the time scale $\mathcal{O}(1)$. This means that the solution of the hybrid system cannot be determined by direct integration of (2.57) and (2.58). In [181] it is shown that partial expectation values of observables $F = F(\boldsymbol{y}, \boldsymbol{c})$ given by

$$\mathbb{E}^\varepsilon_{\boldsymbol{y}}(F, t) := \int_{\mathbb{R}_+^{L - l'}} F(\boldsymbol{y}, \boldsymbol{c}) p^\varepsilon(\boldsymbol{c}, \boldsymbol{y}, t) \, d\boldsymbol{c}$$

can be computed up to an error ε on the $\mathcal{O}(1)$ scale by

$$\mathbb{E}^\varepsilon_{\boldsymbol{y}}(F, t) = F\left(\boldsymbol{y}, \Phi^t_{\boldsymbol{y}} \boldsymbol{c}_0 \right) p^0(\boldsymbol{y}, t) + \mathcal{O}(\varepsilon).$$

This can be used to derive an evolution equation for the continuous dynamics on the time scale $\mathcal{O}(1)$. To this end, we consider the leading-order approximation of $\mathbb{E}^\varepsilon_{\boldsymbol{y}}(F, t)$ for the observable $F(\boldsymbol{c}) = \boldsymbol{c}$

$$\mathbb{E}^0_{\boldsymbol{y}}(\boldsymbol{c}, t) := p^0(\boldsymbol{y}, t) \Phi^t_{\boldsymbol{y}} \boldsymbol{c}_0, \tag{2.59}$$

which can be seen to follow the evolution equation

$$\frac{d}{dt} \mathbb{E}^0_{\boldsymbol{y}}(\boldsymbol{c}, t) = p^0(\boldsymbol{y}, t) \sum_{k \in \mathbb{K}_c} \boldsymbol{\nu}_k^{(z)} \tilde{\alpha}_k(\boldsymbol{y}, \Phi^t_{\boldsymbol{y}} \boldsymbol{c}_0)$$
$$+ \sum_{k \in \mathbb{K}_d} \left[\alpha_k(\boldsymbol{y} - \boldsymbol{\nu}_k^{(y)}, \Phi^t_{\boldsymbol{y} - \boldsymbol{\nu}_k^{(y)}} \boldsymbol{c}_0) \mathbb{E}^0(\boldsymbol{c}, t | \boldsymbol{y} - \boldsymbol{\nu}_k^{(y)}) \right. \tag{2.60}$$
$$\left. - \alpha_k(\boldsymbol{y}, \Phi^t_{\boldsymbol{y}} \boldsymbol{c}_0) \mathbb{E}^0_{\boldsymbol{y}}(\boldsymbol{c}, t) \right],$$

that in fact is valid on time scales of order $\mathcal{O}(1)$. Given the solution of (2.60), one can use definition (2.59) to derive a solution $\Phi^t_{\boldsymbol{y}} \boldsymbol{c}_0$ that is valid for time scales of order $\mathcal{O}(1)$. The combined system of Eqs. (2.58) and (2.60) is called *hybrid master equation*. This hybrid master equation can be used to compute the evolution of expectation values of the full CME up to order ε

$$\mathbb{E}^\varepsilon(F,t) = \sum_{\boldsymbol{y}} \int F(\boldsymbol{y},\boldsymbol{c})p^\varepsilon(\boldsymbol{y},\boldsymbol{c},t)\, d\boldsymbol{c} = \sum_{\boldsymbol{y}} F(\boldsymbol{y},\Phi_{\boldsymbol{y}}^t \boldsymbol{c}_0)p^0(\boldsymbol{y},t) + \mathcal{O}(\varepsilon)$$

for observables F depending on both the discrete component $\boldsymbol{y}$ and the continuous component $\boldsymbol{c} = \varepsilon \boldsymbol{z}$.[11]

This completes our result. Under our scaling assumptions and in case the conditional distributions in the high-abundant species z initially concentrate on one point asymptotically, we have found the following: In leading order, i.e., up to errors of first order in ε, we can replace the full CME by the reduced CME (2.58) for the low-abundant species. The propensities of the reduced CME depend on the (deterministic) trajectories of the evolution Eq. (2.60), which have to be computed in parallel. As the size of the state space is drastically reduced for this hybrid approach, a standard numerical integration of the combined system of Eqs. (2.58) and (2.60) is possible even for complex systems. A numerical algorithm for solving the hybrid master equation is presented in [125].

Numerical Effort

At this stage, we have developed two different hybrid approaches: the approach based on the hybrid master equation (hME) and the one based on hybrid processes (hP) such as the piecewise-deterministic reaction process from Sect. 2.2.1. An interesting question to answer is: Is there a setting in which the hME approach is computationally less demanding than the hP approach? Assume that we are interested in the time period $[0, T]$ and that the reduced CME (2.58) has a finite state space of M discrete states (or can be reduced to M states by truncation). The system (2.60) consists of M coupled ordinary differential equations of dimension N, if we have N high-abundant species, so NM equations in total. Let us proceed under the assumption that the matrix formed out of all state-change vectors is sparse. Then, we can assume that the hME, i.e., the reduced CME (2.58) and the system (2.60) together, can be solved numerically using linear-implicit integrators with computational effort E_{hME} scaling as

$$E_{\mathrm{hME}} = \mathcal{O}(T(M + NM)) = \mathcal{O}(TNM)$$

---

[11]Specifically, it holds $\mathbb{E}^\varepsilon(\boldsymbol{C}(t)) = \sum_{\boldsymbol{y}} \mathbb{E}_{\boldsymbol{y}}^0(\boldsymbol{c},t) + \mathcal{O}(\varepsilon)$ for the first moment of the component $\boldsymbol{C}(t) = \varepsilon \boldsymbol{Z}(t)$ of the multiscale stochastic process.

and gives us $\mathcal{O}(\varepsilon)$ error for expectation values (see Sect. A.5.2). In comparison, the Monte Carlo approximation with n trajectories of the piecewise-deterministic reaction process (hP approach, with dimension $N + M$) will lead to $\mathcal{O}(1/\sqrt{n})$ error. In order to achieve comparable accuracy, we thus need $n = \mathcal{O}(\varepsilon^{-2})$ trajectories. Each of these trajectories will need more computational effort than one of the RRE trajectories, since its jump process part rather requires smaller time steps. Even if we assume that the required effort just scales optimally, that is, like $\mathcal{O}(T(N + M))$, we have a total effort E_{hP} of

$$E_{\mathrm{hP}} = \mathcal{O}(T(M + N)\varepsilon^{-2}).$$

Comparing the two terms demonstrates that, if we aim for similar accuracy regarding expectation values, the CME-based procedure should be more efficient if

$$\min\{M, N\} < C\varepsilon^{-2},$$

with an unknown constant C as a rule of thumb. That is, if the number of discrete states of the reduced CME and of the high-abundant species is not too large, we are to assume that the computation of expectation values is more efficient by means of the CME-based algorithm. Yet, the reader should be aware of our assumption that the stoichiometric matrix is sparse. Without this presumption the comparison would be drastically worse.

Remark 2.13. Similar approaches for model reduction in terms of the chemical Fokker-Planck equation can be found in [77, 173]. The fundamental assumption of these concepts is that the dimension can be reduced by presuming that most of the molecular species have a normally distributed population size with a small variance.

2.3   Application: Autorepressive Genetic System

A set of meaningful examples for multiscale reactive systems is given by gene regulatory networks [22]. The template for transcription, the DNA, is present only in one or two copies which can be active or inactive. Depending on the state of the DNA, the synthesized messenger RNA (mRNA) and protein molecules may reach medium or high population levels. There exist many network models for the interactions between the involved species and the regulation of gene activity (see [41] for a set of more or less complex gene network models). In [85, 250, 251] an *autocatalytic* network of gene expression is investigated as an example for a piecewise-deterministic reaction process. In such a network, the proteins regulate the transcription by a positive feedback law, i.e., the proteins activate the gene by binding to its binding sites. It has been shown that for such a system the proteins become extinct in finite time, and the point mass in zero is the only stationary distribution of the system.

However, many genes are controlled by *negative* self-regulating transcription factors, especially in prokaryotes, for example, *E. coli* organisms [219]. For such autorepressive genetic systems, gene expression typically occurs in bursts. Bursts can only appear when part of the system (at least the DNA) is treated as a discrete random variable. In the limit of describing the whole system by ODEs, the bursts naturally get lost because the activity state of DNA is averaged. Bursting of gene expression can arise at the transcriptional or the translational level [22], or the bursts are modelled as individual stochastic events by which a certain number of products are instantaneously introduced to the system [87]. We will here describe the transcriptional bursting where the gene switches between activity and inactivity, i.e., between its sole states 0 and 1.

2.3.1 The Biochemical Model

We consider the gene expression system depicted in Fig. 2.9. If the DNA is active (species S_1: active DNA), messenger RNA (species S_2: mRNA) is transcribed at rate γ_1 and translated into proteins P (species S_3) at rate γ_2. Each protein can interrupt this stepwise production and repress the gene by binding to it and causing its inactivity at rate γ_3 (species S_4: inactive DNA denoted by DNA_0). The gene is reactivated at a rate γ_4 when the protein becomes detached again. Both within time periods of gene activity and within time periods of gene inactivity, mRNA is degraded at rate γ_5 and proteins are degraded at rate γ_6. All individual reactions are displayed in the following reaction scheme (2.61) with DNA_0 denoting the repressed (inactive) gene

$$
\begin{aligned}
\mathcal{R}_1 &: \ \mathrm{DNA} \xrightarrow{\gamma_1} \mathrm{DNA} + \mathrm{mRNA} \\
\mathcal{R}_2 &: \ \mathrm{mRNA} \xrightarrow{\gamma_2} \mathrm{mRNA} + \mathrm{P} \\
\mathcal{R}_3 &: \ \mathrm{DNA} + \mathrm{P} \xrightarrow{\gamma_3} \mathrm{DNA}_0 \\
\mathcal{R}_4 &: \ \mathrm{DNA}_0 \xrightarrow{\gamma_4} \mathrm{DNA} + \mathrm{P} \\
\mathcal{R}_5 &: \ \mathrm{mRNA} \xrightarrow{\gamma_5} \emptyset \\
\mathcal{R}_6 &: \ \mathrm{P} \xrightarrow{\gamma_6} \emptyset
\end{aligned}
\tag{2.61}
$$

For the stochastic reaction jump process $(\boldsymbol{X}^V(t))_{t \geq 0}$ characterized by the CME, we denote the state of the gene expression system at time t by

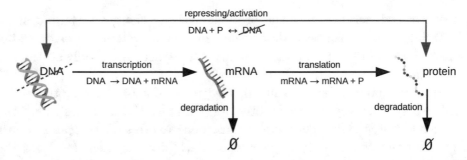

Figure 2.9. Gene expression with negative self-regulation. By transcription, the active gene (DNA) produces messenger RNA (mRNA) which is translated into proteins. Both mRNA and proteins degrade with time. The proteins deactivate the gene by binding to it. This repressing is reversible: The protein can unbind again, thereby reactivating gene

$\boldsymbol{X}^V(t) = (X_1^V(t), X_2^V(t), X_3^V(t)) \in \{0,1\} \times \mathbb{N}_0^2$ with $X_1^V(t) \in \{0,1\}$ referring to the number of active DNA and $X_2^V(t) \in \mathbb{N}_0$ and $X_3^V(t) \in \mathbb{N}_0$ giving the number of mRNA and proteins, respectively. The number $X_4(t)$ for inactive DNA need not be considered, as it is implicitly given by $X_4(t) = 1 - X_1(t)$ for all times t. The state-change vectors of the reactions listed in (2.61) are given by

$$
\boldsymbol{\nu}_1^V = \begin{pmatrix} 0 \\ 1 \\ 0 \end{pmatrix}, \quad \boldsymbol{\nu}_2^V = \begin{pmatrix} 0 \\ 0 \\ 1 \end{pmatrix}, \quad \boldsymbol{\nu}_3^V = \begin{pmatrix} -1 \\ 0 \\ -1 \end{pmatrix},
$$

$$
\boldsymbol{\nu}_4^V = \begin{pmatrix} 1 \\ 0 \\ 1 \end{pmatrix}, \quad \boldsymbol{\nu}_5^V = \begin{pmatrix} 0 \\ -1 \\ 0 \end{pmatrix}, \quad \boldsymbol{\nu}_6^V = \begin{pmatrix} 0 \\ 0 \\ -1 \end{pmatrix},
$$

$$\tag{2.62}$$

and the related propensity functions read

$$
\alpha_1^V(\boldsymbol{x}) = V\gamma_1 x_1, \quad \alpha_2^V(\boldsymbol{x}) = \gamma_2 x_2, \quad \alpha_3^V(\boldsymbol{x}) = \frac{\gamma_3}{V} x_1 x_3,
$$

$$
\alpha_4^V(\boldsymbol{x}) = \gamma_4(1 - x_1), \quad \alpha_5^V(\boldsymbol{x}) = \gamma_5 x_2, \quad \alpha_6^V(\boldsymbol{x}) = \gamma_6 x_3.
$$

$$\tag{2.63}$$

In this case, the population size of both mRNA and proteins scales with the volume V. In the hybrid models, DNA is therefore treated as the stochastic, discrete variable, and mRNA and proteins are treated as continuous variables. The state space of the PDRP is given by

$$
\mathbb{S} = \left\{ \begin{pmatrix} y \\ \boldsymbol{c} \end{pmatrix} : y \in \{0,1\}, \boldsymbol{c} \in \mathbb{R}^2 \right\}.
$$

For each $t \geq 0$, the state of the PDRP is denoted by $(Y(t), C(t)) \in \mathbb{S}$ with $Y(t) \in \{0, 1\}$ defining the state of the DNA ($Y(t) = 1$ referring to active DNA and $Y(t) = 0$ referring to repressed DNA) and $C(t) = (C_1(t), C_2(t))^\mathsf{T} \in \mathbb{R}^2$ defining the concentration of mRNA and proteins at time $t \geq 0$, respectively. The partially scaled state-change vectors are given by $\tilde{\boldsymbol{\nu}}_1^V = (0, 1, 0)^\mathsf{T}$, $\tilde{\boldsymbol{\nu}}_2^V = (0, 0, \frac{1}{V})^\mathsf{T}$, $\tilde{\boldsymbol{\nu}}_3^V = (-1, 0, -\frac{1}{V})^\mathsf{T}$, $\tilde{\boldsymbol{\nu}}_4^V = (1, 0, \frac{1}{V})^\mathsf{T}$, $\tilde{\boldsymbol{\nu}}_5^V = (0, -\frac{1}{V}, 0)^\mathsf{T}$, $\tilde{\boldsymbol{\nu}}_6^V = (0, 0, -\frac{1}{V})^\mathsf{T}$, such that $\mathbb{K}_0 = \mathbb{K}_d = \{3, 4\}$ (see (2.31)) and $\mathbb{K}_1 = \mathbb{K}_c = \{1, 2, 5, 6\}$. The resulting limit vectors are given by

$$\tilde{\boldsymbol{\nu}}_1 = \begin{pmatrix} 0 \\ 1 \\ 0 \end{pmatrix}, \quad \tilde{\boldsymbol{\nu}}_2 = \begin{pmatrix} 0 \\ 0 \\ 1 \end{pmatrix}, \quad \tilde{\boldsymbol{\nu}}_3 = \begin{pmatrix} -1 \\ 0 \\ 0 \end{pmatrix},$$

$$\tilde{\boldsymbol{\nu}}_4 = \begin{pmatrix} 1 \\ 0 \\ 0 \end{pmatrix}, \quad \tilde{\boldsymbol{\nu}}_5 = \begin{pmatrix} 0 \\ -1 \\ 0 \end{pmatrix}, \quad \tilde{\boldsymbol{\nu}}_6 = \begin{pmatrix} 0 \\ 0 \\ -1 \end{pmatrix}.$$

Assumption 2.6 is fulfilled with the limit propensities depending on the state $\boldsymbol{\theta} = (y, \boldsymbol{c})$, $\boldsymbol{c} = (c_1, c_2)^\mathsf{T}$, given as

$$\tilde{\alpha}_1(\boldsymbol{\theta}) = \gamma_1 y, \quad \tilde{\alpha}_2(\boldsymbol{\theta}) = \gamma_2 c_1, \quad \tilde{\alpha}_3(\boldsymbol{\theta}) = \gamma_3 y c_2,$$

$$\tilde{\alpha}_4(\boldsymbol{\theta}) = \gamma_4 (1 - y), \quad \tilde{\alpha}_5(\boldsymbol{\theta}) = \gamma_5 c_1, \quad \tilde{\alpha}_6(\boldsymbol{\theta}) = \gamma_6 c_2.$$

For the vector field μ defined in (2.35), we obtain

$$\mu(\boldsymbol{\theta}) = \gamma_1 y \begin{pmatrix} 0 \\ 1 \\ 0 \end{pmatrix} + \gamma_2 c_1 \begin{pmatrix} 0 \\ 0 \\ 1 \end{pmatrix} + \gamma_5 c_1 \begin{pmatrix} 0 \\ -1 \\ 0 \end{pmatrix} + \gamma_6 c_2 \begin{pmatrix} 0 \\ 0 \\ -1 \end{pmatrix},$$

such that—depending on the state $Y(t) = y \in \{0, 1\}$ of the DNA—the deterministic flow for the mRNA and protein concentration is given by the vector field

$$\mu_y : \mathbb{R}^2 \to \mathbb{R}^2, \quad \mu_y(\boldsymbol{c}) = \begin{pmatrix} \gamma_1 y \\ 0 \end{pmatrix} + \begin{pmatrix} -\gamma_5 & 0 \\ \gamma_2 & -\gamma_6 \end{pmatrix} \boldsymbol{c}. \tag{2.64}$$

The rate function for jumps of the DNA takes the form

$$\tilde{\lambda}(y, \boldsymbol{c}) = \gamma_3 y c_2 + \gamma_4 (1 - y),$$

and the transition kernel is deterministic in the sense of

$$\mathcal{K}((y, \boldsymbol{c}), (y', \boldsymbol{c}')) = \begin{cases} 1 & \text{if } y' = 1 - y \text{ and } \boldsymbol{c}' = \boldsymbol{c}, \\ 0 & \text{otherwise}, \end{cases}$$

for $y \in \{0,1\}, \mathbf{c} \in \mathbb{R}^2$. This means that for each time t the stochastic repressing reaction $\mathcal{R}_3$ takes place at rate $\gamma_3 Y(t) C_2(t)$, while activation $\mathcal{R}_4$ takes place at rate $\gamma_4(1 - Y(t))$, giving

$$\frac{d}{dt}\mathbb{P}(Y(t) = 1) = \gamma_4 \mathbb{P}(Y(t) = 0) - \gamma_3 C_2(t)\mathbb{P}(Y(t) = 1) = -\frac{d}{dt}\mathbb{P}(Y(t) = 0).$$

With $m = |\mathbb{K}_1| = 4$, the noise function of the corresponding piecewise chemical Langevin dynamics turns out to be

$$\sigma_y(\mathbf{c}) = \begin{pmatrix} \sqrt{\gamma_1 y} & 0 & -\sqrt{\gamma_5 c_1} & 0 \\ 0 & \sqrt{\gamma_2 c_1} & 0 & -\sqrt{\gamma_6 c_2} \end{pmatrix},$$

meaning independent noise in the flows of mRNA and proteins. In this special case, the flow for the continuous component $(\mathbf{C}(t))_{t \geq 0}$ can be written as a coupled set of stochastic differential equations

$$dC_1(t) = (\gamma_1 Y(t) - \gamma_5 C_1(t))\, dt + \sqrt{\gamma_1 Y(t) + \gamma_5 C_1(t)}\, dW_1(t)$$

for mRNA and

$$dC_2(t) = (\gamma_2 C_1(t) - \gamma_6 C_2(t))\, dt + \sqrt{\gamma_2 C_1(t) + \gamma_6 C_2(t)}\, dW_2(t) \qquad (2.65)$$

for proteins, with independent Wiener processes W_1, W_2. Here, we used the fact that, given two independent Wiener processes W_1, W_2, the weighted sum $\frac{aW_1 + bW_2}{\sqrt{a^2 + b^2}}$ is another Wiener process.

Hybrid Master Equation

For the hybrid master equation, we accordingly consider $M = 2$ discrete states $y \in \{0,1\}$ and the continuous components $\mathbb{E}_y^0(\mathbf{c}, t)$ of partial expectations for the high-abundant species given by mRNA and proteins. In order to explicitly write down the propensity functions $\tilde{\alpha}_k$ in the following equations, we will use the notation $\Phi_y^t \mathbf{c}_0 = \mathbf{c}(t|y) = (c_1(t|y), c_2(t|y))^\mathsf{T}$. As described in Sect. 2.2.4, we formulate the system of ODEs consisting of the reduced CME (2.58) which here has the form

$$\frac{d}{dt}p^0(0, t) = \gamma_3 c_2(t|1)p^0(1, t) - \gamma_4 p^0(0, t) = -\frac{d}{dt}p^0(1, t) \qquad (2.66)$$

and the evolution Eq. (2.60) for the partial expectations, here given by

$$
\frac{d}{dt}\mathbb{E}_0^0(\boldsymbol{c},t)
$$
$$
= p^0(0,t)\left[\begin{pmatrix} 0 \\ 1 \end{pmatrix}\gamma_2 c_1(t|0) + \begin{pmatrix} -1 \\ 0 \end{pmatrix}\gamma_5 c_1(t|0) + \begin{pmatrix} 0 \\ -1 \end{pmatrix}\gamma_6 c_2(t|0)\right] \quad (2.67)
$$
$$
+ \gamma_3 c_2(t|1)\mathbb{E}_1^0(\boldsymbol{c},t) - \gamma_4\mathbb{E}_0^0(\boldsymbol{c},t)
$$

and

$$
\frac{d}{dt}\mathbb{E}_1^0(\boldsymbol{c},t)
$$
$$
= p^0(1,t)\left[\begin{pmatrix} 1 \\ 0 \end{pmatrix}\gamma_1 + \begin{pmatrix} 0 \\ 1 \end{pmatrix}\gamma_2 c_1(t|1) + \begin{pmatrix} -1 \\ 0 \end{pmatrix}\gamma_5 c_1(t|1) + \begin{pmatrix} 0 \\ -1 \end{pmatrix}\gamma_6 c_2(t|1)\right]
$$
$$
- \gamma_3 c_2(t|1)\mathbb{E}_1^0(\boldsymbol{c},t) + \gamma_4\mathbb{E}_0^0(\boldsymbol{c},t),
$$
$$
(2.68)
$$

where $\mathbb{E}_y^0(\boldsymbol{c},t) = \boldsymbol{c}(t|y)p^0(y,t) \in \mathbb{R}_+^2$. In total, this is a system of ODEs of dimension $M+MN = 6$ which can easily be solved by numerical integration. Its solution approximates the time-dependent first-order moments of the three reactive species.

2.3.2 Simulation Results and Model Comparison

We analyze the dynamics defined by the three path-wise models (RJP, PDRP, PCLE) and compare their empirical averages to the solution of the corresponding hybrid master equation for the self-regulated gene expression system with rates

$$
\gamma_1 = 10^{-2}, \quad \gamma_2 = 0.5, \quad \gamma_3 = 10^{-2}, \quad \gamma_4 = 10^{-2}, \quad \gamma_5 = 5 \cdot 10^{-3}, \quad \gamma_6 = 0.2
$$
$$
(2.69)
$$

and a large volume $V = 100$. Note that these are artificial values without any claim to capture reality. The parameter values of gene expression systems in nature vary a lot and depend on the organism under consideration. The rates proposed here induce high population levels for both mRNA and proteins and comparatively long time periods of active DNA and are thereby well suited to demonstrate the relation between the modeling approaches under consideration—which is the main purpose of this section.

Sample Paths

Sample paths of the RJP are generated computationally by the stochastic simulation algorithm (cf. Sect. 1.3.2). For the PDRP and the PCLE, we applied the stochastic simulation schemes introduced in Sects. 2.2.1 and 2.2.2, respectively (see pages 68 and 72). Trajectories of all three path-wise models are given in Fig. 2.10. One can observe comparatively long periods of time with active DNA (marked by gray areas) where both populations show a positive trend, while a repressed DNA (white areas) induces negative trends. The Gillespie simulation of the RJP shows fluctuations around the trends which are well reproduced by the PCLE. The high population levels of mRNA and proteins induce long run times for the simulations of the RJP, while the approximations by PDRP or PCLE significantly reduce the computational effort.

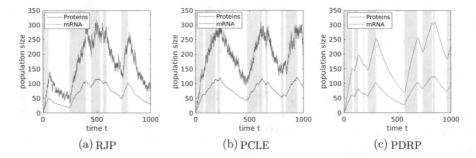

(a) RJP   (b) PCLE   (c) PDRP

Figure 2.10. Simulations of gene expression dynamics. Independent simulations of **(a)** the RJP, **(b)** the hybrid diffusion process, and **(c)** the piecewise-deterministic reaction process for parameter set (2.69) and initial state given by active DNA $(X_1^V(0) = Y(0) = 1)$ and absence of mRNA and proteins $(X_2^V(0) = C_1(0) = 0, X_3^V(0) = C_2(0) = 0)$. The gray areas indicate the time periods of active DNA $(X_1^V(t) = Y(t) = 1)$. In the interest of comparability with the RJP, we plot the abundance $(VC_1(t))_{t\geq 0}$ and $(VC_2(t))_{t\geq 0}$ for the PCLE and PDRP $(V = 100)$

Long-Term Averages and Protein Bursts

Statistic investigations in [247] have shown that the empirical long-term averages and standard deviations of the population size agree very well for all three models. In order to reveal the differences in the approximation properties of PDRP and PCLE, also the local maxima of the population size— referring to protein bursts—have been compared in [247]. Such bursts are typical for gene expression systems with negative self-regulation. One may ask: What exactly do we consider as a burst? In the stochastic systems (RJP and PCLE) "bursts" (large numbers of molecules) can appear by chance at any time. It is the piecewise-deterministic model which clarifies best what kind of bursts we are interested in: the maxima in the protein population after each individual repressing reaction (cf. Fig. 2.10c). This motivates to define the *burst size* as the maximum number of protein molecules within a time interval of inactive DNA. Note that by this definition also small peaks in the molecular population are counted, namely, if repressing takes place although the number of proteins is comparatively small (which can occur in all three models). The results given in Table 2.3 reveal that the size of protein bursts is much better reproduced by the PCLE than by the PDRP, which is due to the lack of noise in the PDRP.

Table 2.3. Protein bursts: results of long-term simulation ($T = 10^6$) for gene expression system with parameter set (2.69)

	RJP	PCLE	PDRP
Mean burst size	250.43	250.08	223.13
Maximum burst size	500	500.55	420.84
Minimum burst size	39	32.16	20.25

Temporal Evolution

Next, we compare the time-dependent evolution of the system starting with initial state given by active DNA ($X_1^V(0) = 1$) and absence of mRNA and proteins ($X_2^V(0) = 0$, $X_3^V(0) = 0$). Based on Monte Carlo simulations (10^4 runs for each model), we draw histograms of the protein population at different points in time (Fig. 2.11) and calculate the time-dependent empirical averages of the three populations as an approximation of the first-order moments in comparison to the solution of the hybrid master equation (Fig. 2.12).

The histogram of the PDRP reveals a clear deviation from the histograms of the other two approaches (see the spike at ~190 in Fig. 2.11a). The reason for this deviation is the following: At time $t = 100$ there are about 35% of trajectories with "non-stop" gene activity (i.e., the DNA has never been repressed within the time interval $[0, 100]$). For the PDRP—due to its determinism within this period of time—all these trajectories are completely consistent in the time interval $[0, 100]$. In particular, the populations of

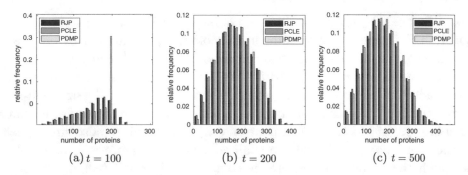

(a) $t = 100$ \hspace{3cm} (b) $t = 200$ \hspace{3cm} (c) $t = 500$

Figure 2.11. Empirical distribution of the number of proteins. Histrograms for the protein population at times **(a)** $= 100$, **(b)** $t = 200$, **(c)** $t = 500$ taken from 10^4 Monte Carlo simulations of each model, all with initial state $X_1^V(0) = 1$, $X_2^V(0) = 0$, $X_3^V(0) = 0$ (or $Y(0) = 1$, $C_1(0) = 0$, $C_2(0) = 0$, respectively)

mRNA and proteins agree with a value of $\sim$190 proteins at time $t = 100$. For the RJP and the PCLE, the proportion of non-stop activity at time $t = 100$ is the same, but the number of molecules varies for these non-stop activity trajectories due to randomness. The stochastic noise is able to create protein abundance up to 250 at time $t = 100$ which is simply impossible in the setting of PDRP. As time goes by, the deviation of the PDRP vanishes because the DNA process $(X_1^V(t))_{t\geq 0}$ (or $(Y(t))_{t\geq 0}$, respectively) reaches an equilibrium.

Although the empirical distribution of the PDRP does not coincide with the one of the two other path-wise modeling approaches on short time scales, the empirical averages agree very well for all times (cf. Fig. 2.12). Also the solution of the hybrid master equation is consistent.

As for computational effort, the two hybrid path-wise methods outperform the Gillespie simulations by several orders of magnitude in case of large volume systems. Simulations of the hybrid diffusion process defined by the PCLE take longer than simulations of the PDRP because random noise terms have to be drawn in each iteration step. In return, the PCLE simulations reproduce the variance in the dynamics and the time-dependent distribution. Naturally, the computational burden is the smallest for the hybrid master equation, which cannot be used to draw sample paths or investigate the burst size and distribution of the high-level molecular population but only approximates its first-order moments.

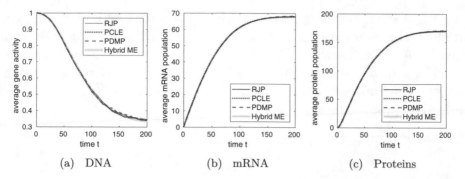

(a) DNA   (b) mRNA   (c) Proteins

Figure 2.12. Average dynamics of gene expression for large volume $V = 100$. RJP/PCLE/PDRP: **(a)** Empirical mean of the DNA state $X_1^V(t)$ or $Y(t)$, respectively, depending on time. This empirical mean also approximates the probability $\mathbb{P}(X_1^V(t) = 1)$ (or $\mathbb{P}(Y(t) = 1)$, respectively), which defines the marginal distribution of the DNA state. **(b)/(c)** Empirical means of the number of mRNA and proteins depending on time. Results from Monte Carlo simulation with 10^4 runs for each model given the initial state $X_1^V(0) = 1$, $X_2^V(0) = 0$, $X_3^V(0) = 0$ (or $Y(0) = 1$, $C_1(0) = 0$, $C_2(0) = 0$, respectively). Hybrid master equation: **(a)** Solution $P^0(1,t)$ of the reduced CME (2.66). **(b)/(c)** First and second component of the rescaled sum $V \cdot [\mathbb{E}_0^0(\boldsymbol{c},t) + \mathbb{E}_1^0(\boldsymbol{c},t)]$ of partial expectations. Results from numerical integration (by explicit Euler) of the coupled ODEs (2.66)–(2.68)

2.3.3 Small Volume Failure

The preceding investigations demonstrate that for high levels of mRNA and protein populations the hybrid models deliver good approximations of the RJP in terms of long-term averages and that the PCLE is also able to properly reproduce bursts and distributions of the number of molecules. In case of parameter values which induce small population levels for mRNA and proteins, however, the approximation methods may fail. This is shown in Fig. 2.13 where the overall empirical averages of the populations disagree in case of $V = 1$ (with propensity functions given by (2.63) and rate constants given by (2.69)). Both PDRP and PCLE show a decreased DNA activity (cf. Fig. 2.13a) combined with lower levels of mRNA and protein abundance (cf. Figs. 2.13b,c). On the other hand, the conditional expectations of the mRNA and protein population given a fixed value of the DNA state are still consistent (cf. Fig. 2.14 where the evolution of the mean number of mRNA molecules and proteins conditioned on $X_1^V(t) = 1$ (or $Y(t) = 1$, respectively) is shown). This consistency directly results from the fact that the conditional dynamics only contain first-order reactions. The only second-order reaction of the system (repressing) induces the switching between the conditions and is ignored in the regime of constant DNA activity. Given the fixed DNA state, the approximation of the mRNA and protein dynamics complies with a one-level approximation by a CLE or RRE such that the corresponding approximation properties apply and first-order moments are perfectly reproduced for any volume.

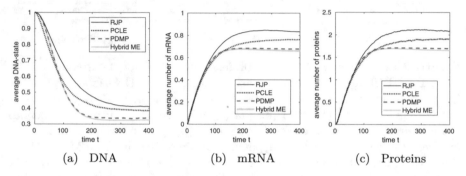

(a) DNA   (b) mRNA   (c) Proteins

Figure 2.13. Average dynamics of gene expression for small volume $V = 1$. RJP/PCLE/PDRP: (**a**) Empirical mean of the DNA state $X_1^V(t)$ or $Y(t)$, respectively, depending on time. This empirical mean also approximates the probability $\mathbb{P}(X_1^V(t) = 1)$ (or $\mathbb{P}(Y(t) = 1)$, respectively), which defines the marginal distribution of the DNA state. (**b**)/(**c**) Empirical means of the number of mRNA and proteins depending on time. Results from Monte Carlo simulation with 10^4 runs for each model given the initial state $X_1^V(0) = 1$, $X_2^V(0) = 0$, $X_3^V(0) = 0$ (or $Y(0) = 1$, $C_1(0) = 0$, $C_2(0) = 0$, respectively). Hybrid master equation: (**a**) Solution $P^0(1, t)$ of the reduced CME (2.66). (**b**)/(**c**) First and second component of the rescaled sum $V \cdot [\mathbb{E}_0^0(\boldsymbol{c}, t) + \mathbb{E}_1^0(\boldsymbol{c}, t)]$ of partial expectations. Results from numerical integration of the coupled ODEs (2.66)–(2.68)

Consequently, the divergence of the overall averages results from the fact that the marginal distributions of the DNA state do not coincide (cf. Fig. 2.13a). This deviation of the marginal distributions is caused by the second-order repressing reaction $DNA + P \to DNA_0$. The following analysis, taken from [247], clarifies this aspect.

Consider at first the PDRP with initial state $Y(0) = 1$ (active DNA) and $\boldsymbol{C}(0) = (C_1(0), C_2(0))^\mathsf{T} = \boldsymbol{c}_0 \in \mathbb{R}_+^2$. Let T_{PDRP} denote the random time of first repression, i.e.

$$T_{\mathrm{PDRP}} := \inf\{t > 0 | Y(t) = 0\}, \tag{2.70}$$

which will be called *repressing time* in the following. The repressing time T_{PDRP} is exponentially distributed with time-dependent rates $\Upsilon_{\mathrm{PDRP}}(t) \geq 0$

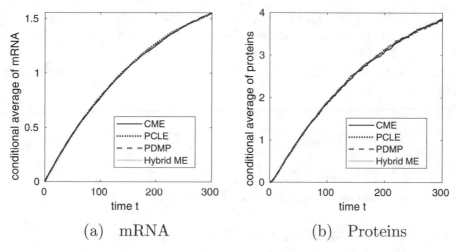

Figure 2.14. Conditional average dynamics for small volume $V = 1$. RJP/PCLE/PDRP: Empirical mean of (**a**) mRNA and (**b**) proteins depending on time *conditioned on* $X_1^V(t) = 1$ (or $Y(t) = 1$, respectively) for all t. Results from Monte Carlo simulation with 10^5 runs for each model (RJP, PCLE, PDRP) given the initial state $X_1^V(0) = 1$, $X_2^V(0) = 0$, $X_3^V(0) = 0$ (or $Y(0) = 1$, $C_1(0) = 0$, $C_2(0) = 0$, respectively). Hybrid ME: Solution of the conditional reaction rate Eq. (2.57) for $y = 1$ and initial state $c_0(1) = (0,0)^\mathsf{T}$

given by $\Upsilon_{\mathrm{PDRP}}(t) := \gamma_3 c_2(t)$ where $(c_2(t))_{t \geq 0}$ is the deterministic trajectory of the protein concentration during the active period of time $[0, T_{\mathrm{PDRP}}]$ given the initial state c_0. More precisely, it holds

$$\mathbb{P}\left(T_{\mathrm{PDRP}} \leq t\right) = 1 - e^{-\int_0^t \Upsilon_{\mathrm{PDRP}}(s)\,ds}.$$

As for the PCLE, the rates for repression are not only time dependent but also random, given by $\Upsilon_{\mathrm{PCLE}}(t) := \gamma_3 C_2(t)$ with $(C_2(t))_{t \geq 0}$ denoting the random trajectory of the protein concentration defined in (2.65) and given some initial state $Y(0) = 1$, $C(0) = c_0 \in \mathbb{R}_+^2$. The corresponding distribution of the first repressing time T_{PCLE} (again defined by (2.70)) is given by

$$\mathbb{P}\left(T_{\mathrm{PCLE}} \leq t\right) = 1 - \mathbb{E}\left(e^{-\int_0^t \Upsilon_{\mathrm{PCLE}}(s)\,ds}\right).$$

Equivalently, for the CME the repressing time $T_{\mathrm{CME}} := \inf\{t > 0 | X_1^V(t) = 0\}$ has the distribution

$$\mathbb{P}\left(T_{\mathrm{CME}} \leq t\right) = 1 - \mathbb{E}\left(e^{-\int_0^t \Upsilon_{\mathrm{CME}}(s)\,ds}\right)$$

with $\Upsilon_{\mathrm{CME}}(t) := \frac{\gamma_3}{V} X_3^V(t)$, where $X_3^V(t) \in \mathbb{N}_0$ denotes the number of proteins at time t in the jump process defined by the CME, again with initially active DNA, i.e., $X_1^V(0) = 1$.

Now the first-order moments of the number of proteins agree for all three models, i.e., it holds

$$c_2(t) = \mathbb{E}(C_2(t)) = \mathbb{E}\left(\frac{X_3^V(t)}{V}\right)$$

for all $t \in [0, T]$. Despite this identity, the distributions of the repressing time are not identical due to the general inequality

$$\exp\left(-\int_0^t \Upsilon_{\mathrm{PDRP}}(s)\,ds\right) \neq \mathbb{E}\left(e^{-\int_0^t \Upsilon_{\mathrm{PCLE}}(s)\,ds}\right) \neq \mathbb{E}\left(e^{-\int_0^t \Upsilon_{\mathrm{CME}}(s)\,ds}\right).$$

Note that in the small volume setting, the PCLE is even likely to produce negative molecule numbers. This is not only infeasible for interpretations of the underlying real-world application but also has a disproportional impact on the repressing propensities.

In summary, the different quality of fluctuations in the number of proteins during DNA activity (no fluctuations for PDRP, white noise for PCLE, Poisson-like variance for the CME) cause the deviations in the repressing times and with it the deviations of the marginal distributions of the DNA state. Although even in this small volume setting the *conditional* dynamics within time periods of constant DNA agree for all three approaches with respect to the first-order moments, the switching times between these periods diverge—and with it the overall population averages. For large population levels of proteins, fluctuations become proportionally small and their impact vanishes—and with it the deviations in the repressing times.

2.3.4 Three-Level Dynamics

So far, we assumed that mRNA and proteins show the same level of abundance. Given the scaling of reaction propensities as defined in (2.63), the population size of both species turns out to be of order V. Another setting (which might even be more realistic) assumes a different scaling for the two species, with the abundance of proteins exceeding the one of the mRNA. More precisely, let us consider the same genetic network as before but replace the propensity for transcription by $\alpha_1^V(\boldsymbol{x}) = \sqrt{V}\gamma_1 x_1$ and the propensity for translation by $\alpha_2^V(\boldsymbol{x}) = \sqrt{V}\gamma_2 x_2$. This induces a mRNA abundance of order $\sqrt{V}$ and a protein abundance of order V (cf. Fig. 2.15a).

In this case, the three-level hybrid process given by the joint equation (2.47) is the optimal combination of the different modeling approaches. It couples stochastic jumps of the DNA state with the Langevin dynamics of the medium-abundant mRNA population and the deterministic dynamics of high-abundant proteins. A simulation of the three-level hybrid process is given in Fig. 2.15b. The dynamics could also be approximated by a PCLE or PDRP, but the joint three-level equation combines both of their advantages: The runtime is reduced in comparison to PCLE simulations (because noise is only calculated for mRNA), and in contrast to the PDRP there is stochastic noise in the mRNA process which—by translation—also induces some local variance in the protein dynamics.

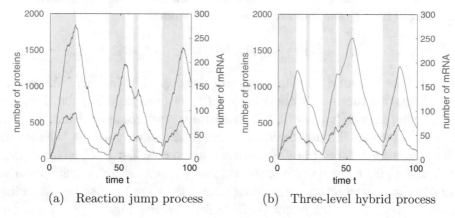

(a) Reaction jump process   (b) Three-level hybrid process

Figure 2.15. Three-level hybrid process for gene expression dynamics. Independent simulations of (**a**) the RJP and (**b**) the joint equation (2.47) for scaled propensity functions $\alpha_1^V(\boldsymbol{x}) = \sqrt{V}\gamma_1 x_1$ and $\alpha_2^V(\boldsymbol{x}) = \sqrt{V}\gamma_2 x_2$ and $V = 100$. Parameter values: $\gamma_1 = 1$, $\gamma_2 = 1$, $\gamma_3 = 10^{-2}$, $\gamma_4 = 0.1$, $\gamma_5 = 0.1$, $\gamma_6 = 0.5$. Initial state given by active DNA and absence of mRNA and proteins. The gray areas indicate the time periods of active DNA

2.4 Dynamic Partitioning

In all the hybrid models considered so far, the split-up of reactions into those treated as stochastic jumps and those approximated by Langevin dynamics or deterministic dynamics was fixed in advance and invariant in time. This is based on the assumption that the existing levels in the multiscale population are independent of the system's evolution: Some species constantly appear in low copy numbers, while others always remain on the level of

high abundance. However, there exist applications (as, e.g., the well-known
Lotka-Volterra dynamics which will be considered below in Example 2.15)
where in the course of time the population size of an individual species may
vary between the levels. In terms of the general multiscale equation (2.47),
this means that the sets $\mathbb{K}_0, \mathbb{K}_1, \mathbb{K}_2$, which classify the reactions, have to be
redefined over time.

2.4.1   Joint Equation for Time-Dependent Partitioning

For each time t, let $\mathbb{K}_0(t)$ be the index set of reactions modeled by stochastic
jumps, while $\mathbb{K}_1(t)$ defines the reactions approximated by Langevin dynamics and $\mathbb{K}_2(t)$ contains the indices of reactions described by deterministic
dynamics. The disjoint sets are chosen such that

$$\{1, \ldots, K\} = \mathbb{K}_0(t) \cup \mathbb{K}_1(t) \cup \mathbb{K}_2(t) \quad \forall t \geq 0.$$

The rules for defining this split-up will be described below. We cannot
directly rewrite the multiscale equation (2.47) (see page 76) for the given
time-varying sets $\mathbb{K}_j(t)$, $j = 0, 1, 2$, because of two reasons: First, we cannot rescale part of the population and consider concentrations because the
scaling now depends on time. We rather have to consider $V \cdot C_l$ as an approximation of X_l for a given value of the volume V. Second, the sums appearing
in (2.47) also become time-dependent such that they have to be taken into
the time integrals, which is not possible for the first line of (2.47) containing the Poisson processes. We therefore switch from integral to differential
notation and define an approximate process $\hat{\boldsymbol{X}} = (\hat{\boldsymbol{X}}(t))_{t \geq 0}$ by

$$\begin{aligned}
d\hat{\boldsymbol{X}}(t) = &\sum_{k \in \mathbb{K}_0(t)} \boldsymbol{\nu}_k \, d\hat{\mathcal{U}}_k(t) \\
&+ \sum_{k \in \mathbb{K}_1(t)} \alpha_k^V(\hat{\boldsymbol{X}}(t)) \boldsymbol{\nu}_k \, dt + \sum_{k \in \mathbb{K}_1(t)} \sqrt{\alpha_k^V(\hat{\boldsymbol{X}}(t))} \boldsymbol{\nu}_k \, dW_k(t) \quad (2.71) \\
&+ \sum_{k \in \mathbb{K}_2(t)} \alpha_k^V(\hat{\boldsymbol{X}}(t)) \boldsymbol{\nu}_k \, dt.
\end{aligned}$$

where again W_k are independent Wiener processes and $d\hat{\mathcal{U}}_k(t) :=$
$\mathcal{U}_k\left(\int_t^{t+dt} \alpha_k^V(\hat{\boldsymbol{X}}(s)) ds\right)$ for independent unit-rate Poisson processes $\mathcal{U}_k$.

Remark 2.14. In this setting, the reactions $\mathcal{R}_k$, $k \in \mathbb{K}_0(t)$ may induce
jumps also in the medium- or high-abundant species. This is because we
consider the original state-change vectors $\boldsymbol{\nu}_k$ instead of the partially scaled

$\tilde{\nu}_k$. An appropriate rescaling of the state-change vectors depending on time (just as the splitting depends on time) would be necessary for such jumps to disappear.

2.4.2 Rules for Partitioning

The fundamental indicator for deciding whether a reaction is treated as a stochastic event or approximated by continuous dynamics is given by the population size of the species involved in the reaction. Different approaches exist for specifying the rules for a split-up [3, 4, 89, 180, 201]. The common idea is to approximate a reaction by continuous dynamics whenever it occurs many times in a small time interval and its effect on the numbers of reactants and products species is small when compared to the total numbers of reactant and product species [201].

In order to make this criterion mathematically more precise, let $X^V(t) = x$ be the actual state of the process and define two thresholds $\tau^*, c^* > 0$. Then, a reaction $\mathcal{R}_k$ is modeled by continuous dynamics if the following two conditions are fulfilled:

(i) The expected time increment $\mathbb{E}(\tau_k | X^V(t) = x) = \frac{1}{\alpha_k^V(x)}$ for the reaction to occur is smaller than τ^*

$$\mathbb{E}(\tau_k | X^V(t) = x) < \tau^*.$$

(ii) The relative change in the population caused by the reaction does not exceed $1/c^*$

$$X_l^V(t) \geq c^* \cdot |\nu_{lk}^V| \quad \forall l = 1, \ldots, L.$$

For three-level systems including approximations in terms of both deterministic dynamics and Langevin dynamics, we can extend this criterion by introducing several threshold values τ_1^*, τ_2^* and c_1^*, c_2^* to define the assignment to the different levels.

For settings where the reaction propensities α_k^V scale with the population size of reactant species (as it is the case for the standard mass-action propensities) and the state-change vectors $\nu_k^V = \nu_k$ are independent of V, both criteria will simultaneously be fulfilled in case of sufficiently large population sizes of involved species. In such a case, it will be enough to directly define thresholds x_1^* and x_2^* for the population size and consider a split-up of reactions given by

$$\mathbb{K}_0(t) = \left\{ k = 1, \ldots K \,\Big|\, \min_{l:\nu_{lk}\neq 0} X_l^V(t) \leq x_1^* \right\},$$

$$\mathbb{K}_2(t) = \left\{ k = 1, \ldots K \,\Big|\, \min_{l:\nu_{lk}\neq 0} X_l^V(t) \geq x_2^* \right\}, \tag{2.72}$$

$$\mathbb{K}_1(t) = \{1, \ldots, K\} \setminus (\mathbb{K}_0(t) \cup \mathbb{K}_2(t)).$$

This means that a reaction is treated as a stochastic jump whenever *any* of its reactant or product species shows low abundance $\leq x_1^*$, while it is described by deterministic dynamics whenever *all* of the affected species occur in high abundance $\geq x_2^*$. Otherwise the reaction is approximated by Langevin dynamics.

Numerical Simulation of the Adaptive Hybrid Process

Within time intervals of constant split-up/splitting (i.e., periods of time where the sets $\mathbb{K}_0(t)$, $\mathbb{K}_1(t)$, $\mathbb{K}_2(t)$ are constant), the dynamics agree with the time-homogeneous three-level setting described in Sect. 2.2.3. Here, one can combine the simulation schemes for PDRPs and PCLEs from page 68 and page 72 to produce suitable trajectories. Of special interest are the points in time at which the population size of any involved species crosses one of the critical values x_1^*, x_2^*, leading to a change in the split-up of the reactions. Given a fixed time-step size for the numerical integration of the RRE and SDE parts, a central question is where to place the trajectory after crossing the critical value. A switch from the CLE level to the discrete level will require some kind of rounding to obtain an integer population state. Vice versa, a jump from the discrete to the continuous level may traverse some of the continuous area. One approach is to simply accept the crossing of the thresholds taking place within an iteration step of the algorithm and to subsequently place the trajectory at the "closest" possible point within the entered level.

Error Analysis

A rigorous error analysis of the hybrid approximation is given in [89] (strong error $\mathbb{E}(|\boldsymbol{X}^V(t) - \hat{\boldsymbol{X}}(t)|)$, only for bounded domain). The upper bound for the error derived in this work takes into account the scaling of the species' abundance as well as the scaling of the rate constants and of time. The error estimate can directly be used to dynamically and objectively partition the reactions into subgroups of those treated as discrete stochastic events and those approximated by ODEs or SDEs [4], as an alternative to the partitioning rule in terms of the thresholds τ^* and c^* that we proposed here.

Example 2.15 (Lotka-Volterra Model). The Lotka-Volterra model is a well-studied system of two interacting species $\mathcal{S}_1$ and $\mathcal{S}_2$ which are considered as *prey* and *predator*, respectively. The model is predominantly used for describing the dynamics of certain biological systems with the species referring to different animal types (e.g., foxes as predators and rabbits as prey). Instead of chemical reactions, one has to think of interactions between these animals. In the stochastic setting, the population changes according to the following reactions (or *interactions*)

$$\mathcal{R}_1 : \mathcal{S}_1 \xrightarrow{\gamma_1} 2\mathcal{S}_1, \qquad \mathcal{R}_2 : \mathcal{S}_1 + \mathcal{S}_2 \xrightarrow{\gamma_2} 2\mathcal{S}_2, \qquad \mathcal{R}_3 : \mathcal{S}_2 \xrightarrow{\gamma_3} \emptyset. \qquad (2.73)$$

These reactions may be interpreted as follows: By $\mathcal{R}_1$, the prey $\mathcal{S}_1$ is able to freely reproduce itself. In contrast, the creation of predators $\mathcal{S}_2$ requires the consumption of $\mathcal{S}_1$-particles as "food" (reaction $\mathcal{R}_2$). An unbounded expansion of the population is avoided by reaction $\mathcal{R}_3$ which refers to a decease of the predator. Actually, the population periodically alternates between low abundance and high abundance (see Fig. 2.16, which contains trajectories of the reaction jump process $(\boldsymbol{X}(t))_{t\geq 0}$ and the deterministic process $(\boldsymbol{C}(t))_{t\geq 0}$).

Let $(\boldsymbol{X}(t))_{t\geq 0}$ denote the corresponding reaction jump process, with $X_1(t)$ referring to the number of $\mathcal{S}_1$-particles and $X_2(t)$ referring to the number of $\mathcal{S}_2$-particles at time t. The corresponding RRE is given by the system of ODEs

$$\begin{aligned} \frac{dC_1(t)}{dt} &= \gamma_1 C_1(t) - \frac{\gamma_2}{V} C_1(t) C_2(t), \\ \frac{dC_2(t)}{dt} &= \frac{\gamma_2}{V} C_1(t) C_2(t) - \gamma_3 C_2(t) \end{aligned} \qquad (2.74)$$

for the concentrations C_1 and C_2 of $\mathcal{S}_1$ and $\mathcal{S}_2$, respectively.

The time-dependent multiscale nature of the dynamics motivates the use of adaptive schemes for simulating the process. Such an adaptive hybrid simulation is shown in Fig. 2.17. We used the simplified rule in terms of population levels described above with the sets $\mathbb{K}_0(t)$, $\mathbb{K}_1(t)$, $\mathbb{K}_2(t)$ chosen according to (2.72) using the thresholds $x_1^* = 10$, $x_2^* = 50$. This means that reaction $\mathcal{R}_1$ (or $\mathcal{R}_3$) appears as a jump whenever there are not more than 10 $\mathcal{S}_1$-particles (or $\mathcal{S}_2$-particles, respectively), while $\mathcal{R}_2$ is treated as a jump whenever either the $\mathcal{S}_1$- *or* the $\mathcal{S}_2$-population (or both) is beneath this threshold, because $\mathcal{R}_2$ affects both species simultaneously. Given that one of the two species is low-abundant while the other is high-abundant, $\mathcal{R}_2$ will thus induce jumps also in the high-abundant species. However, these jumps are not visible in Fig. 2.17, because they are very small in relation to the absolute population level of the high-abundant species.

A comparison of the dynamics is given in Fig. 2.18 where the average population size is shown for the pure jump process and the adaptive hybrid process. In the purely stochastic setting (given by the RJP), it is possible that the prey $\mathcal{S}_1$ or the predator $\mathcal{S}_2$ or both become extinct within finite time. A description by purely deterministic dynamics (given by the RRE (2.74)) precludes such a total extinction, while under the hybrid model (2.71) it can still occur. That is, at least some stochasticity is necessary here to maintain the characteristic features of the dynamics. Figure 2.18c shows the time-dependent proportion of predator extinction for the RJP and the hybrid process in comparison. ◇

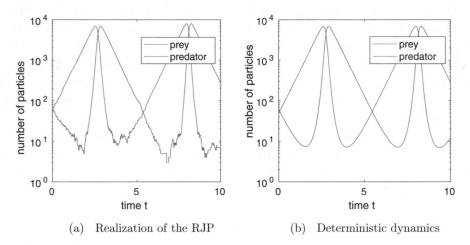

(a)   Realization of the RJP   (b)   Deterministic dynamics

Figure 2.16. Lotka-Volterra dynamics. (a) Gillespie simulation of the reaction jump process $(\boldsymbol{X}(t))_{t\geq 0}$. (b) Solution of the deterministic dynamics given by the RRE (2.74) for $V = 1$. Parameter values: $\gamma_1 = \gamma_3 = 2$, $\gamma_2 = 2 \cdot 10^{-3}$. Initial state: $\boldsymbol{X}(0) = \boldsymbol{C}(0) = (50, 60)$

Remark 2.16 (Blending Region). At the boundaries (given by the critical values x_1^*, x_2^*), one can possibly observe a "flip-flop" behavior of the adaptive hybrid process, meaning that the trajectory of the process quickly switches back and forth between the different levels of description, which is inconvenient for the numerical simulation. This leads to the idea that a splitting by such "strict" thresholds might be questionable because the de-

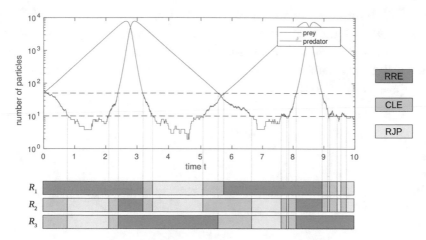

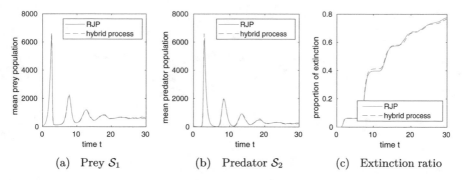

Figure 2.17. Adaptive three-level hybrid process. Trajectory of the three-level hybrid process (2.71) and portion of time where the reactions R_1, R_2, R_3 are treated as discrete stochastic (RJP), continuous stochastic (CLE), or continuous deterministic (RRE)

Figure 2.18. Average Lotka-Volterra dynamics. Empirical averages resulting from $2 \cdot 10^3$ Monte Carlo simulations of the reaction jump process and the three-level adaptive hybrid process, respectively. **(a)**/**(b)**: time-dependent mean of the number of $\mathcal{S}_1$- and $\mathcal{S}_2$-particles, respectively. **(c)**: time-dependent proportion of trajectories with vanished $\mathcal{S}_2$-population (as an estimation of $\mathbb{P}(X_2(t) = 0)$). Note that in the purely deterministic model, the extinction ratio is zero for all times

gree of stochasticity of the dynamics does not change abruptly. Instead, an increasing population should induce a "smooth" shifting from jump dynamics to CLE dynamics and further to deterministic dynamics. In [59], such a smooth transition between jumps and CLE dynamics is achieved by combining both types of dynamics within a "blending region" around the thresholds. The dynamics are weighted by blending functions $\beta_k : \mathbb{R}^L \to [0,1]$ where $\beta_k(\boldsymbol{x}) = 1$ induces pure jump dynamics for the respective reactions, while $\beta_k(\boldsymbol{x}) = 0$ leads to approximative CLE-dynamics, and $0 < \beta_k(\boldsymbol{x}) < 1$ gives a weighted combination of both types of dynamics within the blending region $\{\boldsymbol{x} \in \mathbb{R}^L | 0 < \beta_k(\boldsymbol{x}) < 1\}$.

Chapter 3

Temporal Scaling

For highly reactive systems the stochastic simulation of the reaction jump process $X = (X(t))_{t\geq 0}$ becomes inefficient due to the small time spans between successive reactions. By the structure of the propensity functions α_k, which quantify the systems' reactivity, quickly following reactions can be induced either by high population levels or by large reaction rate constants γ_k (see the definition of α_k in Sect. 1.1). While the case of high population levels has been investigated in Chap. 2, we now assume that some of the reactions $\mathcal{R}_1, \ldots, \mathcal{R}_K$ are "fast" in the sense of having much larger reaction rate constants than the remaining "slow" reactions. As for the population size, in contrast, we assume a uniform low level for all involved species. This means that some components of the process will quickly jump/switch between (comparatively few) low-level population states.

More precisely, we assume that the set of reactions $\mathcal{R}_1, \ldots, \mathcal{R}_K$ can be decomposed into two classes: the class corresponding to the *fast* process components with rate constants γ_k of order $\frac{1}{\varepsilon}$ for some $\varepsilon \ll 1$ and the class corresponding to the *slow* process components with rate constants of order 1. Instead of multiple population scales, as in Sect. 2.2, we are thus concerned with *multiple time scales*. In this case, a separation of species and partial rescaling of the population process (as considered in the previous chapter) is inappropriate, because each species may be affected by both fast and slow reactions. One approach to handle the situation of multiple time scales is to consider the dynamics in terms of reactions extents R_k [109, 110, 116], which leads to a straightforward decomposition of the dynamics. This approach

© The Editor(s) (if applicable) and The Author(s), under exclusive
license to Springer Nature Switzerland AG 2020
S. Winkelmann, C. Schütte, *Stochastic Dynamics in Computational Biology*, Frontiers in Applied Dynamical Systems: Reviews and Tutorials 8, https://doi.org/10.1007/978-3-030-62387-6_3

is presented in Sect. 3.1. An alternative is to find slow *observables* that
are (almost) unaffected by the fast reactions (cf. [244]). The resulting
separation of the dynamics in terms of observables is described in Sect. 3.2.
The existence of slow species which are unaffected by the fast reactions,
as assumed in [28, 29, 105], is included here as a special case. Finally, we
show in Sect. 3.3 how to combine the temporal scaling approaches with the
population scaling approaches from Chap. 2.

3.1   Separation in Terms of Reaction Extents

One key idea to investigate reactive systems with multiple reaction time
scales is to model the state of the system in terms of reaction extents. That
is, we count the number of firings of each reaction up to a certain time in
order to identify the state of the reactive system at this time [109, 116, 117].
The dynamics of the associated counting process are characterized by an
alternative CME that will be presented in Sect. 3.1.1. The advantage of this
alternative representation is that in cases of multiple reaction time scales the
partitioning of the system is quite straightforward: One can directly distinguish between the slowly increasing and the quickly increasing components
of the counting process. Such a partitioning of multiscale systems will be
considered in Sects. 3.1.2–3.1.3, with techniques for approximate solutions
summarized in Sect. 3.1.4.

3.1.1   Alternative Master Equation in Terms of Reaction Extents

Given a chemical reaction network consisting of L species $S_1, \ldots, S_L$ un-
dergoing K reactions $\mathcal{R}_1, \ldots, \mathcal{R}_K$, we consider the stochastic process $\boldsymbol{R} = (\boldsymbol{R}(t))_{t \geq 0}$ of reaction extents

$$\boldsymbol{R}(t) = (R_1(t), \ldots, R_K(t))^\mathsf{T} \in \mathbb{N}_0^K,$$

with $R_k(t)$ denoting the number of $\mathcal{R}_k$-reactions that occurred within the
time interval $[0, t)$ (see Sect. 1.2.1). In the literature, the component $R_k = (R_k(t))_{t \geq 0}$ is also called the *degree of advancement* of the kth reaction (cf.
[109]). Each individual reaction $\mathcal{R}_k$ is considered to be irreversible, and a
reversible reaction complex like, e.g., $S_l \rightleftharpoons S_{l'}$ is always decomposed into
two irreversible reactions $\mathcal{R}_k : S_l \to S_{l'}$ and $\mathcal{R}_{k'} : S_{l'} \to S_l$.

According to Eq. (1.4), the vector $\boldsymbol{R}(t) = (R_k(t))_{k=1,\ldots,K}$ is related to the system's population state $\boldsymbol{X}(t) = (X_l(t))_{l=1,\ldots,L}$ by

$$\boldsymbol{X}(t) = \boldsymbol{x}_0 + \boldsymbol{\nu}\boldsymbol{R}(t), \tag{3.1}$$

where $\boldsymbol{\nu} = (\nu_{lk})$ is the $L \times K$ stoichiometric matrix containing the state-change vectors $\boldsymbol{\nu}_k$ as columns and $\boldsymbol{R}(0) = \boldsymbol{0}$ is the null vector. Note again that by (3.1) $\boldsymbol{X}(t)$ is uniquely determined by $\boldsymbol{R}(t)$, while vice versa this is not true because different combinations of reactions might lead to the same molecular population. In this sense, the counting process $\boldsymbol{R}$ retains more information than the population process $\boldsymbol{X}$.

Parallel to the CME (1.10) in terms of particle numbers, there is an evolution equation for the probability $p(\boldsymbol{r}, t) := \mathbb{P}(\boldsymbol{R}(t) = \boldsymbol{r} | \boldsymbol{R}(0) = \boldsymbol{0}, \boldsymbol{X}(0) = \boldsymbol{x}_0)$ to find the process $\boldsymbol{R}$ in state $\boldsymbol{r} = (r_1, \ldots, r_K)^\mathsf{T}$ at time t given some initial population state $\boldsymbol{x}_0 \in \mathbb{N}_0^L$ and supposing that at initial time $t = 0$ no reactions have occurred yet. This alternative CME will be called *reaction master equation (RME)* and has the form

$$\frac{dp(\boldsymbol{r}, t)}{dt} = \sum_{k=1}^{K} [\beta_k(\boldsymbol{r} - \boldsymbol{1}_k)p(\boldsymbol{r} - \boldsymbol{1}_k, t) - \beta_k(\boldsymbol{r})p(\boldsymbol{r}, t)] \tag{3.2}$$

where $\boldsymbol{1}_k$ is the kth column of the $K \times K$ identity matrix and

$$\beta_k(\boldsymbol{r}) := \alpha_k(\boldsymbol{x}_0 + \boldsymbol{\nu}\boldsymbol{r})$$

is the reaction propensity function in terms of reaction extents.

This reformulation of the dynamics does not decrease the complexity of the system. Solving the RME (3.2) analytically is again a challenging task, and indirect methods have to be applied, like Gillespie's SSA which can directly be used to simulate the counting process $\boldsymbol{R}$. However, in case of highly reactive systems, the reaction extents $R_k(t)$ quickly increase in time, which renders the simulation algorithm inefficient and calls for model reduction methods.

Example 3.1 (Enzyme Kinetics). A well-studied reactive system is given by the Michaelis-Menten system of enzyme kinetics where an enzyme E reversibly binds to a substrate S to form the complex ES which releases a product P and regenerates the enzyme. Thereby, the substrate S is converted into the product P. This system is given by the coupled elementary reactions

$$\mathrm{E} + \mathrm{S} \underset{\gamma_3}{\overset{\gamma_2}{\rightleftharpoons}} \mathrm{ES} \overset{\gamma_1}{\longrightarrow} \mathrm{E} + \mathrm{P}. \tag{3.3}$$

Let $R_1(t)$ denote the number of product release reactions $\mathcal{R}_1 : ES \to E + P$ up to time t, while $R_2(t)$ and $R_3(t)$ refer to the number of binding reactions $\mathcal{R}_2 : E + S \to ES$ and unbinding reactions $\mathcal{R}_3 : ES \to E + S$, respectively. In terms of molecule numbers, the state of the system is given by $\boldsymbol{X}(t) \in \mathbb{N}_0^4$ with $X_l(t)$ for $l = 1, 2, 3, 4$ denoting the number of product molecules $\mathcal{S}_1 =$ P, enzymes $\mathcal{S}_2 = $ E, substrates $\mathcal{S}_3 = $ S, and enzyme-substrate complexes $\mathcal{S}_4 = $ ES at time t, respectively. Let the system start at time $t = 0$ with $e_0 \geq 1$ enzymes and $s_0 \geq 1$ substrates, combined with absence of products and complexes, such that the initial state is given by

$$\boldsymbol{x}_0 = (\, 0 \;\; e_0 \;\; s_0 \;\; 0 \,)^\mathsf{T}.$$

The stoichiometric matrix is given by

$$\nu = \begin{pmatrix} 1 & 0 & 0 \\ 1 & -1 & 1 \\ 0 & -1 & 1 \\ -1 & 1 & -1 \end{pmatrix},$$

and the propensity functions in terms of reaction extents are given by

$$\beta_1(\boldsymbol{r}) = \gamma_1(-r_1 + r_2 - r_3),$$
$$\beta_2(\boldsymbol{r}) = \frac{\gamma_2}{V}(e_0 + r_1 - r_2 + r_3)(s_0 - r_2 + r_3), \qquad (3.4)$$
$$\beta_3(\boldsymbol{r}) = \gamma_3(-r_1 + r_2 - r_3),$$

for a fixed volume $V > 0$ of the system. By (3.1), the population state $\boldsymbol{X}(t)$ at time t is given by

$$\begin{aligned} X_1(t) &= R_1(t), \\ X_2(t) &= e_0 + R_1(t) - R_2(t) + R_3(t), \\ X_3(t) &= s_0 - R_2(t) + R_3(t), \\ X_4(t) &= -R_1(t) + R_2(t) - R_3(t). \end{aligned}$$

Typically, the binding and unbinding of enzyme and substrate are fast reactions, while the release of the product is slow; it thus holds $\gamma_1 \ll \gamma_2, \gamma_3$. Figure 3.1 shows exemplary realizations of both the reaction extent process $\boldsymbol{R}$ and the corresponding population process $\boldsymbol{X}$ for such a setting of time scale separation. This example will be revived in Sect. 3.1.4. $\diamond$

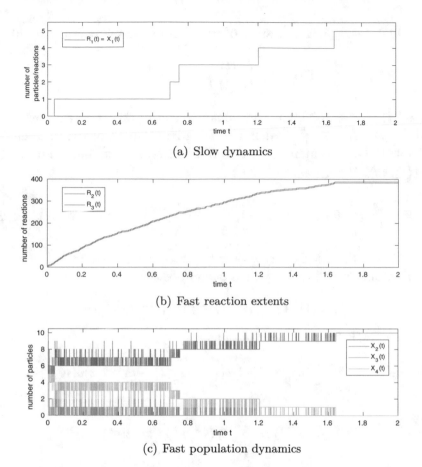

(a) Slow dynamics

(b) Fast reaction extents

(c) Fast population dynamics

Figure 3.1. Multiscale enzyme kinetics. Gillespie simulation of the reaction network (3.3) of Example 3.1 given a time scale separation with $\gamma_1 = 1$ (rate constant for the product release $\mathcal{R}_1$) and $\gamma_2 = \gamma_3 = 10^2$. Initial states: $\boldsymbol{R}(0) = \boldsymbol{0}$ and $\boldsymbol{X}(0) = \boldsymbol{x}_0 = (0\ e_0\ s_0\ 0)^\mathsf{T}$ with $e_0 = 10$ enzymes E and $s_0 = 5$ substrates S. The number $R_1(t)$ of product release reactions (and accordingly the number $X_1(t)$ of products P) increases slowly in time (see (**a**)), while $R_2(t)$ and $R_3(t)$ increase quickly and reach a high level (see (**b**)). Consequently, the other components of the population process $\boldsymbol{X}$ quickly switch between low-level states (see (**c**))

3.1.2 Slow-Scale RME

In order to simplify the notation, we w.l.o.g. assume that the slow reactions are given by $\mathcal{R}_1, \ldots, \mathcal{R}_m$ for some $1 \leq m < K$, while the remaining reactions $\mathcal{R}_{m+1}, \ldots, \mathcal{R}_K$ are fast, and we set $\mathbb{K}_s := \{1, \ldots, m\}$ and

$\mathbb{K}_f := \{m+1, \dots, K\}$. The state of reaction extents is partitioned like

$$\boldsymbol{R}(t) = \begin{pmatrix} \boldsymbol{R}^s(t) \\ \boldsymbol{R}^f(t) \end{pmatrix}, \quad \boldsymbol{r} = \begin{pmatrix} \boldsymbol{r}^s \\ \boldsymbol{r}^f \end{pmatrix},$$

with $\boldsymbol{R}^s(t) = \boldsymbol{r}^s \in \mathbb{N}_0^m$ giving the advancement of the slow reactions at time t and $\boldsymbol{R}^f(t) = \boldsymbol{r}^f \in \mathbb{N}_0^{K-m}$ defining the advancement of the fast reactions. Let $\boldsymbol{1}_k$ be the kth column of the $K \times K$-identity matrix, and define $\boldsymbol{1}_k^s \in \mathbb{N}_0^m$ and $\boldsymbol{1}_k^f \in \mathbb{N}_0^{K-m}$ such that

$$\boldsymbol{1}_k = \begin{pmatrix} \boldsymbol{1}_k^s \\ \boldsymbol{0} \end{pmatrix} \quad \text{for } k = 1, \dots, m, \quad \boldsymbol{1}_k = \begin{pmatrix} \boldsymbol{0} \\ \boldsymbol{1}_k^f \end{pmatrix} \quad \text{for } k = m+1, \dots, K.$$

Summation of the RME (3.2) with respect to $\boldsymbol{r}^f$ and using the fact that the joint distribution has the Bayesian decomposition

$$p(\boldsymbol{r}, t) = p(\boldsymbol{r}^s, \boldsymbol{r}^f, t) = p(\boldsymbol{r}^f, t | \boldsymbol{r}^s) p(\boldsymbol{r}^s, t) \tag{3.5}$$

result in a RME for the slow reactions

$$\frac{dp(\boldsymbol{r}^s, t)}{dt} = \sum_{k \in \mathbb{K}_s} \left[\bar{\beta}_k(\boldsymbol{r}^s - \boldsymbol{1}_k^s, t) p(\boldsymbol{r}^s - \boldsymbol{1}_k^s, t) - \bar{\beta}_k(\boldsymbol{r}^s, t) p(\boldsymbol{r}^s, t) \right], \tag{3.6}$$

where

$$\bar{\beta}_k(\boldsymbol{r}^s, t) := \sum_{\boldsymbol{r}^f} \beta_k(\boldsymbol{r}^s, \boldsymbol{r}^f) p(\boldsymbol{r}^f, t | \boldsymbol{r}^s). \tag{3.7}$$

We call (3.6) the *slow-scale RME* and emphasize the formal analogy to the reduced CME (2.50). Note that, so far, no approximation has been done; the slow-scale RME is exact.

3.1.3   Fast-Scale RME

The crucial point is the determination of the time-dependent rates $\bar{\beta}_k(\boldsymbol{r}^s, t)$ because these contain the unknown conditional probabilities $p(\boldsymbol{r}^f, t | \boldsymbol{r}^s)$. Assuming that within the time scale of the fast reactions the slow reactions do not fire, $\boldsymbol{R}^s(t) = \boldsymbol{r}^s$ can be considered as a constant parameter. Then, $p(\boldsymbol{r}^f, t | \boldsymbol{r}^s)$ approximately satisfies the *virtual fast-scale RME*

$$\begin{aligned} \frac{dp(\boldsymbol{r}^f, t | \boldsymbol{r}^s)}{dt} = \sum_{k \in \mathbb{K}_f} \Big[& \beta_k(\boldsymbol{r}^s, \boldsymbol{r}^f - \boldsymbol{1}_k^f) p(\boldsymbol{r}^f - \boldsymbol{1}_k^f, t | \boldsymbol{r}^s) \\ & - \beta_k(\boldsymbol{r}^s, \boldsymbol{r}^f) p(\boldsymbol{r}^f, t | \boldsymbol{r}^s) \Big]. \end{aligned} \tag{3.8}$$

The following is a rather heuristic justification from [109] that the virtual fast-scale RME may be at least approximately correct: The virtual fast-scale RME concerns the dynamics within the fast time scale where the slow reactions are unlikely to occur. Thereby we may assume $\beta_k(\boldsymbol{r}^s, \boldsymbol{r}^f) \approx 0$ for all $k \in \mathbb{K}_s$ within this time scale, such that the RME (3.2) becomes

$$\frac{dp(\boldsymbol{r}^s, \boldsymbol{r}^f, t)}{dt} \approx \sum_{k \in \mathbb{K}_f} \Big[\beta_k(\boldsymbol{r}^s, \boldsymbol{r}^f - \boldsymbol{1}_k^f) p(\boldsymbol{r}^s, \boldsymbol{r}^f - \boldsymbol{1}_k^f, t) \\ - \beta_k(\boldsymbol{r}^s, \boldsymbol{r}^f) p(\boldsymbol{r}^s, \boldsymbol{r}^f, t) \Big] \tag{3.9}$$

, whereas the slow-scale RME (3.6) gives

$$\frac{dp(\boldsymbol{r}^s, t)}{dt} \approx 0. \tag{3.10}$$

On the other hand, it follows from (3.5) that

$$\frac{dp(\boldsymbol{r}^s, \boldsymbol{r}^f, t)}{dt} = p(\boldsymbol{r}^f, t | \boldsymbol{r}^s) \frac{dp(\boldsymbol{r}^s, t)}{dt} + \frac{dp(\boldsymbol{r}^f, t | \boldsymbol{r}^s)}{dt} p(\boldsymbol{r}^s, t)$$

which, together with (3.10), results in

$$\frac{dp(\boldsymbol{r}^s, \boldsymbol{r}^f, t)}{dt} \approx \frac{dp(\boldsymbol{r}^f, t | \boldsymbol{r}^s)}{dt} p(\boldsymbol{r}^s, t). \tag{3.11}$$

Combining (3.9), (3.11), and (3.5) finally results in the virtual fast-scale RME (3.8).

In order to replace these heuristic arguments by a formal derivation, we have to express the scaling of the reaction rates explicitly. For $k \in \mathbb{K}_f$ we set

$$\beta_k = \frac{1}{\varepsilon} b_k \tag{3.12}$$

where $b_k = \mathcal{O}(1)$. This leads to the following form of the full reaction extent RME (3.2)

$$\frac{dp(\boldsymbol{r}, t)}{dt} \tag{3.13}$$
$$= \frac{1}{\varepsilon} \sum_{k \in \mathbb{K}_f} \Big[b_k(\boldsymbol{r}^s, \boldsymbol{r}^f - \boldsymbol{1}_k^f) p(\boldsymbol{r}^s, \boldsymbol{r}^f - \boldsymbol{1}_k^f, t) - b_k(\boldsymbol{r}^s, \boldsymbol{r}^f) p(\boldsymbol{r}^s, \boldsymbol{r}^f, t) \Big]$$
$$+ \sum_{k \in \mathbb{K}_s} \Big[\beta_k(\boldsymbol{r}^s - \boldsymbol{1}_k^s, \boldsymbol{r}^f) p(\boldsymbol{r}^s - \boldsymbol{1}_k^s, \boldsymbol{r}^f, t) - \beta_k(\boldsymbol{r}^s, \boldsymbol{r}^f) p(\boldsymbol{r}^s, \boldsymbol{r}^f, t) \Big].$$

This form of the RME exhibits its temporal multiscale nature directly. It can be rewritten in a more abstract form as

$$\frac{dp}{dt} = \left(\frac{1}{\varepsilon}\mathcal{G}_f + \mathcal{G}_s\right)p$$

where $\mathcal{G}_f$ denotes the operator acting on p as defined in the second line of (3.13), while $\mathcal{G}_s$ is the operator related to the third line. The part related to $\mathcal{G}_f$ generates a time scale $\tau = \mathcal{O}(\varepsilon)$ in addition to the time scale $t = \mathcal{O}(1)$. Analysis of this setting requires so-called methods of multiple time scales (see [170] or Section 9.8 in [155]). The asymptotic behavior of the fast reaction extents is crucial for this kind of analysis. By definition, each component $R_k = (R_k(t))_{t \geq 0}$ of the counting process is monotonically increasing in time and eventually either reaches a maximum value (if only finitely many reactions of this type can occur) or tends to infinity for large t such that the process $\boldsymbol{R}$ never equilibrates to a steady state but is explosive. In Appendix A.5.3 the application of the method of multiple time scales to the (non-explosive) case of a maximum value is discussed. As a result we get a slow-scale RME of the form (3.6) with an explicit formula for the propensities $\bar{\beta}$ that does not require to solve any virtual fast-scale RME (see Eq. (A.74)). The solution of the slow-scale RME is shown to yield an $\mathcal{O}(\varepsilon)$-approximation of the full solution. The explosive case evades such kind of analysis, i.e., in this case a sound justification of the virtual fast-scale RME by means of multiscale asymptotics is still missing.

3.1.4   Approximative Solutions

Unfortunately, solving the virtual fast RME (3.8) is by itself a difficult issue, and also the evaluation of the propensities $\bar{\beta}_k$ is not trivial as it requires the summation over all possible states $\boldsymbol{r}^f$. Different approaches to address these issues have been developed. For example, $\bar{\beta}_k(\boldsymbol{r}^s, t)$ could be sampled by Monte Carlo simulations, which is again computationally expensive. For the special case where the propensities β_k of the slow reactions depend linearly on the fast variables, it holds

$$\bar{\beta}_k(\boldsymbol{r}^s, t) = \beta_k(\boldsymbol{r}^s, \mathbb{E}(\boldsymbol{R}^f(t)|\boldsymbol{R}^s(t) = \boldsymbol{r}^s)). \tag{3.14}$$

That is, the problem of approximating the fast reactions reduces to the problem of calculating the conditional mean extents of the fast reactions. Finding approximations of these means is feasible for many applications, which will be demonstrated here for the example of enzyme kinetics.

Example 3.1 (continued). Consider again the Example 3.1 of enzyme kinetics given by

$$\text{E} + \text{S} \underset{\gamma_3}{\overset{\gamma_2}{\rightleftharpoons}} \text{ES} \overset{\gamma_1}{\longrightarrow} \text{E} + \text{P}.$$

We assume that the binding reaction $\mathcal{R}_2$ and the unbinding reaction $\mathcal{R}_3$ of the substrate S and the enzyme E are fast reactions with rate constants of order $\frac{1}{\varepsilon}$ (for $\varepsilon \ll 1$), while the release of the product P (reaction $\mathcal{R}_1$) is slow with $\gamma_1 = \mathcal{O}(1)$. The separation of the state $\boldsymbol{R}(t) = \boldsymbol{r} \in \mathbb{N}_0^3$ is then given by $\boldsymbol{r}^s = r_1$ and $\boldsymbol{r}^f = (r_2, r_3)^{\mathsf{T}}$. With the propensity of the slow reaction given by $\beta_1(\boldsymbol{r}) = \gamma_1(-r_1 + r_2 - r_3)$ (see (3.4)), we obtain from (3.14) that

$$\beta_1^{(t)}(r_1) = -\gamma_1 r_1 + \gamma_1 \left[\pi_2(r_1, t) - \pi_3(r_1, t)\right],$$

where $\pi_l(r_1, t) := \mathbb{E}(R_l(t) | R_1(t) = r_1)$ for $l = 2, 3$. In [109], the difference $\pi_2(r_1, t) - \pi_3(r_1, t)$ is approximated by $\min\{s_0, e_0 + r_1\}$ where e_0 is the initial number of enzymes and s_0 is the initial number of substrates, leading to the slow-scale RME

$$\frac{dp(r_1, t)}{dt} = \gamma_1[\min\{s_0, e_0 + r_1 - 1\} - r_1 + 1]p(r_1 - 1, t) \tag{3.15}$$
$$- \gamma_1[\min\{s_0, e_0 + r_1\} - r_1]p(r_1, t).$$

Given that $e_0 \geq s_0$, it holds $\min\{s_0, e_0 + r_1\} = s_0$ and (3.15) becomes

$$\frac{dp(r_1, t)}{dt} = \gamma_1(s_0 - r_1 + 1)p(r_1 - 1, t) - \gamma_1(s_0 - r_1)p(r_1, t).$$

This RME can be solved analytically by means of the binomial distribution

$$p(r_1, t) = \binom{s_0}{r_1}(1 - e^{-\gamma_1 t})^{r_1} e^{-\gamma_1 t(s_0 - r_1)}, \quad r_1 = 0, \dots, s_0. \tag{3.16}$$

With the number of products P given by $X_1(t) = R_1(t)$, it directly follows

$$\mathbb{E}(X_1(t)) = \sum_{r_1=0}^{s_0} r_1 p(r_1, t) = s_0(1 - e^{-\gamma_1 t})$$

and

$$\text{Var}(X_1(t)) = s_0(1 - e^{-\gamma_1 t})e^{-\gamma_1 t}.$$

For $e_0 < s_0$ there exists no analytical solution of (3.15). Instead, the underlying slow-scale process can be sampled by Monte Carlo simulations to estimate its mean and variance. Note that only the slow reaction $\mathcal{R}_1$ occurs in this sampling. Simulations of the fast reactions are completely circumvented which drastically reduces the computational effort. For a comparison of the approximative dynamics to the original dynamics, see Fig. 3.4 on page 122. ◇

3.2   Separation in Terms of Observables

As before, let us consider reaction networks with slow and fast reactions. With the evolution of the reactive system described by particle numbers X_l rather than by reaction extents R_k, a decomposition into slow and fast components is not that straightforward anymore. For settings where some species remain unaffected by the fast reactions, the following separation of species can be reasonable: Any species whose population gets changed by a fast reaction is classified as a "fast species," while all other species are classified as "slow." Such split-ups are discussed in [28, 29, 105]. However, if each species is affected both by slow and by fast reactions, this approach becomes inappropriate. In this case, a more general analysis by means of process "observables" is recommended (cf. [244]) and will be presented now.

3.2.1   Multiscale Observables

As in Sect. 1.3, an *observable* is a function $F : \mathbb{X} \to \mathbb{R}$ of the state variable $\boldsymbol{x} \in \mathbb{X} = \mathbb{N}_0^L$. Such an observable is termed as "slow" if it is not affected by fast reactions, i.e., F is a *slow observable* if it holds

$$F(\boldsymbol{x} + \boldsymbol{\nu}_k) = F(\boldsymbol{x}) \quad \forall k \in \mathbb{K}_f \tag{3.17}$$

for all states $\boldsymbol{x}$. Otherwise it is called *fast observable*. Especially, the separation into slow and fast species is contained as a special case by choosing the observables to be the components x_l of $\boldsymbol{x}$, i.e., by defining $F_l(\boldsymbol{x}) = x_l$.

Any linear observable F given by $F(\boldsymbol{x}) = \boldsymbol{b}^\mathsf{T}\boldsymbol{x}$ for some $\boldsymbol{b} \in \mathbb{R}^L$ is called *variable*. In accordance with (3.17), such a variable is a *slow variable* if

$$\boldsymbol{b}^\mathsf{T}\boldsymbol{\nu}_k = 0 \quad \forall k \in \mathbb{K}_f. \tag{3.18}$$

Given a set of basis vectors $\boldsymbol{b}_1, \ldots, \boldsymbol{b}_J$ that satisfy (3.18), we consider the slow variables v_j defined as $v_j(\boldsymbol{x}) := \boldsymbol{b}_j^\mathsf{T}\boldsymbol{x} \in \mathbb{R}$. Then, all slow observables can be expressed as functions of $v_1, \ldots, v_J$. Let $B \in \mathbb{R}^{J,L}$ be the matrix containing the basis vectors $\boldsymbol{b}_1, \ldots, \boldsymbol{b}_J$ as rows, and let $B^+ \in \mathbb{R}^{L,J}$ denote its pseudoinverse.[1] In addition to the slow variables $\boldsymbol{v}(\boldsymbol{x}) := B\boldsymbol{x} \in \mathbb{R}^J$, we define their orthogonal complement[2]

$$\boldsymbol{w}(\boldsymbol{x}) := \boldsymbol{x} - B^+ B\boldsymbol{x}$$

---

[1] The pseudoinverse B^+ fulfills $BB^+B = B$.

[2] It is $\mathfrak{B} := \mathrm{span}\{\boldsymbol{b}_j | j = 1, \ldots, J\}$ a subspace of $\mathbb{R}^L$, and it holds $B^+B\boldsymbol{x} \in \mathfrak{B}$ and $\boldsymbol{w} \in \mathfrak{B}^\perp$ with respect to the Euclidean scalar product, where $\mathfrak{B}^\perp$ is the orthogonal complement of $\mathfrak{B}$.

such that
$$w(x) + B^+ v(x) = x - B^+ Bx + B^+ Bx = x.$$

Example 3.2. Let four species $S_1, \ldots, S_4$ undergo the reactions

$$S_1 \underset{\gamma_2^f}{\overset{\gamma_1^f}{\rightleftharpoons}} S_2 \underset{\gamma_2^s}{\overset{\gamma_1^s}{\rightleftharpoons}} S_3 \underset{\gamma_4^f}{\overset{\gamma_3^f}{\rightleftharpoons}} S_4 \tag{3.19}$$

with $\gamma_k^f = 10^5$ for $k = 1, \ldots, 4$ (fast reactions) and $\gamma_k^s = 1$ for $k = 1, 2$ (slow reactions). In this example, every species is involved in at least one fast reaction such that a distinction between slow and fast species is impossible. The slow variables, on the other hand, are given by $v_1(x) = x_1 + x_2$ and $v_2(x) = x_3 + x_4$ as these quantities are conserved during the occurrence of the fast reactions $\mathcal{R}_k^f$, $k = 1, \ldots, 4$. The vectors b_j, $j = 1, 2$ are given by $b_1 = (1, 1, 0, 0)$ and $b_2 = (0, 0, 1, 1)$. We obtain

$$B = \begin{pmatrix} 1 & 1 & 0 & 0 \\ 0 & 0 & 1 & 1 \end{pmatrix}, \quad B^+ = \begin{pmatrix} 0.5 & 0 \\ 0.5 & 0 \\ 0 & 0.5 \\ 0 & 0.5 \end{pmatrix},$$

and

$$v(x) = Bx = \begin{pmatrix} x_1 + x_2 \\ x_3 + x_4 \end{pmatrix}, \quad w(x) = x - B^+ Bx = \frac{1}{2} \begin{pmatrix} x_1 - x_2 \\ -x_1 + x_2 \\ x_3 - x_4 \\ -x_3 + x_4 \end{pmatrix}.$$

This multiscale network will be considered again on page 117. $\diamond$

From now on, we represent the state of the system by (v, w) instead of x. The propensity functions $\hat{\alpha}_k$ in terms of the new state variable are defined as

$$\hat{\alpha}_k(v, w) := \alpha_k(w + B^+ v).$$

For the fast reactions we can again, as in (3.12), rescale the propensity functions according to

$$\hat{\alpha}_k(v, w) = \frac{1}{\varepsilon} a_k(v, w), \quad k \in \mathbb{K}_f,$$

with $\mathcal{O}(1)$-functions a_k and the smallness parameter ε measuring the gap in reaction rates as above. The transition of the form $x \mapsto x + \nu_k$ induced by reaction $\mathcal{R}_k$ is replaced by a coupled transition of the form

$$v \mapsto v + B\nu_k \quad \text{and} \quad w \mapsto w + \varphi_k$$

where $\boldsymbol{\varphi}_k := \boldsymbol{\nu}_k - B^+ B \boldsymbol{\nu}_k$. Noting that by construction it holds $B\boldsymbol{\nu}_k = 0$ (and consequently $\boldsymbol{\varphi}_k = \boldsymbol{\nu}_k$) for $k \in \mathbb{K}_f$, this leads to the following form of the full CME

$$\frac{dp(\boldsymbol{v}, \boldsymbol{w}, t)}{dt} \tag{3.20}$$
$$= \frac{1}{\varepsilon} \sum_{k \in \mathbb{K}_f} [a_k(\boldsymbol{v}, \boldsymbol{w} - \boldsymbol{\nu}_k) p(\boldsymbol{v}, \boldsymbol{w} - \boldsymbol{\nu}_k, t) - a_k(\boldsymbol{v}, \boldsymbol{w}) p(\boldsymbol{v}, \boldsymbol{w}, t)]$$
$$+ \sum_{k \in \mathbb{K}_s} [\hat{a}_k(\boldsymbol{v} - B\boldsymbol{\nu}_k, \boldsymbol{w} - \boldsymbol{\varphi}_k) p(\boldsymbol{v} - B\boldsymbol{\nu}_k, \boldsymbol{w} - \boldsymbol{\varphi}_k, t) - \hat{a}_k(\boldsymbol{v}, \boldsymbol{w}) p(\boldsymbol{v}, \boldsymbol{w}, t)],$$

where $\mathbb{K}_f$ and $\mathbb{K}_s$ denote the index sets of the fast and slow reactions, respectively.

3.2.2 Separation of Fast and Slow Scales

For analyzing the multiscale system in terms of observables, the method of multiple time scales can again be applied (see Sect. A.5.3 in the Appendix). In contrast to the setting of reaction extents discussed above, however, the virtual fast process—given a fixed value of the slow variable $\boldsymbol{v}$—might possess a stationary distribution (see Definition A.3 in the Appendix). That is, the conditional distribution $p(\boldsymbol{w}, t | \boldsymbol{v})$ given by

$$p(\boldsymbol{w}, t | \boldsymbol{v}) = \frac{p(\boldsymbol{v}, \boldsymbol{w}, t)}{p(\boldsymbol{v}, t)}$$

possibly converges to a stationary distribution $\pi(\boldsymbol{w} | \boldsymbol{v})$ on the fast time scale. For the following analysis, we assume that for every fixed value of the slow variables $\boldsymbol{v}$ there is a unique stationary distribution $\pi(\boldsymbol{w} | \boldsymbol{v})$ that satisfies

$$\sum_{k \in \mathbb{K}_f} [a_k(\boldsymbol{v}, \boldsymbol{w} - \boldsymbol{\nu}_k) \pi(\boldsymbol{w} - \boldsymbol{\nu}_k | \boldsymbol{v}) - a_k(\boldsymbol{v}, \boldsymbol{w}) \pi(\boldsymbol{w} | \boldsymbol{v})] = 0. \tag{3.21}$$

Moreover, we suppose the associated virtual fast process to be geometrically ergodic with respect to this stationary distribution (see Assumption A.19 on page 232 for a specification). The distribution π will be called *quasi-equilibrium distribution* in the following.

In Appendix A.5.3 the application of the method of multiple time scales to this case is demonstrated. The result (see the paragraph "The fast process is geometrically ergodic" in Appendix A.5.3) is that the solution $p(\boldsymbol{v}, \boldsymbol{w}, t)$ of the full multiscale CME (3.20) has the form

$$p(\boldsymbol{v}, \boldsymbol{w}, t) = \pi(\boldsymbol{w} | \boldsymbol{v}) p^0(\boldsymbol{v}, t) + \mathcal{O}(\varepsilon) + \mathcal{O}(e^{-rt/\varepsilon}), \tag{3.22}$$

where the slow-scale distribution p^0 satisfies

$$\frac{dp^0(\boldsymbol{v}, t)}{dt} = \sum_{k \in \mathbb{K}_s} \left[\bar{\alpha}_k(\boldsymbol{v} - B\boldsymbol{\nu}_k) p^0(\boldsymbol{v} - B\boldsymbol{\nu}_k, t) - \bar{\alpha}_k(\boldsymbol{v}) p(\boldsymbol{v}, t) \right] \qquad (3.23)$$

with *averaged propensities*

$$\bar{\alpha}_k(\boldsymbol{v}) := \sum_{\boldsymbol{w} \in \mathbb{R}^L} \alpha_k(\boldsymbol{v}, \boldsymbol{w}) \pi(\boldsymbol{w}|\boldsymbol{v}). \qquad (3.24)$$

The last term in the identity (3.22) means that there will be an $\mathcal{O}(\varepsilon)$ long phase in which the distribution converges to the given form by relaxation of the fast process. This is a kind of *averaging result* (see also [189]). The interpretation of this result is obvious: Given that the rates of the fast reactions are of order $\frac{1}{\varepsilon}$ for $\varepsilon \ll 1$ while the rates of the slow reactions are of order 1, the time scale for the convergence of $p(\boldsymbol{w}, t|\boldsymbol{v})$ to the quasi-equilibrium distribution is of order ε. The convergence of the original dynamics (3.20) to the effective dynamics (3.23) with an error of order $O(\varepsilon)$ and $\mathcal{O}(1)$ time scales can also be found in [243, 244], but in a simplified setting (see also Remark A.20).

By this approximation, the problem of calculating the conditional probabilities $p(\boldsymbol{w}, t|\boldsymbol{v})$ reduces to the problem of calculating the conditional quasi-equilibrium distribution $\pi(\boldsymbol{w}|\boldsymbol{v})$. Equation (3.21), which characterizes π, is a set of purely algebraic equations which might be easier to solve than the coupled differential equations of the fast-scale CME. As soon as $\pi(\boldsymbol{w}|\boldsymbol{v})$ is known such that the average propensities $\bar{\alpha}_k$ are explicitly given, the effective slow process can directly be simulated by standard methods, applying Gillespie's SSA to the slow-scale CME (3.23). In each iteration step, the slow variable is replaced by $\boldsymbol{v} \rightarrow \boldsymbol{v} + B\boldsymbol{\nu}_k$ for some slow reaction $\mathcal{R}_k$, and the state $\boldsymbol{w}$, which captures also the fast reactions, is sampled according to $\pi(\boldsymbol{w}|\boldsymbol{v})$. In [29], this approach is described in terms of slow and fast species, with the resulting algorithm called *slow-scale SSA*. Of course, such calculations are much faster than simulations of the original system of slow and fast reactions because the fast reactions are not performed at all.

Example 3.2 (continued). For the multiscale system of four species $\mathcal{S}_1, \ldots, \mathcal{S}_4$ interacting by fast and slow reactions according to (3.19), the slow variables are given by the total number $v_1 = x_1 + x_2$ of $\mathcal{S}_1$- and $\mathcal{S}_2$- species and the total number $v_2 = x_3 + x_4$ of $\mathcal{S}_3$- and $\mathcal{S}_4$-species. This means that the processes $(X_1(t) + X_2(t))_{t \geq 0}$ and $(X_3(t) + X_4(t))_{t \geq 0}$ both evolve slowly in time, while each of the summands $(X_l(t))_{t \geq 0}$ changes quickly in time (see Fig. 3.2 for an illustration). Given $v_1 \geq 0$, the distribution of $X_1(t)$ quickly converges with increasing t to a binomial distribution $B(v_1, q_1)$ for $q_1 = \frac{\gamma_2^f}{\gamma_1^f + \gamma_2^f}$ (see Fig. 3.2b), while the number of $\mathcal{S}_2$-particles is given by $X_2(t) = v_1 - X_1(t)$. Equivalently, $X_3(t)$ has a quasi-equilibrium distribution given by $B(v_2, q_2)$ for $q_2 = \frac{\gamma_4^f}{\gamma_3^f + \gamma_4^f}$, and it holds $X_4(t) = v_2 - X_3(t)$.

This means that the quasi-equilibrium distribution $\pi(\boldsymbol{w}|\boldsymbol{v})$ in terms of $\boldsymbol{w} = \boldsymbol{x} - B^+ B \boldsymbol{x}$ and $\boldsymbol{v} = (v_1, v_2)^\mathsf{T}$ is given by

$$
\pi(\boldsymbol{w}|\boldsymbol{v}) = \begin{cases} \pi_1\left(w_1 + \tfrac{1}{2}v_1 \,\middle|\, v_1\right) \pi_2\left(w_3 + \tfrac{1}{2}v_2 \,\middle|\, v_2\right) & \text{if } \boldsymbol{w} = (w_1, -w_1, w_3, -w_3)^\mathsf{T} \\ 0 & \text{otherwise,} \end{cases}
$$

where

$$
\pi_j(x|v) := \begin{cases} \binom{v}{x} q_j^x (1 - q_j)^{v-x} & \text{if } x \in \{0, \dots, v\}, \\ 0 & \text{otherwise,} \end{cases}
$$

for $j = 1, 2$. In this example, we thus obtain an explicit formula/analytical expression for the averaged propensities $\bar{\alpha}_k(\boldsymbol{v})$ defined in (3.24). If interested in the long-term dynamics, we can simulate the slow reactions with the averaged propensities $\bar{\alpha}$ while ignoring the fast reactions. In case of a time scale separation (in the sense of $\gamma_k^f \gg \gamma_{k'}^s$ for all k, k'), this delivers a good approximation and drastically reduces the numerical effort. $\diamond$

3.2.3   Nested Stochastic Simulation Algorithm

For many complex biological systems, the averaged propensities $\bar{\alpha}_k$ or the conditional equilibrium distributions $\pi(\boldsymbol{w}|\boldsymbol{v})$ are not known explicitly. In this case, it is possible to find approximative values for the averaged rates by Monte Carlo simulations of the fast reactions. The following *nested stochastic simulation algorithm (nested SSA)* uses an *outer SSA* to simulate the slow process with rates estimated by an *inner SSA* which simulates the fast process. It uses the fact that the fast process is geometrically ergodic, and thus the average with respect to the conditional equilibrium distribution can be replaced by a running average of the fast process.

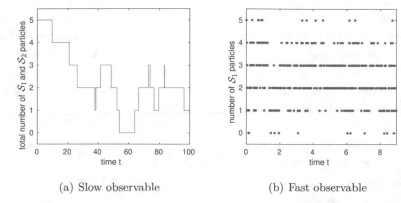

(a) Slow observable   (b) Fast observable

Figure 3.2. Stochastic simulation for multiple time scales. (a) Slow variable v_1 given by the total number $X_1(t) + X_2(t)$ of S_1- and S_2-particles for the multiscale reaction network given in Example 3.2. (b) Quickly changing number $X_1(t)$ of S_1-particles within a time interval of constant $v_1 = 5$ for the same Gillespie simulation as in (a). The quasi-equilibrium distribution of X_1 within this time interval is given by the binomial distribution $B\left(5, \frac{1}{2}\right)$. Parameter values: $\gamma_k^f = 10$ for $k = 1, 2, 3, 4,$, $\gamma_k^s = 0.1$ for $k = 1, 2$. Initial state $X(0) = (5, 0, 0, 0)$

Nested SSA:

1. Initialize time $t = t_0$ and states $x = x_0$, and choose a time horizon $T > t_0$ as well as a time length $T^f > 0$ for the inner SSA.

2. Inner SSA: Generate a trajectory $\tilde{x}(s)$ by stochastic simulation of the fast reactions only, over a time interval of length T^f and with initial state $\tilde{x}(0) = x$. For each slow reaction $k \in \mathbb{K}_s$, compute the modified rate by

$$\tilde{\alpha}_k(x) = \frac{1}{T^f} \int_0^{T^f} \alpha_k(\tilde{x}(s)) \, ds. \qquad (3.25)$$

3. Outer SSA: Run one step of SSA for the slow reactions only, using the modified rates $\tilde{\alpha}_k(x)$. Thereby, determine a reaction time τ and a reaction $\mathcal{R}_k$, $k \in \mathbb{K}_s$, and replace $t \leftarrow t + \tau$ and $x \leftarrow x + \nu_k$.

4. If $t \geq T$, end the simulation. Otherwise return to step 2.

Instead of creating a single trajectory of the fast process in each step 2 of the nested SSA, it is also possible to parallelize the inner SSA to create several independent replicas of the fast process and use the average of their modified rates for the outer SSA (cf. [244]). This approach further increases the efficiency of the algorithm. The nested SSA is very general and straightforward. It is a so-called seamless integration algorithm because it does not require to identify the slow and the fast variables or to determine an analytical approximation of the slow rates.

For the nested SSA to reduce the total computational effort on the one hand and to deliver good approximations on the other hand, a careful choice of the time length T^f for the inner SSA is of crucial significance. By ergodicity, the approximation $\tilde{\alpha}_k$ converges to the true averaged propensity $\bar{\alpha}_k$ with growing T^f in the sense of

$$\tilde{\alpha}_k(\boldsymbol{x}) \to \bar{\alpha}_k(B\boldsymbol{x}) \quad \text{for} \quad T_f \to \infty.$$

For finite T^f one obtains an approximation in the sense of $\tilde{\alpha}_k(\boldsymbol{x}) \approx \bar{\alpha}_k(B\boldsymbol{x})$. An estimate of the quantitative error is given in [244]. With the given time scale separation of reactions, this error becomes small as soon as $T^f/\varepsilon \gg 0$. This means that even small values of T^f give good approximations. Particularly T^f can be chosen smaller than the time scale of the slow reactions, i.e., $T^f \ll 1$, which implies the increase of numerical efficiency in comparison with the original SSA.

Note that even if the exact effective rates $\bar{\alpha}_k$ are known and the nested SSA can be carried out using these exact rates instead of their approximations $\tilde{\alpha}_k$, this is still an approximation of the true dynamics because $\bar{\alpha}_k$ is defined in terms of the virtual fast process which ignores the slow reactions.

The nested SSA algorithm is a special form of a family of algorithms that have become well-known under the name *equation-free* simulation algorithms (cf. [144]). In the fully equation-free version of the nested SSA algorithms, not even the distinction between fast and slow reaction is necessary. If the time scale T_f is selected appropriately, the algorithm will still have the properties described above (cf. [143] for more details).

Example 3.1 (continued). Consider again the example of enzyme kinetics from Sect. 3.1.1 given by

$$E + S \underset{\gamma_3}{\overset{\gamma_2}{\rightleftharpoons}} ES \overset{\gamma_1}{\longrightarrow} E + P \tag{3.26}$$

and set $\gamma_1 = 1$ and $\gamma_2 = \gamma_3 = 10^2$ (i.e., $\varepsilon = 10^{-2}$). Let the initial state be given by $\boldsymbol{x}_0 = (0\ e_0\ s_0\ 0)^\mathsf{T}$ with $e_0 = 10$ enzymes E and $s_0 = 5$ substrates S. The fast reactions are given by $\mathbb{K}_f = \{2, 3\}$. The slow observables are given by the number x_1 of products and the total number $x_3 + x_4$ of free and bound substrates (S and ES together) (cf. Fig. 3.3).

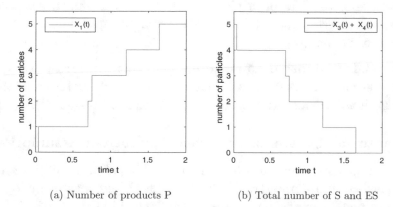

(a) Number of products P   (b) Total number of S and ES

Figure 3.3. Slow observables for enzyme kinetics. Gillespie simulation of the enzyme kinetics of Example 3.1. (**a**) The number $X_1(t)$ of products P and (**b**) the total number $X_3(t) + X_4(t)$ of substrates S and complexes ES evolve slowly in time, which is due to the small reaction rate constant γ_1 for the release reaction $\mathcal{R}_1 : \text{ES} \rightarrow \text{E} + \text{P}$. See also Fig. 3.1, where the fast process components are plotted as well

We compare the dynamics of the following modeling approaches: (a) full stochastic jump process with original SSA, (b) stochastic quasi-equilibrium assumption with nested SSA, and (c) analytic solution (3.16) of the approximative slow-scale CME (3.15). Figure 3.4 shows the mean $\mathbb{E}(X_1(t))$ and standard deviation of the product P for the three approaches.

In Fig. 3.5 the convergence to the quasi-equilibrium distribution is illustrated. Within a time $t \leq 0.01$, this quasi-equilibrium is reached. The time for the slow reaction, however, is of order $1/\gamma_1 = 1$. The solutions depicted in Fig. 3.4 result from a choice of $T^f = 0.05$. For these parameter values, the nested SSA is more than three times as fast as the original SSA and reproduces the dynamics very well. Also the solution defined by the binomial distribution (3.16) delivers a suitable approximation.   ◇

Remark 3.3 (Extension to More Than Two Scales). An extension of the nested SSA to more than two scales is straightforward: Imagine there are three scales given by slow reactions, fast reactions, and ultrafast reactions. A corresponding three-level nested SSA would consist of an outer SSA, an inner SSA, and innermost SSA, with the innermost SSA simulating the ultrafast reactions to determine approximative effective rates for the fast reactions and the inner SSA simulating the fast reactions to approximate the effective rates of the slow reactions. Also the investigations from the previous subsections can be extended to systems with more than two levels of reaction speeds. The principle will be always the same: The effective dynamics of a certain speed level are derived by approximating a given subsystem of higher speed.

Remark 3.4 (Adaptive Partitioning). There exist reactive systems where the levels of reaction speed depend on the time evolution of the system. In such a case, adaptive partitioning is required (cf. [244]).

Relation to the Deterministic Quasi-Equilibrium Assumption

The approximation of the time-dependent conditional probabilities $p(\boldsymbol{w}, t|\boldsymbol{v})$ by the quasi-equilibrium distribution $\pi(\boldsymbol{w}|\boldsymbol{v})$ is the stochastic analog to the well-known *quasi-equilibrium approximation (QEA)* in the setting of deterministic enzyme kinetics. Denote by $C_S(t)$, $C_E(t)$, $C_{ES}(t)$, and $C_P(t)$ the time-dependent concentration of substrates S, enzymes E, complexes ES, and products P, respectively. The deterministic reaction rate equations for the system of reactions (3.26) are given by

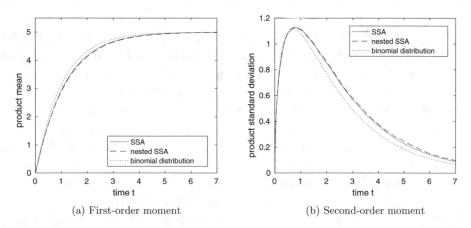

(a) First-order moment   (b) Second-order moment

Figure 3.4. Comparison of (a) first- and (b) second-order moments for enzyme kinetics. Time evolution of the mean and standard deviation of the number $X_1(t)$ of products P in the enzyme kinetics example resulting from 10^4 Monte Carlo simulations of the exact dynamics (solid blue lines) and the nested SSA (black dashed lines) in comparison with the analytical solution of the approximative slow-scale CME (3.15) (black dotted lines). See Example 3.1 on pages 107, 113, and 120

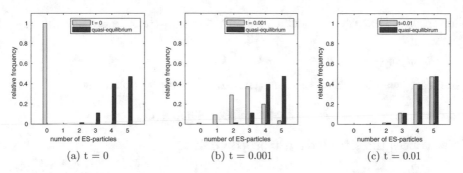

(a) t = 0   (b) t = 0.001   (c) t = 0.01

Figure 3.5. Convergence to the quasi-equilibrium distribution. Empirical distributions (gray bars) of the number $X_4(t)$ of enzyme-substrate complexes ES in the virtual fast system at times **(a)** $t = 0$, **(b)** $t = 0.001$ and **(c)** $t = 0.01$ compared to quasi-equilibrium distribution (dark blue bars) of the virtual fast system, given the initial state of $X_2(0) = 10$ enzymes E, $X_3(0) = 5$ substrates S and $X_4(0) = 0$ complexes ES

$$\frac{d}{dt}C_S = -\gamma_2 C_S C_E + \gamma_3 C_{ES},$$

$$\frac{d}{dt}C_E = -\gamma_2 C_S C_E + (\gamma_1 + \gamma_3)C_{ES},$$

$$\frac{d}{dt}C_{ES} = \gamma_2 C_S C_E - (\gamma_1 + \gamma_3)C_{ES},$$

$$\frac{d}{dt}C_P = \gamma_1 C_{ES}.$$

(3.27)

First of all, we see that the total enzyme concentration $\hat{e}_0 := C_E + C_{ES}$ is constant in time. Again we assume a time scale separation in the sense that the binding and unbinding of the substrate and the enzyme are fast reactions with $\gamma_2, \gamma_3 = \mathcal{O}(\frac{1}{\varepsilon})$ for $\varepsilon \ll 1$, while the release of the product is slow with $\gamma_1 = \mathcal{O}(1)$.

The quasi-equilibrium approximation is based on the so-called geometric limit theorems for singularly perturbed ODEs. In order to see this, we first rewrite $\gamma_i = \tilde{\gamma}_i/\varepsilon$, $i = 2, 3$, and denote $y(t) := C_E(t) - C_P(t)$, and use $C_{ES} = \hat{e}_0 - C_E = \hat{e}_0 - y - C_P$ to bring (3.27) into the following singularly perturbed form

$$\frac{d}{dt}C_S = \frac{1}{\varepsilon}\Big(-\tilde{\gamma}_2 C_S(y + C_P) + \tilde{\gamma}_3(\hat{e}_0 - y - C_P)\Big),$$

$$\frac{d}{dt}y = \frac{1}{\varepsilon}\Big(-\tilde{\gamma}_2 C_S(y + C_P) + \tilde{\gamma}_3(\hat{e}_0 - y - C_P)\Big),$$

$$\frac{d}{dt}C_P = \gamma_1(\hat{e}_0 - y - C_P).$$

This implies that $C_S - y$ is constant in time, i.e., there is a constant $\hat{c}_0$ such that

$$\hat{c}_0 = C_S - y = -C_E + C_P + C_S = C_{ES} + C_P + C_S - \hat{e}_0.$$

Therefore, we also have another (time-independent) constant

$$\hat{s}_0 := \hat{c}_0 + \hat{e}_0 = C_{ES} + C_P + C_S.$$

This reduces the singularly perturbed ODE to

$$\frac{d}{dt}C_S = \frac{1}{\varepsilon}\Big(-\tilde{\gamma}_2 C_S(\hat{e}_0 - \hat{s}_0 + C_S + C_P) + \tilde{\gamma}_3(\hat{s}_0 - C_S - C_P)\Big), \quad (3.28)$$

$$\frac{d}{dt}C_P = \gamma_1(\hat{s}_0 - C_S - C_P). \quad (3.29)$$

Standard theorems of geometric singular perturbation theory (cf. [75]) now imply that for $\varepsilon \to 0$, the solution of the singularly perturbed ODE converges to the solution of the following differential-algebraic equation (DAE)

$$0 = -\tilde{\gamma}_2 C_S(\hat{e}_0 - \hat{s}_0 + C_S + C_P) + \tilde{\gamma}_3(\hat{s}_0 - C_S - C_P),$$

$$\frac{d}{dt}C_P = \gamma_1(\hat{s}_0 - C_S - C_P), \quad (3.30)$$

where the first (algebraic) equation defines an invariant manifold in (C_S, C_P)-space to which the dynamics is geometrically constrained; this is the QEA for this case. Solving it for C_S yields

$$C_S = -\frac{1}{2}(\hat{e}_0 - \hat{s}_0 + C_P + \gamma_d) \pm \sqrt{\frac{1}{4}(\hat{e}_0 - \hat{s}_0 + C_P + \gamma_d)^2 + \gamma_d(\hat{s}_0 - C_P)},$$

where $\gamma_d := \frac{\gamma_3}{\gamma_2}$ is the *dissociation constant*. Closer inspection tells us that only the plus sign gives a consistent solution since only then we have $C_S = 0$ for $C_P = \hat{s}_0$. Putting this into the second equation (3.30) of the DAE, we get the following scalar ODE as the limit equation

$$\frac{d}{dt}C_P = \alpha_{\mathrm{eff}}(C_P), \quad (3.31)$$

where

$$\alpha_{\mathrm{eff}}(c) := \frac{a(c)}{2} - \sqrt{\frac{a(c)^2}{4} - \hat{e}_0(\hat{s}_0 - c)}$$

for $a(c) := -c + \hat{s}_0 + \hat{e}_0 + \frac{\gamma_3}{\gamma_2} \geq 0$.

The translation of these results back to the stochastic setting of a Markov jump process requires the consideration of a reaction jump process $\boldsymbol{X} = (X_S(t), X_P(t))_{t \geq 0}$ for the number $X_S(t)$ of substrate molecules and the number $X_P(t)$ of proteins at time t, undergoing three reactions: a slow reaction $\mathcal{R}_1 : \emptyset \to P$ and two fast reactions $\mathcal{R}_2 : S \to \emptyset$ and $\mathcal{R}_3 : \emptyset \to S$ with the corresponding propensity functions determined by the right-hand side of (3.28) and (3.29), i.e.

$$\alpha_1(x_S, x_P) = \gamma_1(\hat{s}_0 - x_S - x_P),$$

$$\alpha_2(x_S, x_P) = \frac{\tilde{\gamma}_2}{\varepsilon} x_S(\hat{e}_0 - \hat{s}_0 + x_S + x_P),$$

$$\alpha_3(x_S, x_P) = \frac{\tilde{\gamma}_3}{\varepsilon}(\hat{s}_0 - x_S - x_P).$$

The analysis of the corresponding multiscale CME in Appendix A.5.3 tells us that for $\varepsilon \to 0$ the fast reactions give rise to a stationary distribution

$$\pi(x_S|x_P) = \delta_{\alpha_1(x_S, x_P) = \alpha_2(x_S, x_P)}(x_S)$$

that is singularly supported (and corresponds to the invariant manifold discussed above). Our formula (3.24) for the averaged propensity of the slow reaction then yields

$$\bar{\alpha}_2(x_P) = \sum_{x_S=0}^{\infty} \alpha_2(x_S, x_P)\pi(x_S|x_P).$$

That is, we get an effective CME for the number X_P of proteins with the effective reaction

$$\mathcal{R}_{\text{eff}} : \emptyset \to P \tag{3.32}$$

and non mass-action propensity $\alpha_{\text{eff}} : \{0, \ldots, s_0\} \to [0, \infty)$ defined as

$$\alpha_{\text{eff}}(x_P) := \bar{\alpha}_2(x_P) = \frac{a(x_P)}{2} - \sqrt{\frac{a(x_P)^2}{4} - e_0(s_0 - x_P)} \tag{3.33}$$

for $a(x_P) := -x_P + s_0 + e_0 + \frac{\gamma_3}{\gamma_2}$, inspired by the limit result above. Setting $\alpha_{\text{eff}}(x_P) := 0$ for $x_P < 0$, the effective/averaged CME reads

$$\frac{d}{dt}p(x_P, t) = \alpha_{\text{eff}}(x_P - 1)p(x_P - 1, t) - \alpha_{\text{eff}}(x_P)p(x_P, t). \tag{3.34}$$

As x_P takes values only in $\{0, \ldots, s_0\}$, this averaged CME is a finite set of linear equations which can directly be solved (see Sect. 1.3.1) such that stochastic simulations of the reduced process are not necessary.

In Fig. 3.6, we can see that the effective reaction $\mathcal{R}_{\text{eff}}$ well reproduces the dynamics in case of a time scale separation with $\gamma_1 \ll \gamma_2, \gamma_3$.

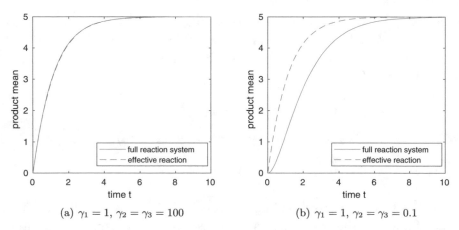

(a) $\gamma_1 = 1$, $\gamma_2 = \gamma_3 = 100$ (b) $\gamma_1 = 1$, $\gamma_2 = \gamma_3 = 0.1$

Figure 3.6. Approximation quality of the averaged CME. Time-dependent average of the number $X_1(t)$ of products P in the enzyme kinetics example (Example 3.1 with $e_0 = 10, s_0 = 5$) resulting from 10^4 Monte Carlo simulations of the exact dynamics (solid blue lines) and direct solution of the averaged CME (3.34), corresponding to the effective reaction $\mathcal{R}_{\text{eff}} : \emptyset \to \text{P}$ with propensity function α_{eff} given in (3.33). In case of a time scale separation with $\gamma_1 \ll \gamma_2, \gamma_3$, the dynamics agree perfectly (see (**a**)) which is consistent with Fig. 3.4a. This is not the case for small γ_2, γ_3 (see (**b**))

3.3   Combination with Population Scaling

In the two preceding sections, we considered the situation of multiple time scales in a reaction network with an entirely low-level population. The setting of multiple population scales, on the other hand, has been addressed separately in Chap. 2. While most of the literature concentrates on either separation of time scales or disparities in the population size, there also exist a few approaches to tackle both challenges simultaneously. In [71], a piecewise-deterministic reaction process (PDRP; see Sect. 2.2.1) with fast jumps between the discrete states is considered, and the convergence to a new PDRP with effective force fields and jump rates is proven. We will call this effective process *averaged PDRP*. A hybrid multiscale algorithm to efficiently simulate such an averaged PDRP is proposed in [202], including computational tools to systematically identify the fast and the slow components and to determine the relaxation times of the fast dynamics on the fly.

We will here demonstrate by means of a simple example how the ideas of Chaps. 2 and 3 can be combined with each other in order to define an effective coarsened model in case of multiple scales in time *and* population.

Example 3.5 (Combined Model for Multiple Scales in Time and Population). Some low copy number particles quickly switch between activity and inactivity by the reversible reactions

$$\mathcal{R}_{1,2}: \quad \mathcal{S}_1 \underset{\gamma_2}{\overset{\gamma_1}{\rightleftharpoons}} \mathcal{S}_2$$

where $\mathcal{S}_1$ and $\mathcal{S}_2$ refer to the active status and the inactive status, respectively. Whenever a particle is active, it produces particles of species $\mathcal{S}_3$ (the product) by

$$\mathcal{R}_3: \quad \mathcal{S}_1 \overset{\gamma_3}{\longrightarrow} \mathcal{S}_1 + \mathcal{S}_3$$

and meanwhile might get eliminated by

$$\mathcal{R}_4: \quad \mathcal{S}_1 \overset{\gamma_4}{\longrightarrow} \emptyset.$$

While $\mathcal{R}_1, \mathcal{R}_2, \mathcal{R}_3$ are assumed to be fast reactions with reaction rate constants of order $1/\varepsilon$ for some $0 < \varepsilon \ll 1$, the elimination $\mathcal{R}_4$ is a slow reaction with a reaction rate constant of order 1. The induced size of the product population is of order $V := 1/\varepsilon$. The fast dynamics of $\mathcal{R}_1$ and $\mathcal{R}_2$ affect the low-abundant species $\mathcal{S}_1$, $\mathcal{S}_2$ and suggest an approximation based on the stochastic quasi-equilibrium assumption. The fast reaction $\mathcal{R}_3$, in contrast, induces a quickly increasing product population $\mathcal{S}_3$. This suggests to treat $\mathcal{S}_3$ as a high-abundant species enabling approximations by deterministic dynamics or Langevin dynamics as described in Chap. 2. Ignoring the fast reactions $\mathcal{R}_1$ and $\mathcal{R}_2$, we would arrive at a PDRP (or PCLE) for the species $\mathcal{S}_1$ (discrete) and $\mathcal{S}_3$ (continuous) with the trends in the product evolution changing at each discrete event of $\mathcal{R}_4$. Considering in addition the quickly equilibrating system of $\mathcal{R}_1$ and $\mathcal{R}_2$, the trends simply have to be averaged.

More precisely, denote by

$$\boldsymbol{X}^V(t) = (X_1^V(t), X_2^V(t), X_3^V(t)) \tag{3.35}$$

the state of the reaction jump process at time t for a given value of $V = 1/\varepsilon$. At first, we ignore reaction $\mathcal{R}_3$, and only consider the process components $X_1^V = (X_1^V(t))_{t \geq 0}$ and $X_2^V = (X_2^V(t))_{t \geq 0}$ of $\boldsymbol{X}^V$ which are affected by the remaining reactions $\mathcal{R}_1$, $\mathcal{R}_2$, and $\mathcal{R}_4$. The slow observable of the system is given by the total number $X_{\text{total}}^V(t) := X_1^V(t) + X_2^V(t)$ of $\mathcal{S}_1$- and $\mathcal{S}_2$-particles which is only affected by the slow reaction $\mathcal{R}_4$. Denoting the possible states of the truncated process (X_1^V, X_2^V) by $(x_1, x_2) \in \mathbb{N}_0^2$, we can thus define the slow variable by $v := x_1 + x_2$ (see Sect. 3.2.1). Due to the fast reactions $\mathcal{R}_1$ and $\mathcal{R}_2$, this process quickly equilibrates to a steady state. Actually, ignoring the slow reaction $\mathcal{R}_4$ and given a value v of the slow variable, the process $X_2^V(t)$ converges in distribution to $X_{2,\infty}^V \sim B(v,q)$ for a binomial distribution $B(v,q)$ with $q = \frac{\gamma_2}{\gamma_1+\gamma_2}$ (cf. [29]), while $X_1^V(t)$ converges to $X_{1,\infty}^V := v - X_{2,\infty}^V$ for $t \to \infty$. By the given scaling of propensities, this convergence is of order $1/\varepsilon$. The quasi-equilibrium distribution $\pi(\cdot|v)$ (see (3.21)) is determined by this binomial distribution, and the *effective mean reaction propensity* for $\mathcal{R}_4$ in the quasi-equilibrium results in

$$\bar{\alpha}_4(v) = \gamma_4 vq = \frac{\gamma_4\gamma_2}{\gamma_1+\gamma_2}v.$$

Further, we now consider the high-abundant species $\mathcal{S}_3$ which is produced by the low-abundant species with effective mean propensity

$$\bar{\alpha}_3(v) = \gamma_3 vq = \frac{\gamma_3\gamma_2}{\gamma_1+\gamma_2}v.$$

In the spirit of Chap. 2, define the process $\Theta^V = (\Theta^V(t))_{t\geq 0}$ by $\Theta^V(t) = (X_{\text{total}}^V(t), C^V(t))$ with $C^V(t) = X_3^V(t)/V$. In the limit of $V \to \infty$ ($\varepsilon \to 0$), we arrive at an averaged PDRP $\Theta = (\Theta(t))_{t\geq 0}$ of the form

$$\Theta(t) = (X_{\text{total}}(t), C(t)), \tag{3.36}$$

with discrete component $(X_{\text{total}}(t))_{t\geq 0}$ and continuous component $(C(t))_{t\geq 0}$. Introducing the "species" $\mathcal{S}_{\text{total}}$ which aggregates the two species $\mathcal{S}_1$ and $\mathcal{S}_2$ of reactants, the PDRP is subject to two effective reactions: the continuous production $\mathcal{S}_{\text{total}} \to \mathcal{S}_{\text{total}} + \mathcal{S}_3$ with propensity $\tilde{\alpha}(\theta) = \bar{\alpha}_3(x)$ for $\theta = (x,c)$ and the discrete decay $\mathcal{S}_{\text{total}} \to \emptyset$ with propensity $\tilde{\alpha}(\theta) = \bar{\alpha}_4(x)$. Reactions $\mathcal{R}_1$ and $\mathcal{R}_2$ do not occur in the PDRP because their effect has been averaged out before.

Realizations of the full reaction jump process $\boldsymbol{X}^V$ and the corresponding averaged PDRP Θ are depicted in Fig. 3.7. For the sake of clarity and comparability, we plot the total number $(X_{\text{total}}^V(t))_{t\geq 0}$ of $\mathcal{S}_1$- and $\mathcal{S}_2$-particles for the reaction jump process and the rescaled continuous process $(VC(t))_{t\geq 0}$ of the averaged PDRP. Compared to the Gillespie simulation of the fully stochastic system, the simulation of the averaged PDRP is extremely time-saving because only very few stochastic events (given by the reaction $\mathcal{S}_1 \to \emptyset$) have to be created. Figure 3.8 compares the distribution of the averaged PDRP and the RJP at the time of extinction. Also a comparison of the time-dependent first- and second-order moments of the processes reveals that the averaged PDRP delivers a high-level approximation of the reaction jump process. For all simulations, the initial state is set to $X_1(0) = 5$, $X_2(0) = 0$, $X_3(0) = 0$ (for the RJP), and $X_{\text{total}}(0) = 5$, $C(0) = 0$ (for the averaged PDRP). For the reaction rates, we choose $\gamma_1 = \gamma_2 = \gamma_3 = \frac{1}{\varepsilon}$ with $\varepsilon = 5 \cdot 10^{-3}$ and $\gamma_4 = 1$. For these parameter values, the coarsening in terms of an averaged PDRP induces a reduction of the runtime by a factor of 10^{-4}. ◇

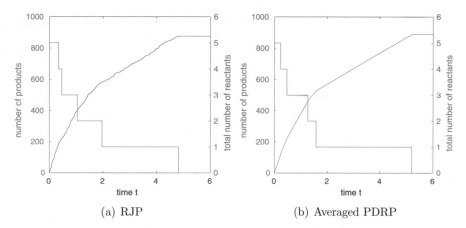

(a) RJP   (b) Averaged PDRP

Figure 3.7. Simulation of the averaged PDRP. Independent simulations of **(a)** the reaction jump process $(\boldsymbol{X}^V(t))_{t\geq 0}$ (3.35) and **(b)** the averaged PDRP $(\Theta(t))_{t\geq 0}$ (3.36) for the system given in Example 3.5. The total number of reactants (i.e., of $\mathcal{S}_1$- and $\mathcal{S}_2$-particles) at time t is given by $X_2(t) + X_1(t)$ for the RJP and by $X_{\text{total}}(t)$ for the averaged PDRP

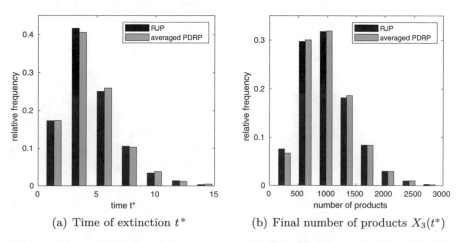

(a) Time of extinction t^* (b) Final number of products $X_3(t^*)$

Figure 3.8. Statistics of the averaged PDRP. See Example 3.5. **(a)** Histogram for the time of extinction $t^* := \min\{t \geq 0 | X_1(t) + X_2(t) = 0\}$. **(b)** Histogram for the final number $X_3(t^*)$ of products at the time of extinction and approximation by $VC(t^*)$. Results from 10^4 Monte Carlo simulations for both models

Chapter 4

Spatial Scaling

All investigations of the preceding chapters rely on the central well-mixed Assumption 1.1. This means that any spatial aspects of a reactive system— like the local positions of reactive particles or spatial inhomogeneities regarding reactivity—are completely ignored. It is supposed that all particles move fast in space, with the overwhelming majority of particle crossings being nonreactive. At any point in time, any pair of existing reactive particles can undergo a second-order reaction, regardless of their current physical distance in space.

However, there exist many applications which do not comply with this well-mixed assumption. The movement of particles in space might actually be quite slow or even limited by local barriers in the environment, such that a frequent crossing of particles seems unrealistic. Instead, the process of two particles finding each other in space can be an essential part of a bimolecular reaction. Moreover, the particles' motion processes are not necessarily independent of each other, but there can be mutual repulsion or attraction forces which lead to spatial exclusion or crowding effects. Also the environment may offer inhomogeneities with respect to the reaction propensities, possibly even restricting the occurrence of certain reactions to specific local subdomains. In such cases, more detailed models are needed, which take into account the spatial aspects of a reactive system.

© The Editor(s) (if applicable) and The Author(s), under exclusive license to Springer Nature Switzerland AG 2020
S. Winkelmann, C. Schütte, *Stochastic Dynamics in Computational Biology*, Frontiers in Applied Dynamical Systems: Reviews and Tutorials 8, https://doi.org/10.1007/978-3-030-62387-6_4

In this chapter, we present different modeling approaches for reaction networks with spatial resolution (see Fig. 4.1 for an overview). We start in Sect. 4.1 with the most detailed models of *particle-based reaction-diffusion* dynamics, where the individual diffusive trajectories of particles are traced in a continuous space and bimolecular reactions can only occur in case of physical proximity of two reactants. In analogy to Chap. 2, we examine the setting of large populations within spatially inhomogeneous environments in Sect. 4.2. The individual trajectories are replaced by space-time concentrations that evolve according to partial differential equations (PDEs). Solving methods for these PDEs typically relies on spatial discretization, which provides a link to another type of modeling framework given by *compartment CMEs*. Here, the continuous diffusive motion of particles is replaced by diffusive jumps between subdomains of space, assuming homogeneous, well-mixed reactive conditions within each compartment. Depending on the type of spatial discretization, these compartment CMEs are called *reaction-diffusion master equation (RDME)* or *spatiotemporal chemical master equation (ST-CME)*—both types are analyzed in Sect. 4.3. Hybrid approaches which combine the models of Sects. 4.1–4.3 are shortly mentioned in Sect. 4.4. The chapter is complemented by an application in Sect. 4.5.

4.1   Particle-Based Reaction-Diffusion

Let us start the chapter by considering the situation of a (comparatively small) population of particles moving in a spatial environment $\mathbb{D} \subset \mathbb{R}^d$ for $d \in \{1, 2, 3\}$. We interpret $\mathbb{D}$ as a biological cell and particles as molecules. Yet, there exist other interpretations, where the individual entities represent persons or organizations called *agents* which move and interact within their physical environment $\mathbb{D}$. In this context, typically the term *agent-based modeling* is used, cf. [53].

As in the preceding chapters, the particles belong to different species $\mathcal{S}_1, \ldots, \mathcal{S}_L$, and the population is affected by chemical reactions. Spatial diffusion is comparatively slow, such that the well-mixed condition (see Assumption 1.1) is violated and the positions of particles in space become fundamentally important. In particular, a second-order reaction can only occur under the condition that two reactive particles cross each other's path in space.

In the following, we give a short description of the detailed models of particle-based reaction-diffusion dynamics where the motion of every individual particle is resolved in time and space and bimolecular reactions

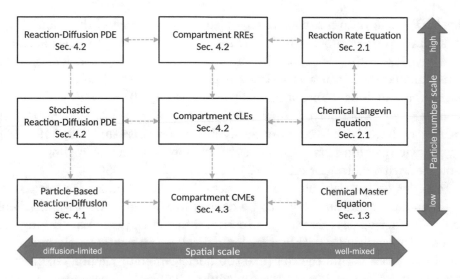

Figure 4.1. Structural overview of modeling approaches. Mathematical models for reaction-diffusion dynamics as presented in Chap. 4 and their relation to the models of the preceding chapters. The vertical axis refers to the population scale; the horizontal axis gives the spatial scale. Note that on the level of well-mixed dynamics (rightmost column), the CME appears as a *microscopic* model with respect to vertical population scale, while it is a *macroscopic* model with respect to the horizontal spatial scale

require spatial proximity. In the literature also the terms "molecular-based model" or "Brownian dynamics" can be found for these types of models [68]. At first, in Sect. 4.1.1, we ignore the chemical reactions and consider different models for the transport of individual particles in space via stochastic dynamics. Reactive interactions of particles will be added in Sect. 4.1.2, and a mathematical model for the combined dynamics will be formulated in Sect. 4.1.3. Numerical methods for particle-based stochastic simulations will be discussed in Sect. 4.1.4.

4.1.1 Modeling Diffusive Motion

The most fundamental approach to model the diffusive motion of particles in space is to consider the particles as points without mass, following independent Brownian motions in free space. The position $Q(t) \in \mathbb{R}^d$ of a particle is then propagated in time by the SDE

$$dQ(t) = \sqrt{2D}\, dW(t), \tag{4.1}$$

where D is the diffusion constant. In the case of *normal* diffusion, W is a standard Wiener process in $\mathbb{R}^d$. Then, the mean squared displacement satisfies $\mathbb{E}\left(Q(t)Q(0)\right) = 2Dt$. In the following, we only consider normal diffusion, mainly because it is the simplest and most discussed case. In contrast, anomalous diffusion means that $\mathbb{E}\left(Q(t)Q(0)\right) = 2Dt^\eta$ with $\eta \neq 1$, where $0 < \eta < 1$ is referred to as *subdiffusion* and $\eta > 1$ as *superdiffusion*. Both cases have been studied extensively in the literature [18, 119, 205]. For cellular reaction networks, the subdiffusion case has been shown to be associated with molecular crowding in the cytoplasm [245].

In order to restrict the dynamics to a bounded domain $\mathbb{D} \subset \mathbb{R}^d$, suitable boundary conditions have to be chosen in addition. Most applications involve further spatial restrictions to the diffusion of particles. Taking the movement of molecules within a biological cell as an example, there can be local boundaries within the cell (like the nuclear membrane in a eukaryote), which inhibit the free flow of the molecules. Typically, there also exist regions which are more attractive than others. Such spatial inhomogeneities in the cellular landscape can be reflected by a potential energy function $U : \mathbb{D} \rightarrow \mathbb{R}$, a smooth function giving the energy depending on the location. A well-known model for diffusive motion within such an energy landscape is given by Langevin dynamics. For a particle of mass $m_0 > 0$, these dynamics are defined by the Langevin equation

$$m_0 \frac{d^2 Q(t)}{dt^2} = -\zeta \frac{dQ(t)}{dt} - \nabla U(Q(t)) + \zeta \sqrt{2D} \xi(t)$$

where $\frac{dQ(t)}{dt}$ is the velocity, $\frac{d^2 Q(t)}{dt^2}$ is the acceleration, ∇U is the gradient of the potential, and ζ denotes the friction coefficient. Here, $(\xi(t))_{t \geq 0}$ is a Gaussian white noise process with mean $\mathbb{E}(\xi(t)) = 0$ and covariance $\mathbb{E}\left(\xi(t)\xi(s)\right) = \delta(t - s)$, with δ being the Dirac delta. In a high friction regime, it holds

$$\zeta \left| \frac{dQ}{dt} \right| \gg m_0 \left| \frac{d^2 Q}{dt^2} \right|,$$

which leads (by setting $m_0 \frac{d^2 Q}{dt^2} = 0$) to *overdamped Langevin dynamics* given by

$$\zeta \frac{dQ(t)}{dt} = -\nabla U(Q(t)) + \zeta \sqrt{2D} \xi(t).$$

Equivalently, we can replace ξ by dW for a standard Wiener process W and obtain the SDE

$$dQ(t) = -\frac{1}{\zeta} \nabla U(Q(t))dt + \sqrt{2D}dW(t). \tag{4.2}$$

By the fluctuation-dissipation relation,

$$D\zeta = k_B T_0,$$

where k_B is the Boltzmann constant and T_0 is the temperature, we can rewrite (4.2) as

$$dQ(t) = -\frac{D}{k_B T_0}\nabla U(Q(t))dt + \sqrt{2D}dW(t), \qquad (4.3)$$

(see [69] and others).[1]

The probability density $\rho(q, t)$ for the process Q given by (4.2) to be in state $q \in \mathbb{R}^d$ at time t satisfies the Fokker-Planck equation

$$\frac{\partial}{\partial t}\rho(q, t) = \frac{D}{k_B T_0}\nabla \cdot (\nabla U(q)\rho(q, t)) + D\nabla^2 \rho(q, t), \qquad (4.4)$$

where $\nabla^2 = \frac{\partial^2}{\partial q_1^2} + \ldots + \frac{\partial^2}{\partial q_d^2}$ is the Laplace operator with respect to q. In case where the potential energy function U is confining in the sense that it holds $\lim_{|q|\to\infty} U(q) = \infty$ and

$$\int_{\mathbb{R}^d} \exp\left(-\frac{U(q)}{k_B T_0}\right) dq < \infty,$$

then a unique equilibrium distribution $\rho^\infty(q)$ exists and has the form

$$\rho^\infty(q) = \frac{1}{A}\exp\left(-\frac{U(q)}{k_B T_0}\right),$$

where A is a normalization constant [190]. Illustrative examples of diffusion processes in potential energy landscapes are given in Figs. 4.2 and 4.3.

Remark 4.1. From now on we use a dimension-free notation for the sake of simplicity.

Example 4.2 (Double-Well Potential in $\mathbb{R}$). Consider the diffusion process $(Q(t))_{t\geq 0}$ in $\mathbb{D} = \mathbb{R}$ following the SDE (4.2) with the double-well potential U defined as

$$U(q) := 0.1((6q - 3)^2 - 1)^2 \qquad (4.5)$$

which has two local minima at $q = \frac{1}{3}$ and $q = \frac{2}{3}$. A trajectory of the process Q as well as the double-well potential U and the related invariant density ρ^∞ are shown in Fig. 4.2. $\diamond$

---

[1] The dimension units are given by $[\zeta] = \frac{Ns}{m}$, $[k_B T_0] =$ Nm and $[D] = \frac{m^2}{s}$ for standard units N (Newton), s (second), and m (meter).

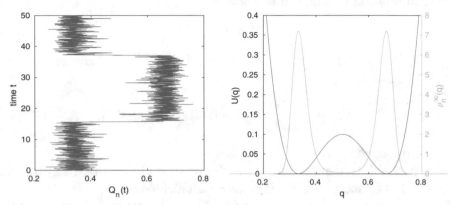

(a) Realization of $(Q(t))_{t\geq 0}$ given by an Euler-Maruyama simulation

(b) Double-well potential $U : \mathbb{D} \to \mathbb{R}$ and equilibrium distribution ρ^∞

Figure 4.2. Diffusion process in double-well potential. Overdamped Langevin dynamics in the double-well potential U defined in (4.5), see Example 4.2. The motion of the particle in space $\mathbb{D} = \mathbb{R}$ is metastable in the sense that the trajectory stays for a comparatively long period of time in one well before a rare transition across the potential barrier to the other well occurs. Note that we switched the axes in (**a**), placing time to the vertical axis, with the purpose of comparability to (**b**)

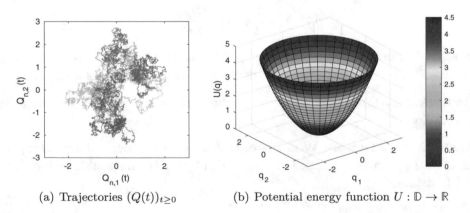

(a) Trajectories $(Q(t))_{t\geq 0}$

(b) Potential energy function $U : \mathbb{D} \to \mathbb{R}$

Figure 4.3. Diffusion process in $\mathbb{D} = \mathbb{R}^2$. (**a**) Four independent trajectories (Euler-Maruyama simulations) of the diffusion process $(Q(t))_{t\geq 0}$ given by (4.2) with the potential energy function given by $U(\boldsymbol{q}) = \frac{1}{2}(q_1^2 + q_2^2)$ and deterministic start in $Q(0) = (0,0)$, $n = 1, \ldots, 4$. (**b**) Potential energy function U. (**a**) Trajectories $(Q(t))_{t\geq 0}$. (**b**) Potential energy function $U : \mathbb{D} \to \mathbb{R}$

Interacting Forces and Crowding

So far, particles were assumed not to influence each other in their movement. When considering particles as massive objects, however, this assumption is not reasonable because the particles mutually exclude each other in space. More sophisticated modeling approaches therefore also take repulsive and attractive forces between nonreactive particles into account, e.g., by defining *interaction potentials* which permit the modeling of effects such as space exclusion, molecular crowding, and aggregation. These particle-particle potentials are functions of the joint *configuration state*

$$\boldsymbol{q} = (q_1, \ldots, q_N) \in \mathbb{D}^N$$

containing the spatial positions of *all* existing particles, with $N \in \mathbb{N}_0$ denoting the (fixed) number of particles. The diffusion process of the n-th individual particle is then given by

$$dQ_n = -\Big(\nabla U(Q_n) + \nabla U_n(\boldsymbol{Q})\Big)dt + \sqrt{2D_n}dW_n$$

where $U_n : \mathbb{D}^N \to \mathbb{R}$ is the interaction potential for particle n, W_n are independent Wiener processes, D_n is the diffusion constant for particle n, and $\boldsymbol{Q} = (\boldsymbol{Q}(t))_{t \geq 0}$,

$$\boldsymbol{Q}(t) = (Q_1(t), \ldots, Q_N(t)),$$

is the configuration process containing the positions of all particles. That is, the dynamics of the particles are not independent anymore but coupled via the interaction potentials.

Molecular Dynamics

Such particle-particle interaction potentials provide a bridge to more detailed *molecular dynamics*, which study, e.g., forces between individual molecules on an atomic scale [86, 199, 214]. In such microscopic molecular dynamics models, the state is defined by the position coordinates and the momenta of all atoms of a molecular system. The objective is to analyze and to understand the geometric structures and dynamical properties of molecular conformations. On this atomic level, a chemical reaction is a whole sequence of atomic movements and interactions. In this sense, the particle-based reaction-diffusion model (which is the most detailed model considered in this book) can itself be interpreted as a coarsening of more in-depth descriptions. We will not touch upon the atomic level of molecular dynamics in more detail, but keep the interpretation of particles as whole molecules and treat each chemical reaction as a single, aggregated event. A coupling of molecular dynamics and reaction-diffusion simulation (by means of Markov state models) is presented in [51].

4.1.2 Diffusion-Controlled Bimolecular Reactions

Given the different types of diffusion equations for the particles' movement in space, we now consider the situation where the diffusing particles may additionally interact with each other by chemical reactions. In contrast to the models of the preceding chapters, we now assume that the occurrence of a bimolecular reaction requires the physical proximity of its reactants. Different modeling approaches for such diffusion-controlled bimolecular reactions have been derived in the last century.

In the theory originally derived by Smoluchowski in 1917, particles are modeled as spheres moving in space by independent Brownian motion, with bimolecular reactions occurring immediately when two reactive particles collide [231, 232]. The precise mathematical formulation is given for a single pair of spherical particles of two species S_1 and S_2 undergoing the bimolecular reaction $S_1 + S_2 \to S_3$. The reaction is not reversible and can therefore happen only once. In order to further reduce the dimensionality of the system, the position of the center of the S_1-particle is fixed, and the *relative* position $q \in \mathbb{R}^d$ of the S_2-particle moving around the S_1-particle is considered. The simultaneous diffusion of both particles is taken care of by setting the diffusion coefficient D to be the sum of the diffusion coefficients D_1 and D_2 of the two particles, i.e., $D := D_1 + D_2$. Given the radii ι_1 and ι_2 of the two spherical particles, the reaction occurs as soon as the S_2-particle reaches by diffusion the distance $\vartheta_1 := \iota_1 + \iota_2$, called *reaction radius* .

Assuming spherical symmetry of the dynamics, the model can be formulated in terms of the relative distance $\vartheta = \|q\|$ between the particles. Let $\rho(\vartheta, t)$ denote the probability density to find the system unreacted at time $t \geq 0$ at a relative distance $\vartheta > 0$, and assuming D to be constant, the system obeys the equation [46]

$$\frac{\partial \rho(\vartheta, t)}{\partial t} = \frac{D}{\vartheta^2} \frac{\partial}{\partial \vartheta} \left(\vartheta^2 \frac{\partial}{\partial \vartheta} \rho(\vartheta, t) \right) \tag{4.6}$$

with boundary condition

$$\rho(\vartheta, t) = 0 \quad \text{for } \vartheta = \vartheta_1. \tag{4.7}$$

The absorbing boundary condition (4.7) reflects the *immediate* reaction in case of proximity. It also means that the S_2-particle cannot be found inside the distance ϑ_1 to the $\mathcal{S}_1$-particle. Equation (4.6) results from a conversion to radial coordinates of the related Smoluchowski equation (Fokker-Planck equation)

$$\frac{\partial}{\partial t}\rho(q,t) = D\nabla^2\rho(q,t), \tag{4.8}$$

which describes the Brownian dynamics of the $\mathcal{S}_2$-particle in Cartesian coordinates, also given by the SDE (4.1). Here, $q \in \mathbb{R}^3$ is the position in $\mathbb{R}^3$ with $\|q\| = \vartheta$, and ∇^2 is the Laplace operator with respect to q.

Later, the model has been extended by Collins and Kimball [38], who introduced a *finite* reactivity, replacing (4.7) by the partially absorbing boundary condition

$$\frac{\partial}{\partial\vartheta}\rho(\vartheta,t) = \gamma_1^{\mathrm{micro}}\rho(\vartheta,t) \quad \text{for } \vartheta = \vartheta_1, \tag{4.9}$$

where $\gamma_1^{\mathrm{micro}} > 0$ is a microscopic reaction rate constant for the binding reaction. For $\gamma_1^{\mathrm{micro}} \to \infty$ this leads back to the Smoluchowski boundary condition (4.7).

The analytical solution of the partial differential equation (PDE) (4.6) depends on the chosen boundary conditions and initial distribution and is expressed in terms of Green's function [222, 235]. An extension to reversible reactions can be found in [1, 146, 194]. We will not carry out any details here. In the original works, the dynamics were described in terms of concentrations rather than by density functions [38, 215, 232], which leads to other solution results for the corresponding boundary value problems.

In the Smoluchowski model (with infinite or finite reactivity), a bimolecular reaction can only occur when the distance between the reactants is *exactly* the reaction radius ϑ_1. In the 1970s, Doi adopted the idea of finite reactivity and introduced a setting where bimolecular reactions between two particles occur with a fixed probability per unit time whenever two reactants are separated by *less* than the reaction radius ϑ_1 [56]. In the *Doi-model*, particles are treated as points (without mass) that undergo independent Brownian motion and can get arbitrarily close to each other in space. The reaction propensity for any pair of reactive particles is of the form

$$\gamma_1^{\mathrm{micro}}\mathbb{1}_{\vartheta_1}(q_1, q_2)$$

where $q_1, q_2 \in \mathbb{D}$ are the spatial positions of the two reactants and

$$\mathbb{1}_{\vartheta_1}(q_1, q_2) = \begin{cases} 1 & \text{if } \|q_1 - q_2\| \leq \vartheta_1, \\ 0 & \text{otherwise,} \end{cases}$$

(see Fig. 4.4 for an illustration). If $\gamma_1^{\mathrm{micro}}$ is very small, the reactants collide many times before a reaction takes place and the system becomes well-mixed, such that the dynamics can be characterized by the CME (1.10). For $\gamma_1^{\mathrm{micro}} \to \infty$, on the other hand, the particles react immediately when getting close enough, which again yields the original Smoluchowski model.

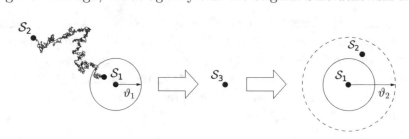

Figure 4.4. Doi-model. Physical model for binding and unbinding $\mathcal{S}_1 + \mathcal{S}_2 \leftrightarrow \mathcal{S}_3$ from the perspective of the $\mathcal{S}_1$-particle. Binding occurs at rate $\gamma_1^{\mathrm{micro}}$ whenever the distance between the $\mathcal{S}_1$- and $\mathcal{S}_2$-particles (black dots) is less than the binding radius $\vartheta_1 > 0$. After unbinding, which occurs at rate $\gamma_2^{\mathrm{micro}}$, the resulting $\mathcal{S}_1$-particle takes the former position of the $\mathcal{S}_3$-particle, and the $\mathcal{S}_2$-particle is placed randomly with a distance drawn from $[0, \vartheta_2]$ where ϑ_2 is called unbinding radius. Choosing $\vartheta_2 > \vartheta_1$ reduces possible rebinding effects

Given that a reaction takes place, also the spatial placement of its products has to be defined. For reactions of type $\mathcal{S}_1 \to \mathcal{S}_2$, product $\mathcal{S}_2$ is naturally placed at the former position of the $\mathcal{S}_1$-particle. If the numbers of reactants and products of a reaction are different, on the other hand, the placement is not that clear. In most particle-based reaction-diffusion models, the products are placed randomly within an environment of its reactants. The placement rule is especially relevant in case of reversible reactions like binding and unbinding. If, after unbinding, the two products are placed with a distance ϑ_2 smaller than the reaction radius of the binding reaction, they immediately get the opportunity to bind again. In the Smoluchowski model, this instant rebinding would definitely occur (due to the infinite reactivity in case of proximity) such that the unbinding distance has to be chosen larger than the binding radius in order to enable unbinding to actually take place [12]. In the more general setting of a finite microscopic binding rate, the particles are able to separate by diffusive motion before rebinding happens. Any choice of the unbinding distance is generally feasible in this case. As rebinding effects are natural phenomena, a complete prevention is not appropriate anyway. Rebinding is even necessary to obtain detailed balance for the overall stochastic process. Figure 4.4 illustrates the binding model of Doi for a reversible reaction and general unbinding distance.

4.1.3 Mathematical Model for Multi-article Systems

Given any of the former rules for the particles' diffusion dynamics and their interactions by chemical reactions, the main goal is to know how the overall system of particles evolves in time. Interesting questions in this context are: How does the population develop in space and time? How often do the reactions occur, and where do they take place? What is the probability to find the total system of reactive particles at a given time in a certain configuration?

The theoretical formulations of Sect. 4.1.2 are limited to isolated pairs of reacting particles. A direct transfer to multi-particle systems with time-varying numbers of particles is rather complicated regarding notation [151]. In contrast to the well-mixed dynamics considered in the previous chapters, a particle-based model needs to label the particles in order to follow their individual diffusive motion. However, given that the particles appear and disappear in the course of time due to chemical reactions, the total number N of particles (and with it the dimension N of the configuration state) is not constant but changes in time, such that already the labeling is ambiguous.

The following notes give a slight impression of how one can tackle the challenge of mathematically formulating a multi-particle system on a spatially resolved scale. The idea is to use Fock space calculus, a rather general approach in which the state space of the systems contains all the state spaces for systems of N particles for all possible numbers of N simultaneously. This approach is often utilized in quantum mechanics and quantum field theory, building on Hilbert space theory, but can also be used in other contexts. Next, we will give a very short introduction to Fock spaces and their use for reacting particle systems. Readers who are mainly interested in the practical implementation of particle-based reaction-diffusion systems might directly proceed with Section 4.1.4.

Fock Space

Let $\mathcal{H}$ denote a Hilbert space of functions on $\mathbb{R}^d$ with scalar product $\langle \cdot, \cdot \rangle$, e.g., $\mathcal{H} = L^2(\mathbb{R}^d) = \left\{ u : \mathbb{R}^d \to \mathbb{R} \, | \int_{\mathbb{R}^d} u(q)^2 \, dq < \infty \right\}$. The space $\mathcal{H}$ is meant to contain the probability distribution functions $u(q)$ for the position $q \in \mathbb{R}^d$ of a single particle and is therefore called the *single particle space*. The tensor product $\mathcal{H} \otimes \mathcal{H}$ of the function space $\mathcal{H}$ with itself is the completion of

$$\text{span}\{u \otimes v | \ u, v \in \mathcal{H}\},$$

where $u \otimes v : (\mathbb{R}^d)^2 \to \mathbb{R}$ is given by $(u \otimes v)(q_1, q_2) := u(q_1)v(q_2)$. It is a Hilbert space with the extended product scalar product

$$\langle u_1 \otimes v_1, u_2 \otimes v_2 \rangle := \langle u_1, v_1 \rangle \langle u_2, v_2 \rangle.$$

If we consider N particles with positions $q_1, \ldots, q_N$, the probability density function generated by this system can be described as a function in

$$\mathcal{H}^{\otimes N} := \bigotimes_{i=1}^{N} \mathcal{H}.$$

If $u_1, u_2, \ldots$ denote an orthonormal basis of $\mathcal{H}$, then

$$u_{i_1, \ldots, i_n} = u_{i_1} \otimes \ldots \otimes u_{i_n}$$

constructs an orthonormal basis of $\mathcal{H}^{\otimes N}$.

In the space $\mathcal{H}^{\otimes N}$, the numbering of the particles matters, which should not be the case if the N particles are indistinguishable. Therefore, we consider the permutation operator U_σ on $\mathcal{H}^{\otimes N}$ induced by a permutation σ on $\{1, \ldots, N\}$

$$U_\sigma(u_1 \otimes \ldots \otimes u_N) := u_{\sigma(1)} \otimes \ldots \otimes u_{\sigma(N)},$$

and define the orthogonal projection $\Pi_N : \mathcal{H}^{\otimes N} \to \mathcal{H}^{\otimes N}$

$$\Pi_N := \frac{1}{N!} \sum_\sigma U_\sigma,$$

and the *symmetrized N-particle space*

$$\Pi_N \mathcal{H}^{\otimes N} = \Pi_N \bigotimes_{i=1}^{N} \mathcal{H}.$$

If the number of particles is not fixed but ensembles of particles with different particle numbers are considered, then the respective function space is the so-called (bosonic) *Fock space*

$$F(\mathcal{H}) := \bigoplus_{N=0}^{\infty} \Pi_N \mathcal{H}^{\otimes N} = \mathbb{C} \oplus \mathcal{H} \oplus (\Pi_2(\mathcal{H} \otimes \mathcal{H})) \oplus \ldots,$$

where the term for $N = 0$ (the 0-particle space) is given by $\mathbb{C}$, the complex scalars. An element of the Fock space $F(\mathcal{H})$ is a tuple $(u^{(0)}, u^{(1)}, u^{(2)}, \ldots)$ with $u^{(N)} \in \Pi_N \mathcal{H}^{\otimes N}$, where $u^{(N)} = u^{(N)}(q_1, \ldots, q_N)$ is the probability distribution function to find N particles at positions $q_1, \ldots, q_N$.

Single-Particle and Two-Particle Operators

If Ψ denotes an operator acting on the single-particle space $\mathcal{H}$, then let $\Psi_j : \mathcal{H}^{\otimes N} \to \mathcal{H}^{\otimes N}$ denote the same operator acting on the jth component of an N-particle state from $\mathcal{H}^{\otimes N}$. Using the definition of Π_N, we have

$$\Pi_N \Psi_1 \Pi_N = \Pi_N \frac{1}{N} \sum_{j=1}^{N} \Psi_j \Pi_N = \frac{1}{N} \sum_{j=1}^{N} \Psi_j \Pi_N, \qquad (4.10)$$

where the first equality means that, after symmetrization, the operator Ψ_1 is indistinguishable from $\frac{1}{N} \sum_{j=1}^{N} \Psi_j$, while the second equality results from the fact that the operator $\frac{1}{N} \sum_{j=1}^{N} \Psi_j$ keeps the state symmetrized. The action of a linear operator Ψ on each particle in the N-particle system is given by the operator

$$\Psi^N := \sum_{j=1}^{N} \Psi_j,$$

which defines a linear map $\Psi^N : \Pi_N \mathcal{H}^{\otimes N} \to \Pi_N \mathcal{H}^{\otimes N}$. An example is given by the diffusion operator that acts component-wise on a N-particle function. On Fock space, the single-particle operator reads

$$\Psi^{F(\mathcal{H})} = \bigoplus_{N=1}^{\infty} \Psi^N = \bigoplus_{N=1}^{\infty} \sum_{j=1}^{N} \Psi_j.$$

Similarly, for a two-particle operator $\hat{\Psi}$ on $\mathcal{H} \otimes \mathcal{H}$ that (simultaneously) acts on two coordinates of any many-particle state, its action on $\Pi_N \mathcal{H}^{\otimes N}$ is given by

$$\hat{\Psi}^N = \sum_{1 \leq i \leq j \leq N} \hat{\Psi}_{ij}$$

where $\hat{\Psi}_{ij}$ denotes the action of $\hat{\Psi}$ on the i and j component of the respective N-particle state. In Fock space $F(\mathcal{H})$, the two-particle operator has the form

$$\hat{\Psi}^{F(\mathcal{H})} = \bigoplus_{N=2}^{\infty} \hat{\Psi}^N = \bigoplus_{N=1}^{\infty} \sum_{1 \leq i \leq j \leq N} \hat{\Psi}_{ij}.$$

Two-Species Fock Space

For systems with two chemical species $\mathcal{S}_1$ and $\mathcal{S}_2$, the Fock space can be composed out of the Fock spaces of each species

$$F(\mathcal{H}_1, \mathcal{H}_2) := \bigoplus_{N,M=0}^{\infty} \Pi_N \mathcal{H}_1^{\otimes N} \otimes \Pi_M \mathcal{H}_2^{\otimes M}.$$

This definition can directly be extended to systems with more than two different species with $\mathcal{H}_1 = \mathcal{H}_2 = \mathcal{H}$ if the species can be considered in the same function space.

Creation and Annihilation Operators

The single-species single-particle and multi-particle operators can be rewritten in terms of the so-called creation and annihilation operators, such that the operators Ψ_j and $\hat{\Psi}_{ij}$ enter (only) through their representation in a basis of the Hilbert space. The operators acting on a multi-species Fock space (including the ones describing reactions between species) can be rewritten using (multi-species) creation and annihilation operators, too. The creation operator maps from $\mathcal{H}^{\otimes N}$ to $\mathcal{H}^{\otimes N+1}$, while the annihilation operator maps $\mathcal{H}^{\otimes N+1}$ to $\mathcal{H}^{\otimes N}$. This way, reaction-diffusion evolution equations for multi-particle systems with several species can be derived and completely expressed by means of creation and annihilation operators (see, e.g., [151] or [55] for a slightly different notation).

We see that the mathematical description of a whole system of interacting particles on a microscopic, particle-based level becomes very complex even for a small number of species and reactions. Analytical solutions of the dynamics exist only for some special reduced settings. Finding more general reaction-diffusion models on the microscopic scale is still an active matter of research. Most of the existing literature on particle-based reaction-diffusion processes, however, concentrates on numerical simulation methods rather than on analytical approaches.

4.1.4 Particle-Based Stochastic Simulation

In most cases of particle-based dynamics, analytical approaches fail because the setting is too complicated to even formulate a mathematical model for the overall dynamics. Instead, the system is typically sampled by stochastic realizations. A basic practical introduction to stochastic simulation methods for reaction-diffusion processes is given in [66]. Many different software tools have been developed in the past, e.g., Smoldyn [11, 12], GFRD (Green's function reaction dynamics) [235, 236], or ReaDDy [210, 211]; reviews are given in [149, 197, 211]. The approaches differ with respect to the level of modeling detail regarding particle diffusion, geometrical confinement, particle volume exclusion, or particle-particle interaction potentials. A higher level of detail increases the model complexity and the computational cost, and it depends on the application at hand in which the level of detail is recommended.

The simulation algorithms are mostly based on a discretization of time. For each time step, the positions of all individual particles are advanced according to a given rule of motion. The simplest possible simulation algorithm can be described as follows: Given the diffusion process (4.3), an Euler-Maruyama discretization with time step $\Delta t > 0$ leads to the recursion

$$q(t + \Delta t) = q(t) - \Delta t \frac{D}{k_B T} \nabla U(q(t)) + \sqrt{2D\Delta t}\, \xi,$$

with ξ being a standard Gaussian random variable in $\mathbb{R}^d$. Each diffusive advance is followed by a checkup for reactive complexes, i.e., sets of reactive particles whose reaction radii overlap. Given such a complex, the reaction can fire; whether this happens or not is decided randomly based on the microscopic reaction rate constant $\gamma_k^{\mathrm{micro}}$. Treating the reactions as independent events, each of them is executed with probability

$$1 - \exp(-\gamma_k^{\mathrm{micro}} \Delta t).$$

For unimolecular reactions no spatial information is required. These reactions occur with a fixed rate independent of the spatial position of the reactant. Reactions of order ≥ 3 are not considered at all; in case of need, they can be decomposed into several second-order reactions.

While both the diffusion coefficients and the macroscopic reaction rate constants γ_1 and γ_2 from the reversible reaction

$$\mathcal{S}_1 + \mathcal{S}_2 \underset{\gamma_2}{\overset{\gamma_1}{\rightleftharpoons}} \mathcal{S}_3$$

are experimentally measurable, the radii ϑ_1, ϑ_2 and the microscopic reaction rate constants $\gamma_k^{\mathrm{micro}}$, $k = 1, 2$ are parameters that have to be chosen by the modeler. There is a certain freedom in choosing these model parameters because different combinations of the binding radius ϑ_1, the unbinding radius ϑ_2, and the microscopic rate constants $\gamma_k^{\mathrm{micro}}$ correspond to the same macroscopic rate constants γ_1, γ_2. A discussion and comparison of different parameter values for such reversible bimolecular reactions can be found in [172].

In detail, the particle-based simulation algorithms differ from each other and can be very complex. Nevertheless, the rough scheme can be summarized as follows see Fig. 4.5.

Scheme for discrete-time particle-based simulation:

1. Initialize time $t \leftarrow t_0$, number of particles $N^l \leftarrow N_0^l$, and configurations $\boldsymbol{q}^l \leftarrow \boldsymbol{q}_0^l \in \mathbb{D}^{N^l}$ for each species $l = 1, \ldots, L$, and choose a time horizon $T > t_0$.

2. Advance the positions of all particles in space by one step of length Δt based on the given diffusion dynamics.

3. Create a list of reactions that can occur and draw random numbers to decide which of them are executed.

4. Place the reaction products according to the defined rules, and set $t \leftarrow t + \Delta t$.

5. If $t \geq T$, end the simulation. Otherwise return to 2.

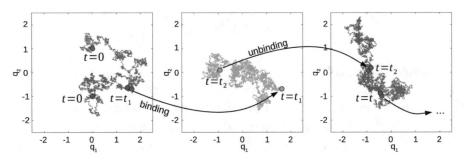

Figure 4.5. Particle-based reaction-diffusion. Particle-based simulation of reaction-diffusion dynamics for binding and unbinding $\mathcal{S}_1 + \mathcal{S}_2 \rightleftharpoons \mathcal{S}_3$ and diffusion in the one-well potential $U(q) = 0.5(q_1^2 + q_2^2)$ (see Fig. 4.3). Left: At time $t = 0$, one particle of species $\mathcal{S}_1$ and one particle of species $\mathcal{S}_2$ start at different positions in $\mathbb{D} = \mathbb{R}^2$ given by $q_0^1 = (0, 1)$ (blue starting point) and $q_0^1 = (0, -1)$ (red starting point). They diffuse in space (blue and red trajectory) and react at the random time point $t = t_1$ after getting close enough in space (Doi-model with binding radius $\vartheta_1 = 0.2$, microscopic binding rate $\gamma_1^{\mathrm{micro}} = 100$). Middle: The resulting particle of species $\mathcal{S}_3$ diffuses in space until it unbinds again at time $t = t_2$ (green trajectory). Right: After unbinding, the two particles restart diffusion from close-by starting points until they eventually bind again at time $t = t_3$. Simulation time step $\Delta t = 10^{-4}$

A central issue for such particle-based simulations is the choice of an appropriate size for the time step Δt. Not only the diffusion speed is relevant in this regard but also aspects like environmental crowding and the size of the reaction radii. With an increasing step size (particularly if the mean spatial advance of a particle is larger than its reaction radius), a "mutual escape" of reactive particles becomes more likely, i.e., particles cross without getting the chance to undergo a reaction. On the other hand, a too small time step obviously renders the simulation inefficient.

Another problem that arises by the discretization of time is the delayed update of the reaction propensities. In fact, two reactions can possibly exclude each other, e.g., if both consume the same reactant particle, such that actually only one of the two reactions can take place. Especially if particles are densely packed in the environment, reaction conflicts are likely to occur. Approaches to handle these reaction conflicts are still under development. A straightforward approach is to redefine the list of possible reactions after each execution of a reaction, discarding those reactions which cannot happen anymore. Another approach is to interpret the dynamics as a time-evolving network, with the vertices given by the particles and edges existing between two particles whenever they are close enough to react. In this context, the *temporal Gillespie algorithm* offers ways for temporally more flexible, system-adapted simulations [53, 237].

A particle-based modeling approach which takes into account the impact of particle collisions onto their velocities is presented in [141, 229]. Positions and velocities are considered as continuous variables, and multi-particle collisions are carried out by partitioning the system into cells and performing special random rotations of the particle velocities. Numerical simulation is again based on temporal discretization.

As an alternative to temporal discretization, there exist modeling approaches that are based on a discretization of space. The continuous diffusive motion of particles is replaced by hopping events between local subdomains, and reaction propensities are updated after each reactive event, such that reaction conflicts are naturally excluded. Such spatially discretized processes will be introduced in Sect. 4.3.

The particle-based resolution is appropriate in cases of low particle concentrations and slow diffusion. Spatial inhomogeneities can be taken into account up to a very detailed level. The simulations, however, can be extremely costly often requiring several CPU years in a single simulation for reaching the biological relevant time scales of seconds. If the population size increases and part of the system is dense, the approach becomes numerically inefficient because reactions would fire in every iteration step. In such a case, a description of the population by concentrations is more appropriate, leading to PDE formulations of the dynamics, which will be investigated in the following section.

4.2 Reaction-Diffusion PDEs

In Sect. 2.1 we considered well-mixed reaction dynamics in terms of particle concentrations rather than particle numbers. We have seen that, by the law of large numbers, randomness vanishes in the large volume limit, such that well-mixed reactive systems with a huge number of particles can be well described by the deterministic reaction rate equation (2.9). In case of a medium-sized population, where the stochasticity of the dynamics cannot be neglected, the fluctuations may be approximated by Gaussian noise, leading to the chemical Langevin equation (2.16). Such modeling approaches by deterministic dynamics or Itô diffusion processes also exist for non-well-mixed reaction-diffusion systems. The underlying condition is that the population is large in the overall space.

As in the setting of particle-based dynamics analyzed in the previous section, the particles are assumed to move within a continuous domain $\mathbb{D} \subset \mathbb{R}^d$. However, instead of considering individual trajectories, we now formulate the dynamics in terms of particle concentrations depending on time and spatial location. The reaction-diffusion dynamics are characterized by partial differential equations (PDEs) or stochastic PDEs (SPDEs), which are formulated in terms of the vector

$$\boldsymbol{C}(q,t) = (C_1(q,t), \ldots, C_L(q,t)),$$

where L is again the number of species and $C_l(q, t)$ denotes the concentration (i.e., number density) of species $\mathcal{S}_l$ depending on the location $q \in \mathbb{D} \subset \mathbb{R}^d$ and time $t \geq 0$ [20, 57, 145]. The relevant equations have the general form

$$\frac{\partial}{\partial t}\boldsymbol{C}(q,t) = \mathcal{L}^{(D)}\boldsymbol{C}(q,t) + \mathcal{L}^{(R)}\boldsymbol{C}(q,t), \tag{4.11}$$

for a diffusion operator $\mathcal{L}^{(D)}$ and a reaction operator $\mathcal{L}^{(R)}$, whose concrete forms will be specified below. In the case where both $\mathcal{L}^{(D)}$ and $\mathcal{L}^{(R)}$ are deterministic operators, Eq. (4.11) is called *reaction-diffusion PDE*. Such deterministic reaction-diffusion dynamics in terms of PDEs will be introduced in Sect. 4.2.1. In case of stochastic operators $\mathcal{L}^{(D)}$ and $\mathcal{L}^{(R)}$, Eq. (4.11) becomes a stochastic PDE, which will analogously be called *stochastic reaction-diffusion PDE* and is the topic of Sect. 4.2.2.

4.2.1 Deterministic Reaction-Diffusion PDEs

If we assume that all particles diffuse in space by independent Brownian motion with diffusion coefficients D_l, the PDE for the diffusive flux in terms of the concentration $C_l(q,t)$ of species $\mathcal{S}_l$ is given by Fick's second law

$$\frac{\partial C_l}{\partial t} = D_l \nabla^2 C_l, \tag{4.12}$$

which is equivalent to the Smoluchowski equation (4.8) for probability densities. This entails a deterministic diffusive operator $\mathcal{L}^{(D)}$ of the form

$$\mathcal{L}^{(D)}\boldsymbol{C}(q,t) = \boldsymbol{D}\nabla^2\boldsymbol{C}(q,t),$$

where $\boldsymbol{D} \in \mathbb{R}^{L,L}$ is a diagonal diffusion coefficient matrix containing the diffusion coefficients D_l on its diagonal, and the Laplace operator ∇^2 acts on $\boldsymbol{C}$ component-wise. Of course, there also exist diffusion operators for the more sophisticated diffusion models from Sect. 4.1.1 and for subdiffusion and superdiffusion processes, but we will stick here with free Brownian motion for the purpose of simplicity. For example, for the dynamics within an energy landscape defined by the external potential U, cf. (4.2), not taking interacting forces between the particles into account, the diffusion operator can be defined component-wise by

$$(\mathcal{L}^{(D)}\boldsymbol{C})_l(q,t) = \nabla \cdot (\nabla U(q)C_l(q,t)) + D_l\nabla^2 C_l(q,t)$$

in analogy to (4.4).[2]

The reaction operator $\mathcal{L}^{(R)}$ can be specified by defining propensity functions $\tilde{\alpha}_k$ in terms of concentrations just as in Sect. 2.1.1. Here, the propensities may additionally depend on the spatial location, that is, they are given by functions of the form $\tilde{\alpha}_k(\boldsymbol{c}, q)$ for each reaction $\mathcal{R}_k$. Letting further $\boldsymbol{\nu}_k$ denote the state-change vector of $\mathcal{R}_k$, we obtain

$$\mathcal{L}^{(R)}\boldsymbol{C}(q,t) = \sum_{k=1}^{K} \tilde{\alpha}_k(\boldsymbol{C}(q,t), q)\boldsymbol{\nu}_k.$$

In total, if we choose Brownian motion for the diffusion processes (i.e., set $U = 0$), the reaction-diffusion PDE (4.11) becomes

$$\frac{\partial}{\partial t}\boldsymbol{C}(q,t) = \boldsymbol{D}\nabla^2\boldsymbol{C}(q,t) + \sum_{k=1}^{K} \tilde{\alpha}_k(\boldsymbol{C}(q,t), q)\boldsymbol{\nu}_k,$$

---

[2]Given the operator $\mathcal{L}^{(D)}$, we can rewrite (4.4) as $\frac{\partial}{\partial t}\rho_n = \mathcal{L}^{(D)}\rho_n$. Note that $\rho_n(q,t)$ is the probability density function for the position $Q(t)$ of a single particle at time t, while $C_l(q,t)$ is the concentration of particles at position q and time t given the whole population.

or, component-wise,

$$\frac{\partial}{\partial t}C_l(q,t) = D_l\nabla^2 C_l(q,t) + \sum_{k=1}^{K} \tilde{\alpha}_k(\boldsymbol{C}(q,t),q)\nu_{lk}. \tag{4.13}$$

Such a system of reaction-diffusion PDEs (4.13) can be regarded as a model independent of others or as an approximation of the more detailed particle-based dynamics. The exact relation between the models is not yet fully understood, mainly because the translation of the reaction parameters is still not completely clear (reaction radius and microscopic rate constant vs. propensity function $\tilde{\alpha}_k$).

Normally, the reaction-diffusion PDEs are too complex to be solved analytically. Instead, standard numerical techniques for solving PDEs are applied: Using finite element or finite-volume methods, the PDE is converted by spatial discretization (method of lines) into coupled sets of *compartment ODEs*, which can then be solved by Runge-Kutta schemes [14, 58, 64]. Depending on the chosen approach, the resulting ODEs may take on the form of reaction rate equations, leading to a system of *RREs in compartments* that are coupled by diffusion terms.

The technique of solving a PDE by discretizing the spatial derivatives but (initially) leaving the time variable continuous is termed *method of lines* [206]. As we have seen, this kind of approach leads from a PDE to a system of ODEs. The counterpart to this approach is given by the so-called *Rothe method* which starts with a discretization of time (e.g., by an implicit Runge-Kutta scheme). By this process, the spatiotemporal PDE problem is turned into an iterative sequence of purely spatial PDE problems which allows to apply the full wealth of finite element methods to its solution, cf. [162].

RREs in Compartments

Instead of spatially continuous dynamics in terms of PDEs, we now consider a discretization of space into compartments $\mathbb{D}_j$, $j \in \mathbb{I}$ for some index set $\mathbb{I}$. The idea is to describe the local reaction kinetics within each compartment by a RRE of type (2.9). Diffusion in space is included by mutual "interaction" of the RREs. Let $\hat{C}_{lj}(t)$ denote the concentration of species $\mathcal{S}_l$ at time t in compartment $\mathbb{D}_j$, $j \in \mathbb{I}$, and set $\hat{\boldsymbol{C}}_j(t) = (\hat{C}_{1j}(t),\ldots,\hat{C}_{Lj}(t))^\mathsf{T}$ for each compartment. The coupled *compartment RREs* are given by a system of equations

$$\frac{d}{dt}\hat{C}_{lj}(t) = \underbrace{\sum_{i\in\mathbb{I}}\lambda_{ij}^l\hat{C}_{li}(t)}_{\text{diffusion}} + \underbrace{\sum_{k=1}^{K}\tilde{\alpha}_k^j(\hat{C}_j(t))\nu_{lk}}_{\text{reactions}}, \tag{4.14}$$

where $\lambda_{ij}^l \geq 0$ is the jump rate constant between compartments $\mathbb{D}_i$ and $\mathbb{D}_j$ for species $\mathcal{S}_l$ and $\tilde{\alpha}_k^j$ which denote the reaction propensities depending on the compartment $\mathbb{D}_j$.[3] This means that there is an exchange between the concentrations of neighboring compartments by diffusive flow, combined with an interior evolution of the concentrations for each compartment due to chemical reactions.

As an example, let the space be discretized by a regular Cartesian grid into mesh cells of constant size $h > 0$. A finite element discretization of the diffusion equation (4.12) leads to jump rate constants given by

$$\lambda_{ij}^l = \begin{cases} \frac{D_l}{h^2} & \mathbb{D}_j \text{ is a nondiagonal neighbor of } \mathbb{D}_i, \\ 0 & \text{otherwise} \end{cases} \tag{4.15}$$

for $i \neq j$ and $\lambda_{ii}^l := -\sum_{j\neq i}\lambda_{ij}^l$ (see Appendix A.2 for the mathematical background). The propensities $\tilde{\alpha}_k^j(\hat{C}_j(t))$, moreover, can be chosen such that they approximate the compartment averages of the spatially continuous propensity functions $\tilde{\alpha}_k(C(q,t),q)$ in (4.13)

$$\tilde{\alpha}_k^j(\hat{C}_j(t)) \approx \frac{1}{|\mathbb{D}_j|}\int_{\mathbb{D}_j}\tilde{\alpha}_k(C(q,t),q)\,dq,$$

which relates the compartment RREs to the reaction-diffusion PDE (4.13) [58].

In comparison to the spatially continuous reaction-diffusion PDE, the compartment RREs reduce spatial information and can therefore be considered as a *mesoscopic* model on the spatial scale (see the overview in Fig. 4.1). On the population scale, both types of systems appear as macroscopic models because stochastic effects are not involved at all. It is worth mentioning that a characterization of the reaction dynamics within each (possibly small) compartment by purely deterministic dynamics may seem inappropriate because the local population can actually be quite small, such that stochastic effects should not be ignored. Stochastic compartment-based models, which tackle this problem, will be discussed in Sect. 4.3 but also appear in the following section.

---

[3]The compartment propensities are functions of the concentration only, i.e., $\tilde{\alpha}_k^j = \tilde{\alpha}_k^j(c)$.

4.2.2   Stochastic Reaction-Diffusion PDEs

The reaction-diffusion PDE model is appropriate if the number of particles is large in the overall space and stochasticity can be neglected. At
levels where spatiotemporal fluctuations in the concentration are significant, the deterministic macroscopic description may fail, and a stochastic
mesoscopic description is needed. In the case of spatially well-mixed dynamics, such a mesoscopic description is given by the chemical Langevin
equation, which is a stochastic differential equation (SDE) (see Sect. 2.1.2).
The analog for spatially inhomogeneous dynamics with comparatively slow
diffusion is given by stochastic PDEs that are formulated in terms of the
concentration $C(q, t)$. The idea is to adapt the approach used for fluctuating hydrodynamics (FHD), where each dissipative flux in the hydrodynamic
equations is augmented with a stochastic flux. This means that the internal
or thermodynamic fluctuations of both reaction and diffusion dynamics are
represented by white noise components that are added to the deterministic
reaction-diffusion PDE (4.13).

The resulting FHD equations for $C(q, t)$ are formally given by the
stochastic partial differential equations (SPDEs) [145]

$$
\begin{aligned}
\frac{\partial}{\partial t} C_l =& \nabla \cdot \left(D_l \nabla C_l + \sqrt{2 D_l C_l} \boldsymbol{Z}_l^{(D)} \right) \\
&+ \sum_{k=1}^{K} \nu_{lk} \left(\tilde{\alpha}_k(\boldsymbol{C}) + \sqrt{\tilde{\alpha}_k(\boldsymbol{C})} Z_k^{(R)} \right),
\end{aligned}
\tag{4.16}
$$

where again D_l is the diffusion coefficient of species $\mathcal{S}_l$, ν_{lk} is the stoichiometric coefficient of species $\mathcal{S}_l$ in reaction $\mathcal{R}_k$, and $\tilde{\alpha}_k$ is the propensity
function in terms of concentrations. For each $l = 1, \ldots, L$, $\boldsymbol{Z}_l^{(D)}(q, t) = (Z_{l,1}^{(D)}(q, t), \ldots, Z_{l,d}^{(D)}(q, t)) \in \mathbb{R}^d$ denotes a vector of independent Gaussian
space-time white noise processes, which are processes with zero mean

$$
\mathbb{E}\left(Z_l^{(D)}(q, t) \right) = 0 \quad \forall q \in \mathbb{R}^d, t \geq 0,
$$

which are uncorrelated in space and time

$$
\mathbb{E}\left(Z_{l,j}^{(D)}(q, t) Z_{l,j'}^{(D)}(q', t') \right) = \delta_{jj'} \delta(q - q') \delta(t - t')
\tag{4.17}
$$

for $j, j' \in \{1, \ldots, d\}$ [20, 145]. Here, $\delta_{jj'}$ is the Kronecker delta, and $\delta(\cdot)$
is the Dirac delta function. Analogously, we define the scalar independent
noise terms $Z_k^{(R)}(q, t) \in \mathbb{R}$ for each of the reactions $\mathcal{R}_k$, $k = 1, \ldots, K$, which
are also assumed to be normalized such that

$$
\mathbb{E}\left(Z_k^{(R)}(q, t), Z_k^{(R)}(q', t') \right) = \delta(q - q') \delta(t - t').
$$

We assume that these processes are all mutually independent.

Formal Derivation

The first line in (4.16) describes the diffusion of particles in space. For a formal derivation of this diffusive part, we follow [45] and consider a microscopic system of particles of species S_l undergoing independent Brownian motion in $\mathbb{R}^d$

$$dQ_n(t) = \sqrt{2D_l}\, dW_n(t),$$

where n is the index of the particle and $(W_n(t))_{t\geq 0}$ is a standard Wiener process in $\mathbb{R}^d$ (see also (4.1)). Applying Itô's lemma (cf. Lemma A.8 in the Appendix) to $f(Q_n(t))$ for a twice-differentiable scalar function $f : \mathbb{R}^d \to \mathbb{R}$ gives

$$df(Q_n(t)) = D_l\nabla^2 f(Q_n(t))\, dt + \nabla f(Q_n(t)) \cdot \sqrt{2D_l}dW_n(t). \qquad (4.18)$$

Define the random field

$$\phi_n(q,t) := \delta(Q_n(t) - q)$$

where δ is again the Dirac delta function, such that

$$f(Q_n(t)) = \int \phi_n(q,t)f(q)\, dq. \qquad (4.19)$$

Using the fact that it holds $\nabla f(Q_n(t)) = \int \phi_n(q,t)\nabla f(q)\, dq$ as well as $\nabla^2 f(Q_n(t)) = \int \phi_n(q,t)\nabla^2 f(q)\, dq$, we can rewrite (4.18) as

$$df(Q_n(t)) = \int \phi_n(q,t)\left(D_l\nabla^2 f(q)\, dt + \nabla f(q) \cdot \sqrt{2D_l}dW_n(t)\right) dq.$$

Integration by parts further gives[4]

$$df(Q_n(t)) = \int f(q)\left(D_l\nabla^2 \phi_n(q,t)\, dt - \nabla \cdot \left(\phi_n(q,t)\sqrt{2D_l}dW_n(t)\right)\right) dy. \qquad (4.20)$$

On the other hand, it follows from (4.19) and the Leipniz integral rule [78] that

$$\frac{d}{dt}f(Q_n(t)) = \int \frac{\partial}{\partial t}\phi_n(q,t)f(q)\, dq. \qquad (4.21)$$

---

[4]Note that ϕ_n is not differentiable with respect to q in the classical sense. The derivative of the distribution ϕ_n is defined by $\int f\nabla\phi_n dq = -\int \nabla f\phi_n dq$ for all test functions $f \in C_0^\infty(\mathbb{D})$. Equivalently, $\nabla^2\phi_n$ is defined by $\int f\nabla^2\phi_n dq = \int \nabla^2 f\phi_n dq$ for all test functions f.

By means of the *fundamental lemma of calculus of variations* [97] and since f is an arbitrary function, a comparison of (4.20) and (4.21) gives

$$\frac{\partial}{\partial t}\phi_n(q,t)\,dt = D_l\nabla^2\phi_n(q,t)\,dt - \nabla\cdot\left(\phi_n(q,t)\sqrt{2D_l}dW_n(t)\right).$$

Now we sum over n and define $C_l(q,t) := \sum_{n=1}^{N_l}\phi_n(q,t)$ in order to obtain

$$\frac{\partial}{\partial t}C_l(q,t)\,dt = D_l\nabla^2 C_l(q,t)\,dt - \sum_n \nabla\cdot\left(\phi_n(q,t)\sqrt{2D_l}dW_n(t)\right)$$

or, equivalently,

$$\frac{\partial}{\partial t}C_l(q,t) = D_l\nabla^2 C_l(q,t) - \sum_n \nabla\cdot\left(\phi_n(q,t)\sqrt{2D_l}\xi_n(t)\right) \qquad (4.22)$$

for some white noise $\xi_n(t)$. The overall (scalar) noise term

$$\chi(q,t) := -\sum_n \nabla\cdot\left(\phi_n(q,t)\sqrt{2D_l}\xi_n(t)\right)$$

appearing in (4.22) cannot be turned into a term that only depends on C_l. However, it can be shown to have the same mean and covariance function as the global noise field $\tilde{\chi}(q,t)$ defined as

$$\tilde{\chi}(q,t) := \nabla\cdot\left(\sqrt{2D_l C_l(q,t)}\mathbf{Z}_l^{(D)}(q,t)\right),$$

where $\mathbf{Z}_l^{(D)}$ is the globally uncorrelated white noise field fulfilling (4.17) [45].[5] Replacing $\chi(q,t)$ by $\tilde{\chi}(q,t)$ in (4.22) finally gives

$$\frac{\partial}{\partial t}C_l(q,t) = D_l\nabla^2 C_l(q,t) + \nabla\cdot\left(\sqrt{2D_l C_l(q,t)}\mathbf{Z}_l^{(D)}(q,t)\right)$$

$$= \nabla\cdot\left(D_l\nabla C_l(q,t) + \sqrt{2D_l C_l(q,t)}\mathbf{Z}_l^{(D)}(q,t)\right), \qquad (4.23)$$

which is the first line of (4.16). That is, the diffusion part of (4.16) refers to free Brownian motion of each particle in space. Within the theory of FHD, there also exist model extensions which take other effects of hydrodynamics into account, like advection, viscous dissipation, thermal conduction, or cross-term effects [20, 57, 145].

---

[5]In [45], it is even claimed that both noise terms $\chi(q,t)$ and $\tilde{\chi}(q,t)$ are statistically identical Gaussian processes.

The multiplicative noise term $\nabla \cdot \sqrt{2D_l C_l} Z_l^{(D)}$ contained in (4.23) is nonlinear and comes with several mathematical problems, such that the question of existence and uniqueness of the solution of (4.23) is still an open problem [39, 152]. First, the divergence form of this term is obscure. One approach to interpret this term is to consider a weak formulation and use partial integration to transfer the divergence operator to test functions [120]. Second, by construction, C_l is still a sum of delta peaks such that its square root is undefined. In a finite element approach for solving (4.23) numerically, the distribution C_l can be approximated by functions for which the square root is well-defined [120]. Furthermore, there exist martingale solutions of (4.23) [152] and path-wise kinetic approaches [74] to solve this problem.

The second line in (4.16) refers to chemical reactions taking place between the particles. Note that by ignoring the diffusion part given by the first line, one obtains

$$\frac{d}{dt} C_l = \sum_{k=1}^{K} \nu_{lk} \left(\tilde{\alpha}_k(\boldsymbol{C}) + \sqrt{\tilde{\alpha}_k(\boldsymbol{C})} Z_k^{(R)} \right),$$

which, after multiplication with $dt \cdot V$ for finite volume V and using $V\tilde{\alpha}_k(\boldsymbol{c}) = \alpha_k^V(V\boldsymbol{c})$, gives a chemical Langevin equation of type (2.16) for each local position $q \in \mathbb{D}$. The local reaction dynamics of the mesoscopic reaction-diffusion model (4.16) are thus consistent with the mesoscopic modeling approach in the well-mixed scenario of Sect. 2.1.2.

Solving SPDEs

The SPDE (4.16) is typically solved by stochastic simulation. Spatial discretization of (4.16) using finite-volume or finite element approaches converts the SPDE into systems of coupled SDEs (possibly of chemical Langevin type) for compartment number densities [145] (see again Fig. 4.1 on page 133 for an overview). The special case where the local reaction kinetics in each compartment are described by a CLE will be considered in the following.

CLEs in Compartments

Given a discretization of space into compartments $\mathbb{D}_i$, one approach is to formulate for each $\mathbb{D}_i$ a chemical Langevin equation of type (2.14), which leads to a system of *compartment CLEs*. In analogy to the compartment

RREs (4.14), the compartment CLEs have the form

$$
\begin{aligned}
dC_{lj}(t) &= \sum_{i\in\mathbb{I}} \lambda_{ij}^l C_{li}\, dt + \sum_{i\in\mathbb{I}} \sqrt{\lambda_{ij}^l C_{li}}\, dW_{ijl}^{(D)}(t) \\
&\quad + \sum_{k=1}^{K} \alpha_k(\boldsymbol{C}_j)\nu_{lk}\, dt + \sum_{k=1}^{K} \sqrt{\alpha_k(\boldsymbol{C}_j)}\nu_{lk}\, dW_{kj}^{(R)}(t)
\end{aligned}
\tag{4.24}
$$

where again C_{lj} denotes the concentration of species $\mathcal{S}_l$ in compartment $\mathbb{D}_j$ and λ_{ij}^l are the jump rates defined in (4.15). Moreover, $W_{ijl}^{(D)}$ and $W_{kj}^{(R)}$ denote independent standard Wiener processes in $\mathbb{R}$. That is, the first line of (4.24) describes the diffusive motion in space, while the second line refers to the reactions within the compartments.

In [96], the compartment CLEs are derived as a spatial extension of the CLE (2.14). Diffusive jumps between compartments are treated as first-order reactions, leading to an extended reaction network that can be approximated by a CLE in the sense of Sect. 2.1.2.

From Chap. 2 we know that the approximation of stochastic reaction kinetics by a CLE requires the reactive population to be large enough such that reactions occur quite often and the randomness may be expressed in terms of Gaussian noise. Given that the compartments $\mathbb{D}_i$ are small, this requirement might actually not be fulfilled, because each compartment can only contain a few particles. In [145], the white noise fluctuations in the reaction part of the compartment CLEs (4.24) are therefore replaced by Poisson fluctuations, which is again more consistent to the discrete nature of chemical reactions. For such a description, the choice of the compartment sizes is a sensitive issue: On the one hand, they have to be small enough such that the dynamics may be considered as homogeneous and well-mixed within each compartment, and on the other hand, there is also a lower bound on the compartment sizes, because bimolecular reactions would become increasingly infrequent as the compartment size decreases. More details regarding this conflict will be given in Sect. 4.3.2.

This partial replacement of continuous stochastic effects by discrete stochastic effects directly leads to the next section. For the compartment CMEs that will be considered therein, also the white noise fluctuations of the *spatial* diffusion dynamics will be replaced by discrete stochastic jump events between compartments.

Example 4.3 (Spreading Dynamics). Let the particles of two species $\mathcal{S}_1$ and $\mathcal{S}_2$ diffuse in space $\mathbb{D} = \mathbb{R}$ with the diffusion dynamics determined by an energy landscape $U : \mathbb{R} \to \mathbb{R}$. The population is affected by the reaction

$$\mathcal{R}_1 : \mathcal{S}_1 + \mathcal{S}_2 \xrightarrow{\gamma_1} 2\mathcal{S}_2$$

which may be interpreted as an "adoption event": A particle of species $\mathcal{S}_1$ may adopt the property of an $\mathcal{S}_2$-particle when they cross in space.[6] For these dynamics, the total number of particles is naturally constant in time. In the long run, the adoption property will spread throughout the whole population, and there will be only $\mathcal{S}_2$-particles/adopters. The reaction-diffusion PDE is given by the coupled equations

$$\frac{\partial}{\partial t} C_1 = \frac{\partial}{\partial q}(U'C_1) + D_1 \frac{\partial^2}{\partial q^2} C_1 - \gamma_1 C_1 C_2,$$

$$\frac{\partial}{\partial t} C_2 = \frac{\partial}{\partial q}(U'C_2) + D_2 \frac{\partial^2}{\partial q^2} C_2 + \gamma_1 C_1 C_2,$$

where $U'(q) = \frac{d}{dq} U(q)$. The corresponding stochastic reaction-diffusion PDE is given by the two equations

$$
\begin{aligned}
\frac{\partial}{\partial t} C_1 = \ & \frac{\partial}{\partial q}(U'C_1) + D_1 \frac{\partial^2}{\partial q^2} C_1 + \frac{\partial}{\partial q}\left(\sqrt{2D_1 C_1} Z_1^{(D)}\right), \\
& \qquad\qquad -\gamma_1 C_1 C_2 + \sqrt{\gamma_1 C_1 C_2} Z_1^{(R)} \\
\frac{\partial}{\partial t} C_2 = \ & \frac{\partial}{\partial q}(U'C_2) + D_2 \frac{\partial^2}{\partial q^2} C_2 + \frac{\partial}{\partial q}\left(\sqrt{2D_2 C_2} Z_2^{(D)}\right), \\
& \qquad\qquad +\gamma_1 C_1 C_2 + \sqrt{\gamma_1 C_1 C_2} Z_1^{(R)}
\end{aligned}
\tag{4.25}
$$

which are again coupled to each other by the reaction terms (second line of each equation). A realization of the spreading dynamics described by this SPDE for a double-well potential U is shown in Fig. 4.6. A comparison of the SPDE dynamics to the corresponding particle-based dynamics can be found in [120]. $\diamond$

4.3   Compartment CMEs

In the previous section, the discretization of space served the purpose of finding an approximate solution to the spatially continuous dynamics characterized by the (deterministic or stochastic) reaction-diffusion PDE. Now

---

[6]Thinking of agents (living beings) instead of chemical particles, such a reaction could refer to the spreading of a disease or an innovation within a population of *non-adopters* $\mathcal{S}_1$ and *adopters* $\mathcal{S}_2$.

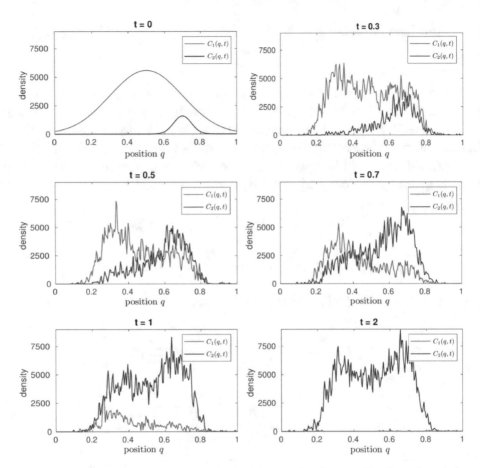

Figure 4.6. Spreading process in a double-well landscape. Realization of the spreading process defined by the SPDE (4.25) given in Example 4.3, with a double-well landscape U on $\mathbb{D} = [0,1]$. We plot the density $C_1(q,t)$ of non-adopters $\mathcal{S}_1$ (in blue) and the density $C_2(q,t)$ of adopters $\mathcal{S}_2$ (in red) at a few time instances. The graphics are taken from [120]

we consider a stochastic reaction-diffusion model that directly relies on spatial discretization and replaces the continuous diffusion by stochastic jumps between spatial compartments. Instead of following individual trajectories or the flow of concentrations, we count the number of particles of each species within each compartment. This is a mesoscopic approach which is less detailed than the microscopic particle-based reaction-diffusion systems of Sect. 4.1. In the mesoscopic approach, particles that are located in the same compartment and belong to the same species are indistinguishable. As we will see, this approach leads to a spatial extension of the fundamental chemical master equation from Sect. 1.2.2.

More precisely, we assume that the diffusion space $\mathbb{D}$ can be split up into a collection of nonoverlapping compartments $\mathbb{D}_i$ ($i \in \mathbb{I}$ for some index set $\mathbb{I}$) in such a way that the reaction-diffusion dynamics can be considered to be well-mixed within each compartment. This implies that particles can react with each other as soon as they are located in the same compartment, regardless of their exact position within the compartment. The spatial movement of particles is incorporated in the model by letting the particles jump between neighboring compartments. Treating these diffusive jumps as first-order reactions, the reaction-diffusion system can be described in terms of a jump process on an extended discrete state space. The resultant system of *compartment CMEs* that characterize the temporal evolution of the corresponding probability distribution will be introduced in Sect. 4.3.1. Different ways to choose the spatial discretization will be presented in Sects. 4.3.2 and 4.3.3 (see Fig. 4.7 for a model hierarchy).

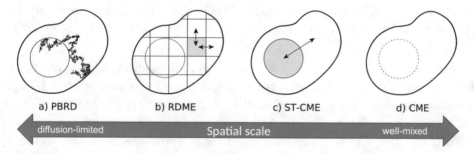

Figure 4.7. Spatial scaling of modeling approaches. The image represents a eukaryotic cell containing a nucleus (circle), with its border giving a natural barrier for diffusion. From left to right: (**a**) On the microscopic level of particle-based reaction-diffusion (PBRD), individual trajectories of particles are traced in continuous space (see Sect. 4.1). On the mesoscopic level, space is discretized into (**b**) mesh compartments or (**c**) metastable compartments, and diffusion is approximated by jumps (see Sect. 4.3). (**d**) In the macroscopic limit, the barriers become entirely permeable, there is no spatial resolution, and dynamics are well-mixed in the overall space (see Chaps. 1–3)

4.3.1 The Reaction-Diffusion Jump Process

Given a discretization of the diffusion space into compartments $\mathbb{D}_i$, set

$$X_{li}(t) := \text{number of particles of species } \mathcal{S}_l \text{ in } \mathbb{D}_i \text{ at time } t \geq 0.$$

We take $\boldsymbol{X}_i(t) = (X_{1i}(t), \dots, X_{Li}(t))^\mathsf{T} \in \mathbb{N}_0^L$ to denote the state of present species in compartment $\mathbb{D}_i$. The total state of the system at some time t is given by the matrix

$$X(t) = (X_i(t))_{i \in \mathbb{I}} \in \mathbb{N}_0^{L,|\mathbb{I}|},$$

where $|\mathbb{I}|$ is the number of compartments. Changes of the state are induced by diffusive transitions (jumps) *between* the compartments and by chemical reactions taking place *within* the compartments.

Given the state $X(t) = x \in \mathbb{N}_0^{L,|\mathbb{I}|}$ of the process, a diffusive jump of a particle of species S_l from compartment $\mathbb{D}_i$ to compartment $\mathbb{D}_j$, $j \neq i$, refers to a transition of the form

$$x \to x + e_{lj} - e_{li}$$

where $e_{li} \in \mathbb{N}_0^{L,|\mathbb{I}|}$ is a matrix whose elements are all zero except the entry (l, i) which is one. Let $\lambda_{ij}^l \geq 0$ denote the rate constant for each individual particle of species S_l to jump from compartment $\mathbb{D}_i$ to $\mathbb{D}_j$. Since all particles are assumed to diffuse independently of each other, the total probability per unit of time for a jump of species S_l from $\mathbb{D}_i$ to $\mathbb{D}_j$ at time t is given by $\lambda_{ij}^l X_{li}(t)$.

For each of the reactions $\mathcal{R}_k$, $k = 1, \ldots, K$, let again $\nu_k = (\nu_{1k}, \ldots, \nu_{Lk})^\mathsf{T}$ describe the change in the number of copies of all species induced by this reaction. Reaction $\mathcal{R}_k$ occurring at time t in the ith compartment refers to the transition $X_i(t) \to X_i(t) + \nu_k$. In order to specify where the reaction takes place, the vector ν_k is multiplied by a $|\mathbb{I}|$-dimensional row vector $\mathbf{1}_i'$ with the value 1 at entry i and zeros elsewhere. This gives a matrix $\nu_k \mathbf{1}_i' \in \mathbb{Z}^{L,|\mathbb{I}|}$ whose ith column is equal to ν_k, while all other entries are zero. With this notation, the change in the total state $X(t) = x$ due to reaction $\mathcal{R}_k$ taking place in compartment $\mathbb{D}_i$ is given by

$$x \to x + \nu_k \mathbf{1}_i'.$$

The propensity for such a reaction to occur is given by the function $\alpha_k^i : \mathbb{N}_0^{L,|\mathbb{I}|} \to [0, \infty)$, where $\alpha_k^i(x)$ denotes the probability per unit of time for reaction $\mathcal{R}_k$ to occur in compartment $\mathbb{D}_i$ given that $X(t) = x$. In analogy to Sect. 1.1, the propensity is considered to be of the form

$$\alpha_k^i(x) = \gamma_k^i \cdot h_k(x_i)$$

for some combinatorial functions h_k and a rate constant $\gamma_k^i > 0$. That is, $\alpha_k^i(x)$ depends only on the column x_i of x referring to the population state of compartment $\mathbb{D}_i$. Note that the rate constant γ_k^i may depend on the compartment, such that—in contrast to the well-mixed scenario of the previous chapters—spatial inhomogeneities in the reaction propensities can be taken into account.

Commonly, the *reaction-diffusion jump process* $\boldsymbol{X} = (\boldsymbol{X}(t))_{t\geq 0}$ is characterized by the corresponding Kolmogorov forward equation. As in the setting of the CME (cf. Sect. 1.2.2), let

$$p(\boldsymbol{x}, t) := \mathbb{P}\left(\boldsymbol{X}(t) = \boldsymbol{x} | \boldsymbol{X}(0) = \boldsymbol{x}_0\right)$$

be the probability that the process is in state $\boldsymbol{x} \in \mathbb{N}^{L, |\mathbb{I}|}$ at time t given an initial state $\boldsymbol{X}(0) = \boldsymbol{x}_0$. The evolution equation for $p(\boldsymbol{x}, t)$ is given by the extended CME (4.26), which in the literature is called *reaction-diffusion master equation (RDME)* [92, 123, 130, 131, 133, 134] or *spatiotemporal chemical master equation (ST-CME)* [246], depending on the underlying spatial discretization. The first line in (4.26) refers to the diffusive part (the transitions between the compartments), while the second line describes the chemical reactions within the compartments.

RDME/ST-CME:

$$\frac{dp(\boldsymbol{x}, t)}{dt} = \sum_{\substack{i,j\in\mathbb{I} \\ i\neq j}} \sum_{l=1}^{L} [\lambda_{ji}^l (x_{lj} + 1) p(\overbrace{\boldsymbol{x} + \boldsymbol{e}_{lj} - \boldsymbol{e}_{li}}^{\text{diffusive jump}}, t) - \lambda_{ij}^l x_{li} p(\boldsymbol{x}, t)]$$

$$+ \sum_{i\in\mathbb{I}} \sum_{k=1}^{K} [\alpha_k^i(\boldsymbol{x} - \boldsymbol{\nu}_k \mathbf{1}_i') p(\underbrace{\boldsymbol{x} - \boldsymbol{\nu}_k \mathbf{1}_i'}_{\text{reaction}}, t) - \alpha_k^i(\boldsymbol{x}) p(\boldsymbol{x}, t)] \tag{4.26}$$

When interpreting the diffusive jumps between the compartments as first-order reactions, the extended CME (4.26) is actually just a CME on the extended state space $\mathbb{N}^{L, |\mathbb{I}|}$. This means that we can directly transfer the path-wise representation (1.6) of a reaction jump process to the reaction-diffusion jump process $(\boldsymbol{X}(t))_{t\geq 0}$. Letting $\boldsymbol{X}(0) = \boldsymbol{x}_0 \in \mathbb{N}^{L, |\mathbb{I}|}$ be the initial state at time $t = 0$, we obtain

$$\boldsymbol{X}(t) = \boldsymbol{x}_0 + \sum_{\substack{i,j\in\mathbb{I} \\ i\neq j}} \sum_{l=1}^{L} \mathcal{U}_{ijl}^{(D)} \left(\int_0^t \lambda_{ij}^l X_{li}(s)\, ds \right) (\boldsymbol{e}_{lj} - \boldsymbol{e}_{li})$$

$$+ \sum_{i\in\mathbb{I}} \sum_{k=1}^{K} \mathcal{U}_{ik}^{(R)} \left(\int_0^t \alpha_k^i(\boldsymbol{X}(s))\, ds \right) \boldsymbol{\nu}_k \mathbf{1}_i',$$

where $\mathcal{U}_{ijl}^{(D)}$ and $\mathcal{U}_{ik}^{(R)}$ $(i, j \in \mathbb{I}, l \in \{1, \ldots, L\}, k \in \{1, \ldots, K\})$ are independent unit-rate Poisson processes which determine the random time points
for the occurrence of diffusive jumps and reactive jumps, respectively. Then,
numerical realizations of the process $(\boldsymbol{X}(t))_{t \geq 0}$ can directly be created by the
stochastic simulation methods presented in Sect. 1.3.2 (leading to a "spatial
Gillespie algorithm"). Of course, the computational complexity increases
in comparison to the well-mixed dynamics, because the first-order diffusive
jumps have to be simulated in addition. How large this increase is depends
on the number of compartments and the diffusion speed.

There exist different approaches for the discretization of space into
compartments $\mathbb{D}_i$. We fundamentally distinguish between *lattice-based* approaches (Sect. 4.3.2), where space is structured by a regular or irregular
grid, and *system-adapted* approaches (Sect. 4.3.3) that make use of natural decompositions of space into metastable areas (e.g., separated by local
barriers) (see Fig. 4.8 for an illustration).

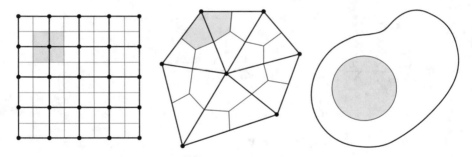

Figure 4.8. Spatial discretizations. Split-up of space by a regular Cartesian
grid (left), by an unstructured triangular mesh (middle), and by a natural
barrier (right). For left and middle: The primal mesh is given by the thick
lines with vertices given by black dots, and the dual mesh is marked by the thin
lines. The compartments are given by the cells of the dual mesh (polygonal
voxels), one of the compartments colored in gray. Right: The circle (thin line)
identifies a local barrier in the space which naturally decomposes the space
into two compartments: the interior (gray) and the exterior of the circle

4.3.2   Lattice-Based Discretization: RDME

One approach for the spatial discretization is to choose a mesh on $\mathbb{D}$ with the
compartments $\mathbb{D}_i$ defined as the dual cells of the mesh (or by the mesh cells
themselves). Given such a mesh discretization, the extended CME (4.26) is
typically called reaction-diffusion master equation (RDME) in the literature
[92, 123, 130, 131, 133, 134]. Most commonly, the mesh is set as a uniform

Cartesian lattice, but there also exist other approaches, e.g., using unstructured triangular meshes [64, 135] which simplify the resolution of curved boundaries. Time-dependent discretizations are considered in [248], and adaptive mesh refinement techniques to improve the approximation quality of molecular diffusion are presented in [17].

The rate constants for the diffusive jumps between the mesh compartments depend on the shape and size of the compartments and on the diffusion properties of the particles. They are generally obtained from a finite element discretization of the diffusion equations [64, 92, 134]. As a simple setting, we again assume the particles to diffuse independently of each other, each following Brownian motion without drift given by $\frac{\partial}{\partial t}\rho(q,t) = D_l \nabla^2 \rho(q,t)$ (see (4.8)), where $\rho(q,t)$ is the probability density for a particle of species S_l to be located at time t in $q \in \mathbb{D}$ and D_l denotes the diffusion coefficient for species S_l. Let the spatial discretization be given by a regular Cartesian mesh on $\mathbb{R}^d$, with the compartments given by hypercubic cells of width $h > 0$. The jump rate from compartment $\mathbb{D}_i$ to $\mathbb{D}_j$ for species S_l is given by Isaacson and Peskin [134] and Gardiner [90]

$$\lambda_{ij}^l := \begin{cases} \frac{D_l}{h^2} & \mathbb{D}_j \text{ is a nondiagonal neighbor of } \mathbb{D}_i, \\ 0 & \text{otherwise,} \end{cases} \tag{4.27}$$

just as in (4.15) (see again Appendix A.2 for further explanation (especially the paragraph *Discretization in space* beginning on page 199)). In the literature, the derivation of the jump rates is also explained for other settings, e.g., diffusion in potential energy [132] or diffusion in complex geometries containing boundaries [134].

Remark 4.4. Note that the diffusive jump rates λ_{ij}^l given in (4.27) are independent of the spatial dimension d. The scaling by h^{-2} results from a discretization of the Laplace operator ∇^2 in the equation $\frac{\partial}{\partial t}\rho(q,t) = D_l \nabla^2 \rho(q,t)$ (see Remark A.9 in the Appendix). In contrast, the rate constants for second-order reactions depend on the spatial dimension (see Eq. (4.28)).

Also the reaction propensities explicitly depend on the mesh size h. In accordance with Sect. 2.1.1, the propensities are scaled by the volume of the compartments, which for d-dimensional hypercubes is given by $V = h^d$. Especially, we obtain

$$\alpha_k^i(\boldsymbol{x}) = \frac{\gamma_k^i}{h^d} x_{li} x_{l'i} \tag{4.28}$$

for a second-order reaction of type $\mathcal{R}_k : S_l + S_{l'} \to \dots$ in compartment $\mathbb{D}_i$.

The meaning of the mesh size and its effect onto the stationary distribution of the process (using the standard rescaling (4.28) of reaction propensities) is illustrated in the following example.

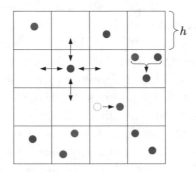

Figure 4.9. Reaction-diffusion master equation. Illustration of the RDME dynamics given in Example 4.5: The spatial domain $\mathbb{D} = [0,1] \times [0,1]$ is discretized by a uniform Cartesian mesh into compartments of equal size. Diffusion of particles in space is described by jumps between nondiagonal neighboring compartments. The second-order reaction $\mathcal{R}_1 : \mathcal{S}_1 + \mathcal{S}_2 \to \mathcal{S}_2$ between a particle of species $\mathcal{S}_1$ (blue circle) and a particle of species $\mathcal{S}_2$ (red circle) can take place when the particles are located in the same compartment. By reaction $\mathcal{R}_2 : \emptyset \to \mathcal{S}_1$, new particles are produced. Within each compartment, the dynamics are considered as well-mixed

Example 4.5. We consider two species $\mathcal{S}_1$ and $\mathcal{S}_2$ moving on $\mathbb{D} = [0,1] \times [0,1]$ and undergoing the chemical reactions

$$\mathcal{R}_1 : \mathcal{S}_1 + \mathcal{S}_2 \xrightarrow{\gamma_1} \mathcal{S}_2, \quad \mathcal{R}_2 : \emptyset \xrightarrow{\gamma_2} \mathcal{S}_1,$$

(see Fig. 4.9 for an illustration). The first reaction refers to the degradation of $\mathcal{S}_1$, which is triggered by $\mathcal{S}_2$. The second reaction induces the production of particles of species $\mathcal{S}_1$. We model the reaction-diffusion system by the RDME (4.26), choosing a regular Cartesian mesh with n_0^2 compartments such that $\mathbb{I} := \{1, \ldots, n_0\}^2$. The state of the stochastic process $\boldsymbol{X}$ at time t is of the form $\boldsymbol{X}(t) = (X_{li}(t))_{l=1,2;i\in\mathbb{I}}$ where $X_{li}(t)$ refers to the number of particles of species $\mathcal{S}_l$, $l \in \{1,2\}$, in compartment $\mathbb{D}_i$, $i \in \mathbb{I} = \{1, \ldots, n_0\}^2$, at time $t \geq 0$.

Let $h^2 = 1/n_0^2$ be the area size of each compartment. Then the rescaled reaction propensities for each compartment $\mathbb{D}_i$ are given by

$$\alpha_1^i(\boldsymbol{x}) = \frac{\gamma_1}{h^2} x_{1i} x_{2i}, \quad \alpha_2^i(\boldsymbol{x}) = h^2 \gamma_2, \tag{4.29}$$

in consistency with the volume-scaled propensities given in (2.1). The diffusive jump rate between nondiagonal neighbor compartments is D_l/h^2, where D_l is the diffusion coefficient of species S_l.

In this system, the total number of S_2-particles is obviously constant, i.e., it holds $\sum_{i \in \mathbb{I}} X_{2i}(t) = X_{2i}(0) \ \forall t$. The total number of S_1-particles depends on time, but in the long run, it will fluctuate around an equilibrium value. More precisely, we define the total number of S_1-particles existing at time t by $X_1^{\text{total}}(t) := \sum_{i \in \mathbb{I}} X_{1i}(t)$ and set $p_1(x,t) := \mathbb{P}(X_1^{\text{total}}(t) = x)$. Then $(X_1^{\text{total}}(t))_{t \geq 0}$ is an ergodic process, and we can define its stationary distribution by

$$\pi(x) := \lim_{t \to \infty} p_1(x,t) \qquad (4.30)$$

for $x \in \mathbb{N}_0$. Figure 4.10 shows that this equilibrium distribution of the overall S_1-population depends on the mesh size. A reduction of the compartment size h^2 shifts the stationary distribution to the right, meaning that the size of the S_1-population increases. This is due to the fact that the degradation $\mathcal{R}_1$ occurs less often. Actually, the encounter of an S_1- and an S_2-particle within the same compartment, which is necessary for the bimolecular reaction $\mathcal{R}_1$ to occur, becomes less likely, as illustrated in Fig. 4.11. The chosen rescaling (4.29) of the propensity α_1^i cannot compensate this rareness of encounters. Modifications of the rescaling (4.29), which are able to improve the consistency of the stationary distributions, are given in [67] (for equivalent dynamics in three-dimensional space). $\diamond$

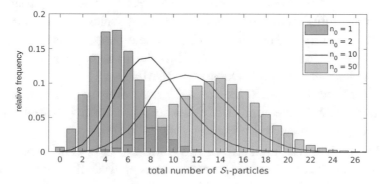

Figure 4.10. Stationary distribution depending on grid size. Stationary distribution defined in (4.30) for the total number of S_1-particles in Example 4.5, estimated by long-term Gillespie simulations of the standard RDME (4.26) for Cartesian meshes with n_0^2 compartments, where $n_0 = 1, 2, 10, 50$. Parameter values: $\gamma_1 = 0.2$, $\gamma_2 = 1$, $D_1 = D_2 = 0.1$, $B_0 = 1$. Rescaling of the reaction propensities given by (4.29)

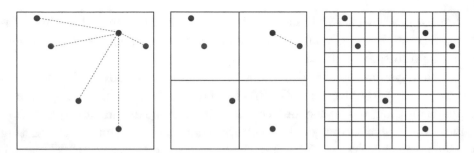

Figure 4.11. Reactivity depending on grid size. Visualization of the differently sized spatial discretizations of the domain $\mathbb{D} = [0,1] \times [0,1]$ considered in Example 4.5. The blue circle represents the $\mathcal{S}_2$-particle, and the red circles refer to the $\mathcal{S}_1$-particles. The number of compartments is given by n_0^2 for $n_0 = 1$ (left), $n_0 = 2$ (middle), and $n_0 = 10$ (right). An $\mathcal{S}_1$-particle can be destroyed by reaction $\mathcal{R}_1 : \mathcal{S}_1 + \mathcal{S}_2 \to \mathcal{S}_2$ if it is located in the same compartment as the $\mathcal{S}_2$-particle. In the well-mixed scenario ($n_0 = 1$, left), the $\mathcal{S}_2$-particle can react with all present $\mathcal{S}_1$-particles (which is illustrated by the dotted lines). With a finer discretization, it becomes less likely for two particles to meet in the same compartment. This illustration is inspired by Smith and Grima [217]

The Microscopic Limit

The RDME can be interpreted as an independent physical model or as an approximation to the spatially continuous stochastic reaction-diffusion model (i.e., the Doi- or Smoluchowski model) from Sect. 4.1 where particles move by continuous diffusion and react when sufficiently close. When interpreting the RDME as an approximation of the particle-based dynamics, the naturally arising questions are as follows: How does the approximation quality depend on the mesh size h? Does the RDME (4.26) with jump rates defined by (4.15) and propensities rescaled according to (4.28) converge for $h \to 0$, and what is the limit? Much effort has been invested to answer these questions. The central points are the following:

On the one hand, a small size of the compartments is necessary to guarantee a good approximation of the continuous spatial diffusion dynamics of each particle. Also the assumption that each compartment is homogeneous and that bimolecular reactions can occur within a compartment entails an upper bound for the mesh size. On the other hand, it appears unphysical to choose subvolumes that are smaller than the size of the particles themselves. Moreover, a fine discretization of a two- or three-dimensional space does not naturally lead to a high approximation quality because encounters of molecules within the same (very small) compartment become unlikely, which reduces the chances for bimolecular reactions to take place. It has been shown that the conventional rescaling (4.28) of the reaction propensities is not able to compensate this reduction of molecular crossings, such that bimolecular reactions are effectively suppressed [130]. Actually, in the continuum limit of $h \to 0$, the RDME breaks down in the sense that it becomes increasingly inaccurate compared to the Smoluchowski model [122].

One idea to handle this problem is to redefine the scaling of the binding propensities and to let them explicitly depend not only on the mesh size but also on the diffusion coefficients [67, 73, 130]. Still, it has been proven that below a certain critical size of the mesh it will be impossible to make the classical RDME (which only allows reactions between particles located in the same compartment) consistent with the Smoluchowski model for any choice of the binding rate [122]. This means that for systems with bimolecular reactions there is always a lower bound on the mesh size, for which the rescaled binding propensity is infinite such that two particles react immediately after entering the same compartment and the error in approximation of the particle-based model cannot be made arbitrarily small. Adapted models to overcome this lower error bound have been developed in [124, 131]. In these adapted models, particles are allowed to react also when occupying neighboring (different but nearby) compartments. In this case, the solution to the resulting modified RDME converges to the solution of the particle-based model as the lattice spacing goes to zero [131]. This *convergent RDME* results from a finite-volume discretization of the Doi-model (see Sect. 4.1.2).

In summary, it is reasonable to interpret the RDME as an independent modeling approach arising from a spatial extension of the CME and not as an approximation of the more complex microscopic particle-based reaction-diffusion model.

Limit of Fast Diffusion/Relation to CME

After having clarified the meaning of the mesh size and the relation between the RDME and the spatially continuous particle-based dynamics from Sect. 4.1, another interesting aspect is the connection to the well-mixed scenario underlying the CME (1.10). Actually, it has been shown that in the limit of fast diffusion (given a constant mesh size), the RDME converges to a chemical master equation [216]. This CME, however, may not be the "right" CME (in the sense that it refers to the same reaction network). Whether the RDME converges to the CME of the same system depends on the propensity functions: For elementary reactions of mass-action type, the convergence is given; for more complex reactions (e.g., Michaelis-Menten kinetics), the convergence cannot be guaranteed, cf. [216, 217].

Reaction-Diffusion in Crowded Media

The standard RDME approach neglects the effects of volume exclusion where each particle occupies a significant subvolume of the available space making this subvolume inaccessible to other particles. Especially biochemical processes in living cells often show a large fraction of occupied volume, a fact which motivates to redefine the rates occurring in the RDME and to let them depend on the occupancy state of the compartments.

On the level of particle-based modeling (cf. Sect. 4.1), volume exclusion effects are incorporated by interaction potentials or, more indirectly, by choosing subdiffusion instead of normal diffusion. For lattice-based random walks, effects of mutual spatial exclusion can be modeled by letting the diffusive jump rates depend on the occupancy of neighboring compartments [72, 161, 224]. While these works do not consider any interaction of particles by chemical reaction events, a special type of reaction-diffusion jump process including volume exclusion effects is presented in [37]. In this modeling approach, each RDME compartment can contain at most one molecule, and obstacles are represented as already occupied compartments. The jump length coincides with the size of a molecule, which induces high simulation costs with many small jumps, and bimolecular reactions occur between particles located in neighboring compartments. Such dynamics have been simulated and applied to different types of diffusion-limited dynamics in crowded environments, cf. [19, 113, 207]. Further methods to model reaction-diffusion systems of particles among stationary or moving obstacles are given in [65, 179]. Here, crowding effects are upscaled from microscopic motion of particles to mesoscopic jump dynamics and further to a macroscopic level given by a deterministic diffusion equation with space-dependent diffusion and reaction rates.

Numerical Simulation and Complexity

There exist many software tools for the compartment-based simulation of reaction-diffusion dynamics, e.g., MesoRD [118] or SmartCell [5]. With an increasing diffusive speed or a decreasing mesh size, the diffusive transition events become more and more frequent. For an event-driven stochastic simulation of the RDME, this leads to very small time steps. Simulations of an RDME are therefore much more complex than simulations of the analog

CME. Finding high-performance stochastic simulation methods for RDMEs is an actual topic of research [198]. Based on the *next reaction method* (cf. Sect. 1.3.2), the *next subvolume method* has been developed in [62] and implemented for Cartesian meshes in [118] and for unstructured meshes in [42]. An adaptive multilevel algorithm combining stochastic simulation, τ-leaping, and deterministic, macroscopic advance for the diffusion processes is presented in [76]. A comparison of RDME simulations and particle-based reaction-diffusion simulations is given in [54].

The numerical effort can be drastically reduced in case of metastable diffusion dynamics, which are subject of the following section.

4.3.3 Metastable Dynamics: ST-CME

In the lattice-based approaches, discretization is essentially uniform over the whole space, not taking into account any prior knowledge about the spatial structure. In some applications, however, the characteristics of the system suggest a more natural, system-adapted split-up, e.g., if there exist local barriers which reduce the diffusive flow and obviously split the space into areas of metastability. The central example that will be investigated in Sect. 4.5 considers diffusion in a eukaryotic cell which, by the nuclear membrane, naturally decomposes into two compartments, the nucleus and the cytoplasm. In such environments, the dynamics of a moving particle are *metastable* in the sense that the particle remains for a comparatively long period of time within one of the compartments, before it eventually switches to another area of space.

Such situations, in which the space $\mathbb{D}$ exhibits a particular structure with respect to the diffusion and reaction properties, motivate a completely different type of coarsening. The basic assumption is the following [246]:

Assumption 4.6. There is a decomposition of $\mathbb{D}$ into compartments $\mathbb{D}_i$, $i \in \mathbb{I}$, such that:

1. Transitions between compartments are rare (metastability),.

2. Within each compartment $\mathbb{D}_i$ diffusion is rapid compared to reaction (local well-mixed property),.

3. Within each compartment $\mathbb{D}_i$ the reaction rates are constant, i.e., independent of the position (local homogeneity).

The first point of Assumption 4.6 delivers the basis for a construction of a *Markov state model (MSM)* for the diffusive part of the reaction-diffusion network. Assuming that the diffusion processes of all particles are metastable with respect to the same space decomposition, they can be approximated by Markov jump processes on the fixed set of compartments $\{\mathbb{D}_i | i \in \mathbb{I}\}$. By the second and third point of Assumption 4.6, we make sure that within each compartment the reaction dynamics can accurately be described by a chemical master equation.

Markov State Modeling

The theory of Markov state models is a well-established tool to approximate complex molecular kinetics by Markov chains on a discrete partition of the configuration space [24, 129, 193, 203, 204, 213, 214]. The underlying process is assumed to exhibit a number of metastable sets in which the system remains for a comparatively long period of time before it switches to another metastable set. A MSM represents the effective dynamics of such a process by a Markov chain that jumps between the metastable sets with the transition rates of the original process. Two main steps are essential for this approach: the identification of metastable compartments and the determination of jump rates for transitions between them. Both steps are extensively studied in the literature, providing concrete practical instructions to implement them [23, 24, 35]. Although usually interpreted in terms of conformational dynamics (i.e., intrinsic dynamics of the atoms of a single molecule), the approaches can directly be transferred to the kinetic dynamics of particles within a reaction-diffusion system. Within the ST-CME (4.26), jumps and reactions are uncoupled from each other and memoryless in the sense that a reaction taking place does not influence the jump propensity of any species, and the reaction rates within each compartment are independent of the previous evolution of the system. Thereby, a separate analysis of the diffusion properties is justified.

Spatial Coarsening

Assuming that the space of motion consists of metastable areas where the reaction-diffusion dynamics can be considered being well-mixed, the first task is to identify the location and shape of these areas. While in certain applications the spatial clustering might be obvious (like for the process of gene expression), others will require an analysis of available data (either from simulations or experiment) to define the number and geometry of compartments.

The MSM approach to perform this step of space decomposition is based on a two-stage process exploiting both geometry and kinetics [24, 35, 212]. Starting with a geometric clustering of space into small volume elements, the kinetic relevance of this clustering is checked for the given simulation data, followed by a merging of elements which are kinetically strongly connected. To illustrate the relevance of both geometric and kinetic aspects for the space discretization, one can consider the example of nuclear membrane transport: By the reduced permeability of the nuclear membrane, the cell decomposes into two metastable regions (nucleus and cytoplasm). A small geometric distance between two points is a first indication for belonging to the same compartment, though this would also group together close points which are located on different sides of the nuclear membrane and are thereby kinetically distant. While other clustering methods might fail to capture the kinetic properties of the system under consideration, the MSM approach incorporates the relevant kinetic information and allocates such points into different compartments [23].

The algorithms used for spatial clustering by MSMs require the availability of a large amount of data. Transitions between metastable sets are infrequent events which even in long (but finite) trajectories might be small in number. Fortunately, for MSMs there exist methods to validate the model and to quantify the statistical error, allowing the rerun of simulations in case of high uncertainty (so-called adaptive sampling) [24, 193, 204]. The development of standards for the number and length of simulations needed for a statistically reliable modeling is in progress [35].

Estimation of Jump Rates λ_{ij}

Once a suitable space partitioning for the problem under consideration is given (known in advance or determined by Markov state modeling as described before), the rates for transitions between the compartments have to be estimated. The theory of MSM also delivers concrete practical techniques to determine the diffusive jump rates out of trajectory data. Applied to the diffusion dynamics of a particle in space, the rate estimation can shortly be summarized as follows.

Let the space of motion $\mathbb{D}$ be decomposed according to Assumption 4.6 into compartments $\mathbb{D}_i$, $i \in \mathbb{I}$ for some finite index set $\mathbb{I}$, with $\bigcup_{i \in \mathbb{I}} \mathbb{D}_i = \mathbb{D}$ and $(\text{int}_{\mathbb{D}_i}) \cap (\text{int}_{\mathbb{D}_j}) = \emptyset$ for all $i \neq j$ where $\text{int}_{\mathbb{D}_i}$ denotes the interior of $\mathbb{D}_i$. The diffusion of a single particle in $\mathbb{D}$ is given by a homogeneous, time-continuous Markov process $(Q(t))_{t \geq 0}$ with $Q(t)$ denoting the position of the particle in $\mathbb{D}$ at time $t \geq 0$. We assume that the process $(Q(t))_{t \geq 0}$ has a unique, positive invariant probability measure π on $\mathbb{D}$. Fixing a lag time $\tau > 0$, the *MSM process* $(\tilde{Q}_n)_{n\in\mathbb{N}}$ is defined as a discrete-time Markov chain on $\mathbb{I}$ with transition probabilities

$$a(\tau, i, j) := \mathbb{P}_\pi(Q(\tau) \in \mathbb{D}_j | Q(0) \in \mathbb{D}_i), \qquad (4.31)$$

(see the definition of the stochastic transition function in Eq. (A.2)), where the subscript π indicates that $Q(0) \sim \pi$. The Markov process $(\tilde{Q}_n)_{n\in\mathbb{N}}$ serves as an approximation of the process $(\hat{Q}_n)_{n\in\mathbb{N}}$ defined by $\hat{Q}_n = i \Leftrightarrow Q(n \cdot \tau) \in \mathbb{D}_i$ which, in general, is itself *not* Markovian but has a memory. For details about the approximation quality, see [204].

Given a trajectory $(q_0, q_1, \ldots, q_N)$ (resulting from experimental data or a separated simulation) of the diffusion process $(Q(t))_{t\geq 0}$ with $q_n := Q(n \cdot \Delta t)$ for a fixed time step $\Delta t > 0$, the transition matrix

$$A(\tau) := (a(\tau, i, j))_{i,j=1,\ldots,|\mathbb{I}|}$$

is estimated by counting the transitions between the compartments: Let the lag time $\tau = l\Delta t$ ($l \in \mathbb{N}$) be a multiple of Δt and set

$$\hat{a}(\tau, i, j) := \frac{a_{ij}}{\sum_{j'} a_{ij'}}, \quad a_{ij} := \sum_{n=0}^{N-l} \mathbb{1}_{\mathbb{D}_i}(q_n)\mathbb{1}_{\mathbb{D}_j}(q_{n+l}) \qquad (4.32)$$

with $\mathbb{1}_{\mathbb{D}_i}$ denoting the indicator function of $\mathbb{D}_i$. Then $\hat{A}(\tau) := (\hat{a}(\tau, i, j))_{i,j=1,\ldots,|\mathbb{I}|}$ is a maximum likelihood estimator for the transition matrix $A(\tau)$ [193, 203].

In order to turn from discrete time back to continuous time within the coarsened setting, the matrix estimation is repeated for a range of lag times τ, and the resulting transition matrices $\hat{A}(\tau)$ are used to determine appropriate jump rates $\lambda_{ij} \geq 0$ between the compartments in the setting of a continuous-time Markov jump process. More precisely, we aim for a rate matrix $\Lambda = (\lambda_{ij})_{i,j=1,\ldots,|\mathbb{I}|}$ with $\lambda_{ij} \geq 0$ for $i \neq j$ and $\lambda_{ii} = -\sum_{j\neq i} \lambda_{ij}$ which describes the time evolution of a memoryless system by the master equation

$$\frac{d\boldsymbol{p}^\mathsf{T}(t)}{dt} = \boldsymbol{p}^\mathsf{T}(t)\Lambda \qquad (4.33)$$

with $\boldsymbol{p}(t) := (p(i, t))_{i=1,\ldots,|\mathbb{I}|} \in \mathbb{R}_+^{|\mathbb{I}|}$ and $p(i, t)$ denoting the probability for the diffusion process $(Q(t))_{t\geq 0}$ of a given particle to be in compartment $\mathbb{D}_i$ at time t [234]. It refers λ_{ij} to the jump rate for transitions from compartment $\mathbb{D}_i$ to compartment $\mathbb{D}_j$. Letting $\boldsymbol{p}_0$ be the initial distribution at time $t = 0$, the solution to (4.33) is given by $\boldsymbol{p}^{\mathsf{T}}(t) = \boldsymbol{p}_0^{\mathsf{T}} \exp(\Lambda t)$ which suggests the relation $A(\tau) \approx \exp(\Lambda \tau)$ and motivates to set

$$\Lambda := \frac{1}{\tau} \log(A(\tau)) \tag{4.34}$$

in case this term does not depend on τ. On short time scales, the "true" process $(Q(t))_{t \geq 0}$ cannot be accurately approximated by such a memory-less system because back-and-forth transitions between the compartments induce memory effects [36]. For τ not too small, on the other hand, $\frac{1}{\tau} \log(A(\tau)) \approx$ constant holds indeed [24, 214] such that (4.34) becomes reasonable. The entries of the matrix Λ are the jump rates appearing in the ST-CME. The discretization error arising with the coarsening can even be quantitatively bounded [24].

Remark 4.7. The estimation of the rate matrix can be made more accurate by forcing it to fulfill certain side constraints like detailed balance or the compliance with an a priori known equilibrium distribution [24].

Beside the practical and algorithmic details, the theory of MSMs also provides a well-understood mathematical background, deducing the coarsened dynamics by Galerkin projections of the original process [214]. Based on this theory, it has been shown in [246] that the ST-CME (4.26) results from a Galerkin projection of the particle-based reaction-diffusion dynamics.

Example 4.8 (Binding and Unbinding in Two Compartments). Let the domain $\mathbb{D}$ be given by a disk of radius ι,

$$\mathbb{D} := \{q \in \mathbb{R}^2 | \ \|q\| \leq \iota\}.$$

A circle of radius $\iota' < \iota$ splits the domain into two compartments $\mathbb{D}_1, \mathbb{D}_2$ given by

$$\mathbb{D}_1 := \{q \in \mathbb{D} | \ \|q\| \leq \iota'\}, \quad \mathbb{D}_2 := \mathbb{D} \setminus \mathbb{D}_1.$$

The diffusion of particles within $\mathbb{D}$ is modeled by independent Brownian motions $dQ(t) = \sqrt{2D} dW(t)$ (where D is the diffusion constant), with reflecting boundary conditions at the outer boundary $\partial \mathbb{D} = \{q \in \mathbb{R}^2 | \|q\| = \iota\}$ and limited flux through the border $\partial \mathbb{D}_1 = \{q \in \mathbb{R}^2 | \|q\| = \iota'\}$ of the interior disk (see Fig. 4.12 for an illustration). That is, inside each of the compartments, the diffusion process of a single particle is given by $dQ(t) = \sqrt{2D} dW(t)$ as in (4.1). Together with the given boundary conditions, this leads to the following set of equations for the probability density $\rho(q,t)$ of finding the process $(Q(t))_{t\geq 0}$ in $q \in \mathbb{D}$ at time t:

$$\frac{\partial \rho}{\partial t} = D\Delta\rho \quad \text{in } \mathbb{D},$$

$$\frac{\partial \rho}{\partial \eta} = 0 \quad \text{on } \partial\mathbb{D}, \tag{4.35}$$

$$-D\frac{\partial \rho}{\partial \eta} = -\kappa[\rho] \quad \text{on } \partial\mathbb{D}_1,$$

where $\kappa \in (0,1)$ is the permeability of the border $\partial\mathbb{D}_1$, $[\rho]$ denotes the jump in $\rho(q,t)$ across this border, and η is the outward pointing normal to a given surface [134].

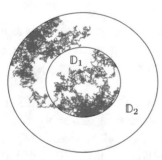

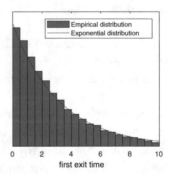

(a) Brownian dynamics on $\mathbb{D}$ (b) Distribution of first exit time from $\mathbb{D}_1$

Figure 4.12. Brownian motion on a disk with metastable compartments. **(a)** The border of the interior disk $\mathbb{D}_1$ exhibits a reduced permeability, rejecting most of the crossing trials of the diffusive trajectory. The particle cannot leave the total domain $\mathbb{D} = \mathbb{D}_1 \cup \mathbb{D}_2$. **(b)** The exit times from compartment $\mathbb{D}_1$ may be approximated by exponential random variables. The empirical distribution is the result of $2 \cdot 10^4$ Monte Carlo simulations. **(a)** Brownian dynamics on $\mathbb{D}$. **(b)** Distribution of first exit time from $\mathbb{D}_1$

Now we consider three species $\mathcal{S}_1, \mathcal{S}_2, \mathcal{S}_3$ which diffuse within $\mathbb{D}$ according to (4.35) and undergo the binding and unbinding reactions

$$\mathcal{R}_1: \ \mathcal{S}_1 + \mathcal{S}_2 \longrightarrow \mathcal{S}_3, \qquad \mathcal{R}_2: \ \mathcal{S}_3 \longrightarrow \mathcal{S}_1 + \mathcal{S}_2. \tag{4.36}$$

We assume that the binding reaction $\mathcal{R}_1$ takes place only in $\mathbb{D}_2$, while the unbinding reaction $\mathcal{R}_2$ is restricted to $\mathbb{D}_1$ (spatially inhomogeneous reaction propensities). This separation of reactions, combined with the metastability of diffusion due to a reduced permeability of $\partial\mathbb{D}_1$, obviously keeps the process from being well-mixed in total space, such that a description by the CME is not appropriate. A minimum spatial information, as given by a ST-CME, is needed for a congruence of the dynamics (see Fig. 4.13 for a quantitative comparison of the modeling approaches). In this figure, we can see that for the CME the number of S_3-particles grows faster (compared to the two other approaches) which results from the fact that in a totally well-mixed scenario two particles of species S_1 and S_2, which emerge by an unbinding reaction R_2, can immediately rebind again. In contrast, both in the particle-based scenario and in the ST-CME approach, the particles can only rebind after having moved from $\mathbb{D}_1$ to $\mathbb{D}_2$, which is due to the assumed restrictions.

The diffusive jump rates of the ST-CME are estimated by simulating the pure diffusion process characterized by (4.35) and applying the techniques from MSM theory described above. For the reaction rates, the diffusion of two particles S_1 and S_2 within $\mathbb{D}_2$ is simulated until they meet and react (for given reaction radii and a microscopic binding rate), and the inverse of the mean reaction time is chosen as an approximation for the binding rate in $\mathbb{D}_2$. For the CME, we average the reaction propensity $\alpha_k^i(x)$ over $\mathbb{D}$ by the equilibrium distribution of the diffusion process in order to get an "overall" reaction propensity for each reaction R_k. $\diamond$

4.4 Hybrid Reaction-Diffusion Approaches

In the previous sections, different modeling approaches for spatiotemporal reaction dynamics have been summarized. The most detailed model of particle-based reaction-diffusion dynamics given in Sect. 4.1 is suitable in case of a comparatively small molecular population. For large populations, the simulation of the corresponding processes becomes numerically too complex. Approximations in terms of PDE formulations for dynamics with large populations have been introduced in Sect. 4.2. A reduction of spatial resolution, on the other hand, leads to the compartment-based approaches of Sect. 4.3.

In application to real-world dynamics, it might not be clear which of the three modeling approaches is best suited, because spatial inhomogenities can appear: Some areas in space require a high spatial resolution, while others do not; or high population levels in some regions are combined with low abundance in other regions.

Similar to Sect. 2.2, where hybrid approaches for well-mixed dynamics with multiple population scales have been investigated, also recombinations of the approaches from Sects. 4.1–4.3 can lead to efficient hybrid models for

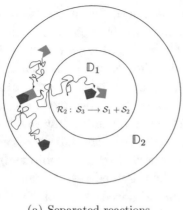

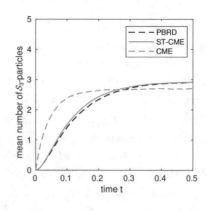

(a) Separated reactions

(b) Time-dependent first-order moments

Figure 4.13. Spatial model for binding and unbinding. **(a)** Separation of reactions with the binding reaction restricted to $\mathbb{D}_2$ and the unbinding reaction restricted to $\mathbb{D}_1$. **(b)** Mean total number of $\mathcal{S}_3$-particles (in both compartments together) for separated reactions, estimated by 10^4 Monte Carlo simulations of the particle-based dynamics (as reference), the ST-CME and the CME (with averaged reaction rates). Initial state: 5 $\mathcal{S}_1$-particles uniformly distributed in $\mathbb{D}_1$ and 5 $\mathcal{S}_2$-particles uniformly distributed in $\mathbb{D}_2$, no $\mathcal{S}_3$-particles

reaction-diffusion dynamics, using different levels of detail in different parts of the computational domain.

A coupling of the mesoscopic lattice-based approach (for regions requiring less detail) and the microscopic particle-based approach (for regions where accuracy is crucial to the model) is presented in [68, 79–81, 121, 148]. A crucial point is the inter-regime transfer of particles at the interface between the regions. Here, different approaches have been proposed in literature: the so-called *two regime method* [79, 80], the *compartment-placement method* [121], and the *ghost cell method* [81]. These micro-meso hybrid reaction-diffusion approaches have been applied to several biological systems such as intracellular calcium dynamics [82] or actin dynamics [68].

Other hybrid approaches couple reaction-diffusion PDEs for regions of high particle abundance with the mesoscopic compartment-based model for regions of low abundance [83, 184, 249] or with the microscopic particle-based model [84, 95].

4.5 Application

For an illustration we again consider the process of gene expression from Sect. 2.3, this time as a reaction-diffusion process within a eukaryotic cell. The cell naturally decomposes into two compartments: the nucleus and the cytoplasm. The nuclear membrane exhibits a reduced permeability which induces a metastability of the diffusion dynamics within the cell. Particles are assumed to move independently of each other and not to influence each other in their transport across the membrane. The cellular membrane is assumed to be impermeable such that the particles cannot reach the cell's exterior.

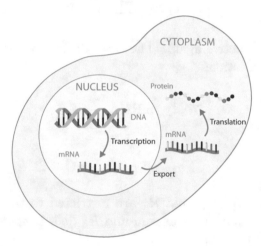

Figure 4.14. Gene expression in a eukaryotic cell. The DNA is located in the nucleus, where the transcription takes place. Before the mRNA molecule can be translated into proteins, it has to diffuse to the nuclear membrane and enter the cytoplasm

We consider the same coarsened system of reactions as in Sect. 2.3 and ignore more detailed steps that may underly each of these effective reactions. The DNA (species $\mathcal{S}_1$) is located at some fixed place within the nucleus, where it is transcribed into messenger RNA (species $\mathcal{S}_2 = \text{mRNA}$) by

$$\mathcal{R}_1: \quad \text{DNA} \xrightarrow{\gamma_1^i} \text{DNA} + \text{mRNA}.$$

After its diffusive export to the cytoplasm, the mRNA is translated into proteins (species $\mathcal{S}_3 = \text{P}$) by

$$\mathcal{R}_2: \quad \text{mRNA} \xrightarrow{\gamma_2^i} \text{mRNA} + \text{P}.$$

For an illustration of these first steps of gene expression given by transcription, export, and translation, see Fig. 4.14. The proteins produced in the cytoplasm then move back to the nucleus, where they can repress the gene by attaching themselves,

$$\mathcal{R}_3: \quad \mathrm{DNA} + \mathrm{P} \xrightarrow{\gamma_3^i} \mathrm{DNA}_0,$$

meaning that the DNA's functionality in transcription is interrupted. The repressing can be reversed by detachment of the (attached) protein,

$$\mathcal{R}_4: \quad \mathrm{DNA}_0 \xrightarrow{\gamma_4^i} \mathrm{DNA} + \mathrm{P},$$

rendering the DNA active again. The degradation of mRNA and proteins is induced by the reactions

$$\mathcal{R}_5: \quad \mathrm{mRNA} \xrightarrow{\gamma_5^i} \emptyset$$

and

$$\mathcal{R}_6: \quad \mathrm{P} \xrightarrow{\gamma_6^i} \emptyset,$$

respectively. While $\mathcal{R}_1, \mathcal{R}_3$, and $\mathcal{R}_4$ are restricted to the nucleus, $\mathcal{R}_2$ and $\mathcal{R}_5$ solely take place in the cytoplasm, and $\mathcal{R}_6$ does not underlie any spatial restrictions (see Table 4.1).

In order to specify the environment, we define the biological cell and its nucleus as concentric circles in $\mathbb{R}^2$ with radius ι_2 and ι_1, respectively. That is, we let the nucleus be given by

$$\mathbb{D}_{\mathrm{nuc}} = \left\{ q \in \mathbb{R}^2 \,\middle|\, \|q\| \leq \iota_1 \right\},$$

while the cytoplasm is defined as

$$\mathbb{D}_{\mathrm{cyt}} = \left\{ q \in \mathbb{R}^2 \,\middle|\, \iota_1 < \|q\| \leq \iota_2 \right\}$$

and the exterior of the eukaryotic cell is set to

$$\mathbb{D}_{\mathrm{ext}} = \left\{ q \in \mathbb{R}^2 \,\middle|\, \|q\| > \iota_2 \right\}.$$

The reaction-diffusion dynamics take place in $\mathbb{D} = \mathbb{D}_{\mathrm{nuc}} \cup \mathbb{D}_{\mathrm{cyt}}$.

Modeling for the RDME

For a modeling by the RDME, we choose a regular Cartesian grid on $[-\iota_2, \iota_2] \times [-\iota_2, \iota_2]$ with $n_0 \times n_0$ compartments, which covers the domain $\mathbb{D}$. For each $i \in \mathbb{I} = \{i = (i_1, i_2) | i_1, i_2 = 1, \ldots, n_0\}$, the compartment $\mathbb{D}_i$ is given by

$$\mathbb{D}_i = \left\{ q = (q_1, q_2) \in \mathbb{R}^2 \,\middle|\, (i_1 - 1)h < q_1 \le i_1 h, (i_2 - 1)h < q_2 \le i_2 h \right\}$$

with $h := \frac{2\iota_2}{n_0}$. Each of these compartments is assigned to one of the three domains $\mathbb{D}_{\mathrm{nuc}}$, $\mathbb{D}_{\mathrm{cyt}}$, or $\mathbb{D}_{\mathrm{ext}}$, depending on its major proportion within these domains. That is, we split up the index set $\mathbb{I}$ according to

$$\mathbb{I} = \mathbb{I}_{\mathrm{nuc}} \cup \mathbb{I}_{\mathrm{cyt}} \cup \mathbb{I}_{\mathrm{ext}}$$

where

$$\mathbb{I}_{\mathrm{nuc}} := \left\{ i \in \mathbb{I} \,\middle|\, |\mathbb{D}_i \cap \mathbb{D}_{\mathrm{nuc}}| > |\mathbb{D}_i \cap \mathbb{D}_{\mathrm{cyt}}| \right\},$$

$$\mathbb{I}_{\mathrm{ext}} := \left\{ i \in \mathbb{I} \,\middle|\, |\mathbb{D}_i \cap \mathbb{D}_{\mathrm{ext}}| > |\mathbb{D}_i \cap \mathbb{D}_{\mathrm{cyt}}| \right\},$$

and $\mathbb{I}_{\mathrm{cyt}} = \mathbb{I} \setminus (\mathbb{I}_{\mathrm{nuc}} \cup \mathbb{I}_{\mathrm{ext}})$. Here, $|A|$ denotes the area size of a given domain A.

Remark 4.9. An unstructured mesh consisting of triangles would better approximate the curved boundaries, but here we want to consider the most basic approach.

For a simulation, we choose the parameter values $\iota_2 = 0.2$, $\iota_1 = 0.1$ (skipping units for simplicity), and $n_0 = 20$, giving $h = 0.02$. The diffusion coefficient is set to $D = 0.001$ for both mRNA and proteins, and the permeability of the nuclear membrane is chosen as $\kappa = 0.1$. This implies jump rates $\lambda_{ij}^l = D/h^2 = 2.5$ ($l = 2, 3$) between neighboring compartments of the same domain, while $\lambda_{ij}^l = \kappa D/h^2 = 0.25$ for neighboring compartments with one of them assigned to the cytoplasm and the other one assigned to the nucleus. In order to exclude the exterior, we set $\lambda_{ij}^l = 0$ for neighboring compartments with $i \in \mathbb{I}_{\mathrm{ext}}$ or $j \in \mathbb{I}_{\mathrm{ext}}$. The DNA (species $\mathcal{S}_1$) is located close to the center in compartment $\mathbb{D}_{(10,10)}$, where it stays permanently, i.e., it holds $\lambda_{ij}^1 = 0 \; \forall i, j \in \mathbb{I}$. The second-order repressing reaction $\mathcal{R}_3$ between a protein and the DNA requires the protein particle to be located in $\mathbb{D}_{(10,10)}$, as well. The reaction rate constants are given in Table 4.1. An illustration of the RDME process can be found in Fig. 4.15.

Table 4.1. Reaction rate constants for RDME

		Nucleus $i \in \mathbb{I}_{\mathrm{nuc}}$	Cytoplasm $i \in \mathbb{I}_{\mathrm{cyt}}$
Transcription	γ_1^i	0.1	0
Translation	γ_2^i	0	0.5
Repressing of DNA	γ_3^i	0.001	0
Activation of DNA	γ_4^i	0.02	0
Degradation of mRNA	γ_5^i	0	0.1
Degradation of proteins	γ_6^i	0.05	0.05

Modeling for the ST-CME

For a comparison with a spatiotemporal CME that considers only two metastable compartments, namely, nucleus $\mathbb{D}_{\mathrm{nuc}}$ and cytoplasm $\mathbb{D}_{\mathrm{cyt}}$, we determine suitable jump rates between these compartments by Markov state model estimation. That is, we simulate the pure diffusive jump process of a single particle on the Cartesian grid of the RDME (or the continuous Brownian motion process) and apply the methods from Sect. 4.3.3 to find the effective jump rates for diffusive transitions between the compartments. For the parameter values given above, we obtain $\lambda_{12}^l = 1/10$ for jumps from nucleus to cytoplasm and $\lambda_{21}^l = 1/30$ for jumps in the opposite direction.

For all first-order reactions, the rate constants of the ST-CME agree with those of the RDME from Table 4.1. The rate for the second-order reaction $\mathcal{R}_3$ (repressing of DNA) is estimated as follows. We simulate the dynamics of a single protein in the nucleus, consisting of jumps between the RDME compartments and binding/unbinding to the DNA, and deduce the mean long-term activity of the DNA from this trajectory. A two-state Markov model between activity and inactivity is derived, with the activation rate given by $\gamma_4^i = 0.02$.

Comparing ST-CME to RDME

Figure 4.16 shows the total number of molecules depending on time for two independent simulations of the reaction-diffusion processes defined by the RDME and the ST-CME. We can observe the same characteristics, with the number of proteins and mRNA increasing and decreasing depending on the DNA activity. The time-dependent averages of the DNA activity, and the two populations are compared in Fig. 4.17. The agreement of RDME and ST-CME can be improved by decreasing the permeability κ (thereby

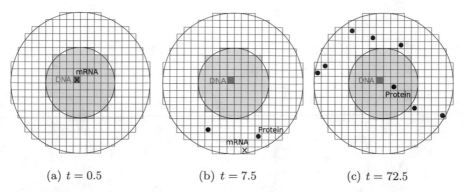

(a) $t = 0.5$	(b) $t = 7.5$	(c) $t = 72.5$

Figure 4.15. Snapshots of gene expression simulation. State of the RDME process at different points in time. (**a**) At time $t = 0.5$, the first mRNA-particle (symbolized by a black cross) has been transcribed in compartment $(10, 10)$ where the DNA is located. (**b**) At time $t = 7.5$, the mRNA-particle has left the nucleus and has been translated into several proteins (black dots). (**c**) At time $t = 72.5$, all mRNA-particles have been degraded, but several proteins exist in the cytoplasm. One of the proteins has entered the nucleus, where it can possibly repress the DNA

increasing the metastability of the diffusion process).

In comparison to the RDME, simulations of the ST-CME reduce the computational effort by a factor of 10^3 in the considered example. This reduction of numerical complexity happens at the expense of spatial resolution and is only justified if the diffusion within each compartment of the ST-CME is fast compared to the reactions, that is, if the dynamics behave well-mixed within each of these metastable compartments. For slowly diffusing particles, the RDME approach is more appropriate. In case of small population levels with the particles not densely packed in their environment, a particle-based approach is even more suitable (see Sect. 4.1).

By this comparison we finalize Chap. 4 and conclude the main part of the book.

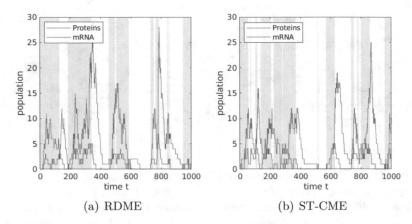

(a) RDME   (b) ST-CME

Figure 4.16. Comparison of RDME and ST-CME: Trajectories. Independent simulations of (**a**) the RDME and (**b**) the ST-CME for the process of gene expression with reaction rates given in Table 4.1. The figures show the total number of protein and mRNA molecules (summed over all compartments) depending on time. For the RDME a 20×20 grid is chosen (see Fig. 4.15), with jump rates defined on page 179. As for the ST-CME, the two metastable compartments are given by nucleus and cytoplasm. As in Sect. 2.3.2, the gray areas indicate the time periods of active DNA

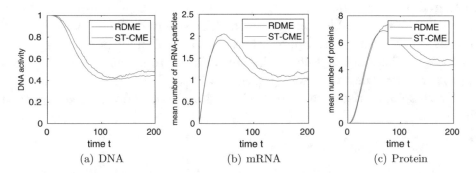

(a) DNA   (b) mRNA   (c) Protein

Figure 4.17. Comparison of RDME and ST-CME: First-order moments. Time-dependent mean of DNA activity and of the total number of mRNA and proteins (in the overall environment), estimated by 10^3 Monte Carlo simulations (10^4 for ST-CME). Each simulation starting with active DNA and total absence of mRNA and proteins

Summary and Outlook

Modeling, simulation, and analysis of interacting particle systems constitute a very active field of research. In computational biology, the particles are typically representing molecules, while the main interactions are chemical reactions and association processes. Typically, the dynamics of such systems exhibit cascades of different scales in time and in most cases also in space. In general, this multiscale behavior renders numerical simulation on the finest level (or resolution) infeasible. Therefore, existing mathematical models are extended, refined, or recombined in order to accurately describe biological or biochemical processes while keeping the model complexity within limits. Simultaneously, the construction of efficient numerical schemes for the different models plays a fundamental role. This book delivers a snapshot of the current state of knowledge. It presents different modeling approaches in a congruent framework, uncovers the interrelation between the models, and discusses the basic numerical schemes used today.

Of course, there are related topics and branches of literature which are not included in our review, for example, the mathematical modeling of dynamics in crowded environments or membrane-associated processes. Such topics reveal promising directions for future research, like a reconstruction of the RDME to include crowding effects or the development of efficient hybrid models combining well-mixed dynamics with spatially resolved processes. Here, the book can serve for orientation and provides a basis for further model creation and application, not only in the context of biochemical reaction-diffusion processes but also for other systems of interacting entities appearing, e.g., in economic, medical, or social sciences.

S. Winkelmann, C. Schütte, *Stochastic Dynamics in Computational Biology*, Frontiers in Applied Dynamical Systems: Reviews and Tutorials 8, https://doi.org/10.1007/978-3-030-62387-6

Appendix A

Mathematical Background

A.1 Markov Jump Processes

Markovian dynamics are dynamics without memory: The future evolution of the process solely depends on the current state and not on the past – a feature which is named *Markov property*. The reaction jump process $\boldsymbol{X} = (\boldsymbol{X}(t))_{t \geq 0}$ introduced in Sect. 1.2.1 fulfills the Markov property and thereby belongs to the class of *Markov jump processes*. In order to provide the background theory for the investigations of the preceding chapters, we will summarize here some well-known results about Markov jump processes [25, 147, 164, 228].

A.1.1 Setting the Scene

Let $(\Omega, \mathcal{E}, \mathbb{P})$ be a probability space, where Ω denotes the sample space, $\mathcal{E}$ is a σ-algebra on Ω containing all possible events, and $\mathbb{P}$ is a probability measure on $(\Omega, \mathcal{E})$. Given a state space $\mathbb{X}$ and a σ-algebra $\mathcal{E}'$ on $\mathbb{X}$, a family $X = (X(t))_{t \geq 0}$ of random variables $X(t) : (\Omega, \mathcal{E}, \mathbb{P}) \to (\mathbb{X}, \mathcal{E}')$ is called *continuous-time stochastic process* on $\mathbb{X}$. If $X(t) = x \in \mathbb{X}$, the process is said to be in state x at time t. For each $\omega \in \Omega$, the set $(X(t, \omega))_{t \geq 0}$ is called *realization* (or *trajectory* or *sample path*) of the process $X = (X(t))_{t \geq 0}$. In the following, we consider the state space $\mathbb{X}$ to be a countable discrete set.

S. Winkelmann, C. Schütte, *Stochastic Dynamics in Computational Biology*, Frontiers in Applied Dynamical Systems: Reviews and Tutorials 8, https://doi.org/10.1007/978-3-030-62387-6

Definition A.1 (Markov Process). A continuous-time stochastic process $(X(t))_{t \geq 0}$ on a countable set $\mathbb{X}$ is called *Markov process*, if for any $t_{j+1} > t_j > \ldots > t_0$ and any $B \subset \mathbb{X}$ the *Markov property*

$$\mathbb{P}(X(t_{j+1}) \in B | X(t_j), \ldots, X(t_0)) = \mathbb{P}(X(t_{j+1}) \in B | X(t_j)) \qquad \text{(A.1)}$$

holds. The Markov process is called *homogeneous*, if the right-hand side of (A.1) only depends on the time increment $\Delta t = t_{j+1} - t_j$, i.e.,

$$\mathbb{P}(X(t_{j+1}) \in B | X(t_j)) = \mathbb{P}(X(\Delta t) \in B | X(0)).$$

Given a homogeneous Markov process, the function $a : \mathbb{R}_+ \times \mathbb{X} \times \mathbb{X} \to [0,1]$ defined as

$$a(t, x, y) := \mathbb{P}(X(t) = y | X(0) = x) \qquad \text{(A.2)}$$

is called *stochastic transition function*. Its values $a(t, x, y)$ give the conditional probabilities for the process to move from x to y within time t. The probability distribution p_0 satisfying

$$p_0(x) = \mathbb{P}(X(0) = x)$$

is called *initial distribution*.

In order to avoid transitions within zero time, we require

$$a(0, x, y) = \delta_{xy} \qquad \text{(A.3)}$$

with $\delta_{xy} = 1$ for $x = y$ and zero otherwise. We further assume the transition function to be continuous at 0

$$\lim_{t \searrow 0} a(t, x, y) = \delta_{xy}, \qquad \text{(A.4)}$$

which guaranties that the realizations of the process are almost surely right-continuous. As the state space $\mathbb{X}$ is considered to be a discrete set, the continuity from the right implies that the realizations are step functions, that is, are piecewise constant in time, with jumps occurring at random jump times (see Fig. A.1 for an illustration). This motivates the name *Markov jump process*.

A.1.2   Infinitesimal Generator and Master Equation

Given the time-dependent transition function $a(t, x, y)$ defined in (A.2), we can introduce a family $\{A(t) | t \geq 0\}$ of (possibly infinite dimensional) transition matrices by

$$A(t) := (a(t, x, y))_{x,y \in \mathbb{X}}.$$

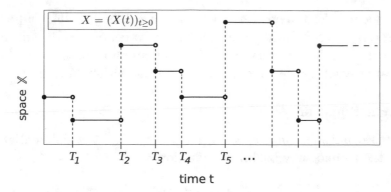

Figure A.1. Schematic illustration of a Markov jump process. The realization is piecewise constant in time. The process switches at random times T_j between the states of a discrete state space $\mathbb{X}$. The trajectory is right-continuous

For every $t \geq 0$, $A(t) \in [0,1]^{|\mathbb{X}|,|\mathbb{X}|}$ is a stochastic matrix, i.e., it contains non-negative real numbers representing probabilities and the entries of each row sum to 1. Utilizing the regularity conditions (A.3) and (A.4), it holds

$$\lim_{t \searrow 0} A(t) = \mathrm{Id} = A(0),$$

where $\mathrm{Id}(x,y) := \delta_{xy}$, i.e., if $\mathbb{X}$ is finite, Id is the $|\mathbb{X}| \times |\mathbb{X}|$-identity matrix. The family of transition matrices is a semigroup, as it satisfies the Chapman-Kolmogorov equation

$$A(t+s) = A(t)A(s) \quad \forall s, t \geq 0.$$

It is fully characterized by its *infinitesimal generator* $G = (g(x,y))_{x,y \in \mathbb{X}}$ given by

$$G := \lim_{t \searrow 0} \frac{A(t) - \mathrm{Id}}{t} \tag{A.5}$$

where the limit is taken entrywise. By definition, the off-diagonal entries of the generator are non-negative and finite

$$0 \leq g(x,y) < \infty \quad \text{for } x \neq y,$$

while the diagonal entries are non-positive with

$$-\infty \leq g(x,x) = -\sum_{y \neq x} g(x,y) \leq 0. \tag{A.6}$$

For $x \neq y$, one can interpret $g(x,y)$ as the *jump rate*, i.e., the average number of jumps per unit time, from state x to state y. Likewise, $|g(x,x)|$ is the total escape rate for the process to leave state x. The case $|g(x, x)| = \infty$ can occur in case of an infinite state space if the rates $g(x, y)$ do not decrease fast enough. For the process this would mean an immediate escape from the state x after entering, i.e., a zero residence time within state x.

Evolution Equations

Under the condition $-g(x, x) < \infty$ for all $x \in \mathbb{X}$, A is differentiable and satisfies the Kolmogorov backward equation

$$\frac{d}{dt} A(t) = GA(t). \tag{A.7}$$

In case of a finite state space $\mathbb{X}$, it holds

$$A(t) = \exp(tG),$$

where $\exp(tG) := \sum_{n=0}^{\infty} \frac{t^n}{n!} G^n$.

Supposing that $-\sum_{y \in \mathbb{X}} a(t, x, y) g(y, y) < \infty$ is satisfied for all $t \geq 0$ and all $x \in \mathbb{X}$ (which is always true in case of a finite state space), then the Kolmogorov forward equation

$$\frac{d}{dt} A(t) = A(t) G \tag{A.8}$$

holds, as well. From the Kolmogorov forward equation (A.8), one can deduce an evolution equation for the distribution vector

$$\boldsymbol{p}(t) = (p(x, t))_{x \in \mathbb{X}},$$

where $p(x, t) := \mathbb{P}(X(t) = x | X(0) \sim \boldsymbol{p}_0)$ is the probability to find the process in state x at time t given the initial distribution $\boldsymbol{p}_0$. It holds $\boldsymbol{p}(t) = A^{\mathsf{T}}(t) \boldsymbol{p}_0$ or $\boldsymbol{p}^{\mathsf{T}}(t) = \boldsymbol{p}_0^{\mathsf{T}} A(t)$, such that multiplying (A.8) with the vector $\boldsymbol{p}_0^{\mathsf{T}}$ from the left gives the *master equation* (in matrix-vector notation)

$$\frac{d}{dt} \boldsymbol{p}^{\mathsf{T}}(t) = \boldsymbol{p}^{\mathsf{T}}(t) G \tag{A.9}$$

with initial condition $\boldsymbol{p}(0) = \boldsymbol{p}_0$. Using (A.6), an alternative ("state-wise") formulation of the master equation is given by

$$\begin{aligned}
\frac{d}{dt} p(x, t) &= \sum_{y \in \mathbb{X}} p(y, t) g(y, x) \\
&= \sum_{y \neq x} [p(y, t) g(y, x) - p(x, t) g(x, y)].
\end{aligned} \tag{A.10}$$

Operator Notation

For each $t \geq 0$, the matrix $A^{\mathsf{T}}(t)$ is the matrix representation of the time-dependent forward transfer operator $\mathcal{A}^t : \ell_1^1 \to \ell_1^1$ of the process (also called *propagator*), which acts on the space

$$\ell_1^1 := \left\{ v : \mathbb{X} \to \mathbb{R}_0^+ \,\middle|\, \sum_{x \in \mathbb{X}} v(x) = 1 \right\}$$

of probability distribution functions on $\mathbb{X}$ and propagates these functions in time by

$$(\mathcal{A}^t v)(x) = \sum_{y \in \mathbb{X}} a(t, y, x) v(y),$$

such that $p(x, t) = (\mathcal{A}^t p_0)(x)$ for $p_0 \in \ell_1^1$ given by $p_0(x) := p(x, 0)$. Parallel to (A.5) one can define the operator $\mathcal{G} : \ell_1^1 \to \ell_1^1$ by

$$\mathcal{G}v := \lim_{t \searrow 0} \frac{\mathcal{A}^t v - v}{t},$$

which is also referred to as the *infinitesimal generator* corresponding to the semigroup $\mathcal{A}^t$. In terms of the operator $\mathcal{G}$, the master equation reads

$$\frac{dp}{dt} = \mathcal{G}p,$$

as an analog of the matrix-vector notation given in (A.9) which can also be written as $\frac{d}{dt}\boldsymbol{p}(t) = G^{\mathsf{T}}\boldsymbol{p}(t)$. That is, G^{T} is the matrix representation of $\mathcal{G}$.[1]

Remark A.2. The matrix G itself is the matrix representation of the adjoint operator $\mathcal{G}^{\mathrm{adj}}$ of $\mathcal{G}$ defined as

$$\mathcal{G}^{\mathrm{adj}}u := \lim_{t \searrow 0} \frac{\mathcal{T}^t u - u}{t},$$

where $\mathcal{T}^t : \ell^1 \to \ell^1$ for $t \geq 0$ is the *backward transfer operator* of the process acting on functions $u \in \ell^1 = \{u : \mathbb{X} \to \mathbb{R} \mid \sum_{x \in \mathbb{X}} |u(x)| < \infty\}$ by

$$\mathcal{T}^t u(x) := \mathbb{E}\Big(u(X(t)) \,\Big|\, X(0) = x\Big) = \sum_{y \in \mathbb{X}} a(t, x, y) u(y).$$

Commonly, the term *infinitesimal generator* is also used for this adjoint operator $\mathcal{G}^{\mathrm{adj}}$.

---

[1] Note that p is a function of time t and state x, while $\boldsymbol{p}$ is a vector-valued function of time.

Stationary Distribution and Steady State

A stationary distribution is a distribution that does not change in time, i.e., it fulfills $\frac{d}{dt}p = 0$. Thus, by setting the left-hand side in the master equation (A.9) to 0, the following definition becomes reasonable:

Definition A.3 (Stationary Distribution). A probability distribution function $\pi \in \ell_1^1$ with

$$0 = \mathcal{G}\pi$$

is called *stationary distribution* or *equilibrium distribution* or *steady-state distribution*. In matrix-vector notation, this means $0 = \pi^\mathsf{T} G$ for a vector $\pi = (\pi(x))_{x \in \mathbb{X}}$.

The dynamical system is said to be in *steady state* if it is distributed according to the stationary distribution π, i.e., if $\mathbb{P}(X(t) = x) = \pi(x)$.

A.1.3   Jump Times, Embedded Markov Chain, and Simulation

Beside the characterization of a Markov jump process in terms of its generator, another way to describe the dynamics is a two-step formalism: For each state, define first the waiting time for the next jump to occur and second the transition probabilities to the other states given that a jump occurs.

The random times $T_0 < T_1 < T_2, \ldots$ at which the Markov jump process $(X(t))_{t \geq 0}$ performs its jumps are recursively defined by $T_0 := 0$, $T_{j+1} := T_j + \tau(T_j)$ where $\tau(t) : \mathbb{X} \to [0, \infty]$ with

$$\tau(t) = \inf\{s > 0 | X(t+s) \neq X(t)\}$$

is the *residual lifetime* in state $X(t)$. In case of $\tau(t) = \infty$, the process will never leave the state $X(t)$. It can be shown (for a time-homogeneous Markov jump process) that this residual lifetime follows an exponential distribution which only depends on the state $X(t)$ while being independent of t. That is, for every $x \in \mathbb{X}$, there exists a *jump rate* $\lambda(x) \geq 0$ such that

$$\mathbb{P}(\tau(t) > s | X(t) = x) = \exp(-\lambda(x)s)$$

for all $s \geq 0$. In terms of the generator entries, this jump rate is given by $\lambda(x) = -g(x, x)$. By the right continuity of the process X, it holds $\lim_{t \searrow T_j} X(t) = X(T_j)$ almost surely for all jump times T_j, i.e.,

$$\mathbb{P}\left(\lim_{t \searrow T_j} X(t) = X(T_j)\right) = 1 \quad \forall T_j.$$

Restricting the investigation to the states of the process at the jump times, one obtains a discrete-time Markov chain that switches between the states according to rules determined by the Markov jump process.

Definition A.4 (Embedded Markov Chain). Given the Markov jump process $(X(t))_{t \geq 0}$ with jump times $(T_j)_{j \in \mathbb{N}_0}$, the *embedded Markov chain* $(\hat{X}_j)_{j \in \mathbb{N}_0}$ is defined as

$$\hat{X}_j := X(T_j), \quad j = 0, 1, 2, \ldots$$

The embedded Markov chain is a discrete-time Markov process with a transition matrix $\hat{P} = (\hat{p}(x, y))_{x,y \in \mathbb{X}}$ given by

$$\hat{p}(x, y) := \mathbb{P}\Big(X(T_1) = y \,\Big|\, X(0) = x\Big) = \begin{cases} \frac{g(x,y)}{-g(x,x)} & \text{if } y \neq x, \\ 0 & \text{if } y = x, \end{cases}$$

for states with $\lambda(x) = -g(x, x) > 0$. In case of $\lambda(x) = 0$, on the other hand, one can set $\hat{p}(x, y) := 0$ for $y \neq x$ and $\hat{p}(x, x) := 1$. In any case, it holds $0 \leq \hat{p}(x, y) \leq 1$ for all $x, y \in \mathbb{X}$ and $\sum_{y \in \mathbb{X}} \hat{p}(x, y) = 1$.

The rates $\lambda(x)$ together with the transition function $\hat{P} = (\hat{p}(x, y))_{x,y \in \mathbb{X}}$ of the embedded Markov chain completely characterize the Markov jump process by the following recursion: Suppose that the process is in state $x \in \mathbb{X}$ at time t. If $\lambda(x) = 0$, the process will stay in state x for all times, i.e., $X(s) = x$ for all $s \geq t$. If $\lambda(x) > 0$, the process will stay in x for a random time that is exponentially distributed with parameter $\lambda(x)$. Then, it jumps to another state $y \neq x$ with probability $\hat{p}(x, y)$. This alternative characterization is the basis for numerical simulations of the Markov jump process. In contrast to approximative numerical integration schemes (like Runge-Kutta methods) for solving ODEs, the following algorithm produces statistically exact trajectories of the underlying Markov jump process.

Simulation scheme for Markov jump processes:

1. Initialize time $t \leftarrow t_0$ and state $x \leftarrow x_0$, and choose a time horizon $T > t_0$.

2. Draw the residual lifetime $\tau \sim \exp(\lambda(x))$ by setting

$$\tau = \frac{1}{\lambda(x)} \ln \left(\frac{1}{r} \right)$$

   with r drawn from a standard uniform distribution $U(0,1)$.

3. Draw the next state y according to the transition function $\hat{p}(x,y)$.

4. Replace $t \leftarrow t + \tau$ and $x \leftarrow y$.

5. End the simulation in case of $t \geq T$. Otherwise, return to 2.

A.1.4 Poisson Processes

One type of Markov jump processes are so-called *counting processes*. These are nondecreasing Markov processes on $\mathbb{X} = \mathbb{N}_0$ which count the number of random events occurring in time. Of special interest are *Poisson processes*, which model the increments by Poisson-distributed random variables. Depending on whether the parameters of these Poisson distributions are constant or depend on time, one distinguishes between *homogeneous* and *inhomogeneous* Poisson processes.

Definition A.5 (Time-Homogeneous Poisson Process). A right-continuous stochastic process $\mathcal{P}_v = (\mathcal{P}_v(t))_{t \geq 0}$ on the state space $\mathbb{N}_0$ is called *time-homogeneous Poisson process with rate constant $v > 0$* if it fulfills the following conditions:

(i) $\mathcal{P}_v(0) = 0$ almost surely.

(ii) $(\mathcal{P}_v(t))_{t \geq 0}$ has independent increments. That is, for any sequence $0 \leq t_1 < t_2 < \ldots < t_n$ of given (nonrandom) time points, the increments $\mathcal{P}_v(t_{j+1}) - \mathcal{P}_v(t_j)$, $j = 1, \ldots, n-1$, are stochastically independent of each other.

(iii) For $t \geq s$, the increment $\mathcal{P}_v(t) - \mathcal{P}_v(s)$ follows a Poisson distribution with mean $v(t-s)$, i.e., it holds

$$\mathbb{P}\big(\mathcal{P}_v(t) - \mathcal{P}_v(s) = n\big) = \frac{(v(t-s))^n}{n!} e^{-v(t-s)} \quad \forall n \in \mathbb{N}_0. \tag{A.11}$$

For $v = 1$ the process is called *unit-rate Poisson process* and denoted by $\mathcal{U} = (\mathcal{U}(t))_{t \geq 0}$ with $\mathcal{U}(t) := \mathcal{P}_1(t)$.

A homogeneous Poisson process with rate $v > 0$ is a monotonically increasing Markov jump process with all jumps having size 1 (see Fig. A.2). Its generator matrix is of the form

$$
G = \begin{pmatrix}
-v & v & 0 & 0 & \cdots \\
0 & -v & v & 0 & \cdots \\
0 & 0 & -v & v & \cdots \\
\vdots & \vdots & \ddots & \ddots & \ddots
\end{pmatrix}.
$$

The residual lifetime fulfills $\tau(t) \sim \exp(v)$ for all t, such that the waiting times between the jumps are independent of each other and fulfill

$$
\mathbb{P}(T_{j+1} - T_j > t) = \exp(-vt), \tag{A.12}
$$

where $(T_j)_{j \in \mathbb{N}_0}$ are the random jump times of the Poisson process. The transition probabilities of the embedded Markov chain are given by

$$
\hat{p}(x, y) = \begin{cases} 1 & \text{if } y = x + 1. \\ 0 & \text{otherwise.} \end{cases}
$$

Consequently, given a sequence $(\tau_j)_{j \in \mathbb{N}_0}$ of exponential random variables with parameter $v > 0$, the Poisson process can inversely be defined as

$$
\mathcal{P}_v(t) := \max \{ j \geq 0 | T_j \leq t \}
$$

where $T_0 = 0$ and $T_{j+1} = T_j + \tau_j$ for all $j \in \mathbb{N}_0$.

The mean and the variance of the Poisson process $\mathcal{P}_v = (\mathcal{P}_v(t))_{t \geq 0}$ are given by

$$
\mathbb{E}(\mathcal{P}_v(t)) = vt, \quad \text{Var}(\mathcal{P}_v(t)) = vt.
$$

From (A.11) it follows that the probability for exactly one jump to occur within a fixed time interval $[t, t + dt)$ for $dt > 0$ fulfills

$$
\mathbb{P}\left(\mathcal{P}_v(t + dt) - \mathcal{P}_v(t) = 1\right) = vdt \cdot \exp(-vdt) = vdt + o(dt) \tag{A.13}
$$

while the probability for no jump to occur is given by $1 - vdt + o(dt)$, and the probability for more than one jump is of order $o(dt)$ for $dt \to 0$.

The definition can be extended to time-varying rates $\Upsilon(t)$ given by a measurable *intensity function* $\Upsilon : \mathbb{R}_+ \to \mathbb{R}_{>0}$, leading to *time-inhomogeneous* Poisson processes. The third condition (iii) in Definition A.5 then has to be replaced by the following formulation:

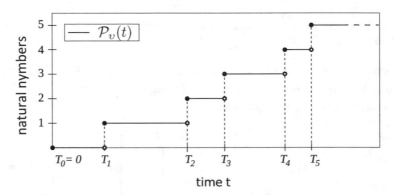

Figure A.2. Trajectory of a homogeneous Poisson process. The time increments $T_{j+1} - T_j$ between the jumps of the homogeneous Poisson process $(\mathcal{P}_v(t))_{t \geq 0}$ with rate constant $v > 0$ follow independent exponential distributions with parameter v. The process is monotonically increasing in time with jumps always of size 1

(iii)' For all $t \geq s$, the increment $\mathcal{P}_\Upsilon(t) - \mathcal{P}_\Upsilon(s)$ follows a Poisson distribution with mean

$$\int_s^t \Upsilon(s')\,ds'.$$

An inhomogeneous Poisson process can be expressed in terms of a homogeneous unit-rate Poisson process by setting

$$\mathcal{P}_\Upsilon(t) = \mathcal{U}\left(\int_0^t \Upsilon(s)\,ds\right). \tag{A.14}$$

Here, two different time frames are linked/related to each other: At the absolute time t of the Poisson process $\mathcal{P}_\Upsilon$, the amount of "internal time" that passed for $\mathcal{U}$ is given by the cumulative intensity $\int_0^t \Upsilon(s)ds$. This internal time does actually not have units of time, but is a dimensionless indicator for the time advance in $\mathcal{U}$ [7].

More general types of counting processes can be obtained by letting the intensity depend on the state (or the past) of the process itself and/or on other stochastic inputs. For example, consider

$$N(t) := \mathcal{U}\left(\int_0^t \Upsilon(s, N(s))\,ds\right),$$

where the intensity is a function $\Upsilon : \mathbb{R}_+ \times \mathbb{N}_0 \to \mathbb{R}_{>0}$ of the time and the actual state of the process. The process $(N(t))_{t \geq 0}$ is a Markovian counting process on $\mathbb{N}_0$; however, it is not a Poisson process because it does not have independent increments.

A.2 Diffusion Process and Random Walk

Beside the Markov jump processes which are described in Sect. A.1, also diffusion processes given by stochastic differential equations play an important role in this book. In the following, we summarize some basic concepts in order to provide a background for readers who are not familiar with the related mathematical theory.

A.2.1 Stochastic Differential Equations

A process of crucial importance for the investigations in this book is the *Wiener process*, also called *Brownian motion*, with the two terms used as synonyms throughout the chapters. The common definition is the following:

Definition A.6 (Wiener Process/Brownian Motion). A *(standard) Wiener process* $W = (W(t))_{t \geq 0}$ on $\mathbb{R}$, also called *Brownian motion process*, is a stochastic real-valued process with the following properties:

1. It holds $W(0) = 0$ almost surely.

2. The process W has independent increments: For all time points $0 \leq t_0 < t_1 < t_2$, the increments $W(t_1) - W(t_0)$ and $W(t_2) - W(t_1)$ are stochastically independent of each other.

3. The increments are normally distributed: For all $0 \leq s < t$, it holds $W(t) - W(s) \sim \mathcal{N}(0, t - s)$.

4. The paths $(W(t))_{t \geq 0}$ on $\mathbb{R}$ are almost surely continuous in time.

A typical trajectory of the one-dimensional Wiener process is depicted in Fig. A.3.

The generalization to the multidimensional case is the following: A stochastic process $(\boldsymbol{W}(t))_{t \geq 0} = (W_1(t), \ldots, W_K(t))_{t \geq 0}$ on $\mathbb{R}^K$ is called (standard) Wiener process if its components $W_k = (W_k(t))_{t \geq 0}$ are independent standard Wiener processes on $\mathbb{R}$ as defined in Definition A.6. In this case, the increments $\boldsymbol{W}(t) - \boldsymbol{W}(s)$ are again independent and normally distributed with $\boldsymbol{W}(t) - \boldsymbol{W}(s) \sim \mathcal{N}(\boldsymbol{0}, (t - s)\mathrm{Id})$, where $\boldsymbol{0} \in \mathbb{R}^K$ is a vector with zero entries and $\mathrm{Id} \in \mathbb{R}^{K,K}$ is the identity matrix.

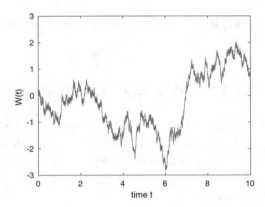

Figure A.3. Wiener process/Brownian motion. Single realization of the one-dimensional Wiener process defined in Definition A.6

The Wiener process is the key process to describe more complicated stochastic processes in the form of stochastic differential equations (SDEs). In this book, only Itô diffusion processes as solution to specific SDEs are relevant. Their formal definition is based on the concept of the Itô integral, which uses stochastic processes as integrands and integrators and can be seen as a stochastic generalization of the Riemann-Stieltjes integral. In the standard setting, which is the only relevant in this book, the Wiener process is employed as integrator: Given the Wiener process W and another stochastic process $(X(t))_{t\geq 0}$, the *Itô integral*

$$I(t) := \int_0^t X(s)\, dW(s)$$

is another stochastic process defined by

$$\int_0^t X(s)\, dW(s) := \lim_{n\to\infty} \sum_{i=1}^n X\left(t_{i-1}\right)\left(W\left(t_i\right) - W\left(t_{i-1}\right)\right) \qquad \text{(A.15)}$$

for $t_i := i \cdot \frac{t}{n}$ and fixed t. Note that for each summand in (A.15), the process X is evaluated at the left-hand endpoint of the respective finite time interval $[t_i, t_{i+1}]$. Thereby, the Itô integral differentiates from the *Stratonovich integral* as another fundamental stochastic integral definition [187].

By means of the Itô integral, we can define the class of *Itô stochastic processes*.

Definition A.7 (Itô Process). A stochastic process $(X(t))_{t\geq 0}$ is called *Itô process* or *Itô diffusion* if it satisfies

$$X(t) = X(0) + \int_0^t \mu(X(s), s)\, ds + \int_0^t \sigma(X(s), s)\, dW(s) \qquad \text{(A.16)}$$

for a standard Wiener process $(W(t))_{t \geq 0}$ and suitable functions $\mu, \sigma : \mathbb{R} \times [0, \infty) \to \mathbb{R}$.

The integral equation (A.16) can be reformulated in terms of the SDE

$$dX(t) = \mu(X(t), t)dt + \sigma(X(t), t)\,dW(t), \qquad (A.17)$$

which may also be used directly to define the corresponding Itô process. Actually, the SDE (A.17) has to be interpreted as a reformulation of the (more formal) integral equation (A.16). The functions μ and σ are called *drift coefficient* and *diffusion coefficient*, respectively. Note that for $\mu = 0$ and $\sigma = 1$ one obtains again a standard Wiener process.

The Itô integral does not obey the ordinary chain rule; instead there is the slightly more complicated *Itô lemma* [187], which we will quote now, thereby completing our short summary of Itô calculus.

Lemma A.8 (Itô's Lemma). *Let $(X(t))_{t \geq 0}$ be the Itô diffusion process given by (A.16) or (A.17), and let $f : \mathbb{R} \to \mathbb{R}$ be a twice differentiable function. Then, the process $Y(t) := f(X(t))$ is again an Itô process fulfilling*

$$dY(t) = \frac{\partial}{\partial x}f(X(t))dX(t) + \frac{1}{2}\frac{\partial^2}{\partial x^2}f(X(t))dX(t)^2. \qquad (A.18)$$

Using $dX(t)^2 = \sigma^2(X(t), t)dt$ and aggregating the dt- and $dW(t)$-terms in Eq. (A.18), the explicit SDE for the process Y turns out to be

$$dY(t) = \left[\frac{\partial}{\partial x}f(X(t), t)\mu(X(t), t) + \frac{1}{2}\frac{\partial^2}{\partial x^2}f(X(t), t)\sigma^2(X(t), t)\right] dt$$
$$+ \frac{\partial}{\partial x}f(X(t), t)\sigma(X(t), t)dW(t).$$

Itô processes appear in several contexts within the preceding chapters: The chemical Langevin equation (2.16) has the form of an Itô SDE (actually with each summand containing a drift and a diffusion coefficient), the hybrid diffusion processes from Sect. 2.2.2 involve SDE-components, and also the spatial movement of particles considered in Sect. 4.1 is given by an Itô diffusion process. In the following section, we describe this last type of spatial diffusion process in more detail for bounded spatial domains, giving some basics about spatial discretization and approximations by jump processes. This serves as a background for Sect. 4.3.

A.2.2 Diffusion in a Bounded Domain

For a start, let us consider diffusion-type stochastic differential equations of the form

$$dQ(t) = \mu(Q(t))dt + \sqrt{2D}dW(t), \tag{A.19}$$

where $Q(t)$ denotes the position of the moving particle in $\mathbb{D} = \mathbb{R}^d$ at time t, $D > 0$ is the diffusion coefficient, $\mu : \mathbb{R}^d \to \mathbb{R}^d$ is a sufficiently smooth function, and $(W(t))_{t\geq0}$ is a standard Wiener process in $\mathbb{R}^d$. The diffusion process Q is a special case of an Itô process given by the SDE (A.17).

The associated Fokker-Planck equation, i.e., the Kolmogorov forward equation associated with (A.19), then has the form

$$\frac{\partial}{\partial t}\rho(q,t) = D\nabla^2\rho(q,t) - \nabla \cdot (\mu(q)\nabla\rho(q,t)), \tag{A.20}$$

where $\rho(q,t)$ is the probability density function to find the particle in position q at time t and ∇ denotes the nabla operator with respect to q, i.e.,

$$\nabla = \left(\frac{\partial}{\partial q_1}, \ldots, \frac{\partial}{\partial q_d}\right)^\mathsf{T}.$$

For the sake of simplicity, we restrict our consideration to the case of pure unbounded diffusion, i.e., we consider $\mu = 0$. Note that the subsequent explanation can be extended to a large class of choices of μ.

In the case of $\mu = 0$, (A.19) yields a scaled Brownian motion

$$Q(t) = \sqrt{2D}W(t), \tag{A.21}$$

and (A.20) reduces to the simple diffusion equation

$$\frac{\partial}{\partial t}\rho(q,t) = D\nabla^2\rho(q,t), \qquad \rho(q,0) = \rho_0(q), \tag{A.22}$$

where we added initial conditions in order to get a Cauchy problem. The connection between (A.21) and (A.22) is the following: The solution of (A.22) can be written in terms of the process $(Q(t))_{t\geq0}$ given by (A.21) by setting

$$\rho(q,t) = \mathbb{E}_q(\rho_0(Q(t))), \tag{A.23}$$

where $\mathbb{E}_q$ denotes the expectation conditioned on the initial value of Q

$$\mathbb{E}_q\big(\rho_0(Q(t))\big) := \mathbb{E}\big(\rho_0(Q(t))|Q(0) = q\big).$$

If we now consider a bounded domain $\mathbb{D} \subset \mathbb{R}^d$ instead of $\mathbb{D} = \mathbb{R}^d$ (e.g., given by a biological cell where the particles move inside), we have to add boundary conditions to the PDE (A.22). In the simplest case where the domain $\mathbb{D}$ is open and the boundary $\partial \mathbb{D}$ of $\mathbb{D}$ can never be overcome by the moving particle but is absorbing, we need Dirichlet boundary conditions, that is, we have to consider the boundary value problem

$$\frac{\partial}{\partial t}\rho(q,t) = D\nabla^2\rho(q,t), \qquad q \in \mathbb{D}, \ t \geq 0, \qquad (A.24)$$

$$\rho(q,0) = \rho_0(q),$$

$$\rho(q,t) = 0, \qquad q \in \partial\mathbb{D}, \ t \geq 0.$$

The solution of (A.24) can also be written in terms of the process $Q = (Q(t))_{t \geq 0}$ that moves within $\mathbb{D}$ according to (A.21) with absorbing boundary conditions. In order to provide the respective identity, we have to introduce the first exit time τ of Q from $\mathbb{D}$

$$\tau := \inf\{t \geq 0 | Q(t) \in \partial\mathbb{D}\}.$$

The first exit time τ is path-dependent and thus is a random variable that gives the first time the respective path of Q that touches the boundary $\partial\mathbb{D}$ of $\mathbb{D}$ (if started within $\mathbb{D}$). With this preparation the solution of (A.24) is given by

$$\rho(q,t) = \mathbb{E}_q\Big(\rho_0(Q(t))\mathbb{1}_{t \leq \tau}\Big), \qquad (A.25)$$

where $\mathbb{1}$ denotes the indicator function, i.e.,

$$\mathbb{1}_{t \leq \tau} = \begin{cases} 1 & \text{if } t \leq \tau, \\ 0 & \text{if } t > \tau. \end{cases}$$

That is, (A.25) tells us that we have to *eliminate* the respective solution path of Q as soon as it exits from $\mathbb{D}$.

Discretization in Space

Our next step is to discretize (A.24) in space (only). This procedure is called *the method of lines* for parabolic PDEs of the form of (A.24). For the sake of simplicity, we will consider the case of particle diffusion in $d = 2$ dimensions and the unit square domain $\mathbb{D} = [0,1] \times [0,1]$ such that

$$\partial\mathbb{D} = \Big\{(q_1,q_2) \in [0,1]^2 \,\big|\, q_1 \in \{0,1\} \text{ or } q_2 \in \{0,1\}\Big\}.$$

For the spatial discretization, we choose finite differences with uniform discretization points $q_{(i_1,i_2)}$

$$q_{(i_1,i_2)} := (i_1 h, i_2 h) \in [0,1]^2, \qquad (i_1, i_2) \in \mathbb{I} := \{0, \dots, n_0\}^2$$

with step size $h = 1/n_0$ for some $n_0 \in \mathbb{N}$. Defining

$$\mathbb{I}_{\text{int}} := \{1, \dots, n_0 - 1\}^2,$$

all points $q_{(i_1,i_2)}$ with $(i_1, i_2) \in \mathbb{I}_{\text{int}}$ lie inside of $\mathbb{D}$, and all points $q_{(i_1,i_2)}$ with $i_1 \in \{0, n_0\}$ or $i_2 \in \{0, n_0\}$ lie on the boundary $\partial\mathbb{D}$. Finite difference discretization means that we seek a discrete time-dependent solution $\hat{\rho}_{(i_1,i_2)}(t)$ (serving as an approximation $\hat{\rho}_{(i_1,i_2)}(t) \approx \rho(q_{(i_1,i_2)}, t)$), being defined at the discretization points only that satisfies

$$\frac{d}{dt}\hat{\rho}_{(i_1,i_2)} = \frac{D}{h^2}\left(\hat{\rho}_{(i_1-1,i_2)} + \hat{\rho}_{(i_1+1,i_2)} + \hat{\rho}_{(i_1,i_2-1)} + \hat{\rho}_{(i_1,i_2+1)} - 4\hat{\rho}_{(i_1,i_2)},\right)$$
$$\text{for all } (i_1, i_2) \in \mathbb{I}_{\text{int}},$$
$$\hat{\rho}_{(i_1,i_2)}(0) = \rho_0\left(q_{(i_1,i_2)}\right) \quad \text{for all } i_1, i_2 = 0, \dots, n_0,$$
$$\hat{\rho}_{(i_1,i_2)} = 0 \quad \text{for all } i_1 \in \{0, n_0\} \text{ or } i_2 \in \{0, n_0\}.$$

That is, the second derivatives in (A.24) given by the Laplace operator ∇^2 have been replaced by second-order central differences in two dimensions at all discretization points that lie within $\mathbb{D}$, while the points on the boundary inherit the boundary condition. In this way the infinite dimensional parabolic problem (A.24) has been replaced by a set of ordinary differential equations.

Remark A.9. In two dimensions ($d = 2$), an approximation of the Laplace operator $\nabla^2 = \frac{\partial^2}{\partial q_1^2} + \frac{\partial^2}{\partial q_2^2}$ by means of the finite-element method results in

$$\nabla^2 u(q_1, q_2)$$
$$\approx \frac{u(q_1 - h, q_2) + u(q_1 + h, q_2) + u(q_1, q_2 - h) + u(q_1, q_2 + h) - 4u(q_1, q_2)}{h^2}$$

for functions $u = u(q_1, q_2)$. Equivalent calculations in $d = 1$ or $d = 3$ also result in the factor h^{-2}, such that the diffusive jump constant given by $\frac{D}{h^2}$ is independent of the spatial dimension d.

We may bring this equation for $\hat{\rho}$ into matrix-vector form by enumeration of the discretization points $q_{(i_1,i_2)}$ and by introducing the generator matrix

$$G_{(i_1,i_2),(j_1,j_2)} := \frac{D}{h^2} \begin{cases} -4 & \text{if } i_1 = j_1, i_2 = j_2, \\ 1 & \text{if } i_1 - 1 = j_1, i_2 = j_2, \\ 1 & \text{if } i_1 + 1 = j_1, i_2 = j_2, \\ 1 & \text{if } i_1 = j_1, i_2 - 1 = j_2, \\ 1 & \text{if } i_1 = j_1, i_2 + 1 = j_2, \\ 0 & \text{otherwise} \end{cases}$$

for $(i_1, i_2) \in \mathbb{I}_{\text{int}}$, $(j_1, j_2) \in \mathbb{I}$. Obviously, G takes the form of a matrix by enumeration of the set $\{0, \ldots, n_0\}^2$ with the interior points being enumerated first. Then, it is of dimension $(n_0 - 1)^2 \times (n_0 + 1)^2$. Using the same enumeration for bringing $\hat{\rho}$ into vector form, it becomes a row vector of dimension $(n_0 + 1)^2$. By clipping the entries that belong to the discretization points on the boundary, we get the $(n_0 - 1)^2$-dimensional row vector $\hat{\rho}^o$. Keeping only the entries for the boundary points yields the $4n_0$-dimensional row vector $\hat{\rho}^b$. That is, we have

$$\hat{\rho} = \left(\hat{\rho}^o, \hat{\rho}^b \right).$$

With this notation, the discretized equation becomes

$$\frac{d}{dt}\hat{\rho} = \hat{\rho}^o G,$$
$$\hat{\rho}(t = 0) = \hat{\rho}_0, \tag{A.26}$$

for an accordingly defined initial distribution vector $\hat{\rho}_0$. Now G is a matrix with non-negative off-diagonal entries and

$$\sum_{(j_1,j_2)\in\mathbb{I}} G_{(i_1,i_2),(j_1,j_2)} = 0.$$

That is, it has the form of a generator (at least in a general sense, since it is not a square matrix). Thus, Eq. (A.26) has the form of a boundary value *master equation*. More precisely, the solution of (A.26) satisfies the preservation of non-negativity (i.e., if $\hat{\rho}_0 \geq 0$ and then $\hat{\rho}(t) \geq 0$ for all $t \geq 0$) and conservation of 1-norm conditions, i.e.,

$$\sum_{(i_1,i_2)\in\mathbb{I}} \hat{\rho}_{(i_1,i_2)}(t) = \sum_{(i_1,i_2)\in\mathbb{I}} \hat{\rho}_{(i_1,i_2)}(0), \qquad \forall t \geq 0.$$

This means that there will be an associated discrete-space, continuous-time Markov process.

In order to write the solution of (A.26) as an expectation value of a Markov process, let us introduce the following continuous-time *random walk* on a two-dimensional infinite lattice with lattice constant h, i.e., on the state space $(h\mathbb{Z})^2$: When being in state $(hi_1, hi_2) \in (h\mathbb{Z})^2$, jumps are possible only to the four states $h(i_1 \pm 1, i_2)$ or $h(i_1, i_2 \pm 1)$, and for each of these states, the exponentially distributed rate of jumping is given by D/h^2. This defines a Markov jump process that we denote by $\hat{Q} = (\hat{Q}(t))_{t \geq 0}$. With this process in hand, we have that the solution of (A.26) is given by

$$\hat{\rho}_{(i_1, i_2)}(t) = \mathbb{E}_{(i_1, i_2)}\Big(\hat{\rho}_0(\hat{Q}(t))\mathbb{1}_{t \leq \tau_{\text{int}}}\Big), \qquad (A.27)$$

where τ_{int} denotes the first exit time of $\hat{Q}$ from $h\mathbb{I}_{\text{int}}$.

The classical theory of finite difference discretization of parabolic equations like (A.24) provides insight into the convergence of the solution of the space-discretized equation (A.26) to the solution of (A.24) for increasingly finer discretization $h \to 0$. In order to state one of the many results, let us denote the set of all discretization points with step size h by

$$\mathbb{D}_h = \Big\{q_{(i_1, i_2)}\Big| \ (i_1, i_2) \in \mathbb{I}_{\text{int}}\Big\} = h\mathbb{I}_{\text{int}} \subset \mathbb{D},$$

and the corresponding solution of (A.26) by $\hat{\rho}_h(q_{(i_1, i_2)}, t) = \hat{\rho}_{(i_1, i_2)}(t)$ where $\hat{\rho}$ is the solution of (A.26) given the step size h.

Provided that (A.24) has a classical solution that is twice differentiable in time and three times continuously differentiable in $\mathbb{D}$ and toward the boundary, then for every $T > 0$ there exists a constant $C > 0$, independent of h such that for small enough $h > 0$ we have

$$\max_{q \in \mathbb{D}_h} |\rho(q, t) - \hat{\rho}_h(q, t)| \leq Ch, \qquad \forall t \leq T.$$

That is, for $h \to 0$ the discretized solution converges (point-wise) to the classical solution of (A.24). Because of the relation of $\hat{\rho}_h$ to the random walk $\hat{Q}$ on $(h\mathbb{Z})^2$ and of ρ to the diffusion process Q, this also implies that the laws of Q and $\hat{Q}$ are close for small enough h. In fact, $\hat{Q}(t)$ on $(h\mathbb{Z})^2$ converges in distribution to $Q(t)$ for all t.

Remark A.10. For the spatial discretization of (A.24), we have many different options beyond finite difference with uniform discretization points. For example, one usually applies finite element [252] (or finite volume) discretizations instead of finite difference ones, and we can use problem-adapted (nonuniform) grids instead of uniform discretization points. It is important to note that the resulting discretization matrix will always exhibit the generator property (as long as the boundary is appropriately represented). Therefore, the discretized form of (A.24) has the form of a master equation, and we can always associate a random walk with it (in the sense that its variant with appropriate elimination at the boundary allows us to give the solution of the discretized diffusion equation in the form of an expectation value). This random walk always converges in distribution to the diffusion process $(Q(t))_{t \geq 0}$.

A.3 Convergence of Stochastic Processes

The convergence of stochastic processes is a subject of central importance in the main part of the book. Different types of convergence are mentioned throughout the chapters. We will here explain the basic concepts and deliver the fundamental definitions that are needed in order to understand the convergence statements.

In general, a stochastic process on the state space $\mathbb{X}$ is a collection $(X(t))_{t \in \mathbb{T}}$ of random variables $X(t) : (\Omega, \mathcal{E}, \mathbb{P}) \to (\mathbb{X}, \mathcal{E}')$, for a probability space $(\Omega, \mathcal{E}, \mathbb{P})$ and a measurable space $(\mathbb{X}, \mathcal{E}')$. Here, we will consider a metric space $(\mathbb{X}, d_{\mathbb{X}})$ such that $\mathcal{E}'$ can be the Borel σ-algebra. The index set $\mathbb{T}$ is mostly given by an interval of the real line, but also other index sets are possible. We here restrict the investigations to $\mathbb{T} = [0, \infty)$ or $\mathbb{T} = [0, T]$ for $T > 0$. For each $t \in \mathbb{T}$, the random variable $X(t)$ is a function of $\omega \in \Omega$. This is emphasized by writing $X(\omega, t)$. For fixed $\omega \in \Omega$, the function $X(\omega, \cdot) : \mathbb{T} \to \mathbb{X}$ is called *sample path* of the stochastic process. Let the space of all possible sample paths be denoted by $\mathcal{X}$, i.e., for each $\omega \in \Omega$, the associated sample path $X(\omega, \cdot)$ is contained in the set $\mathcal{X}$. For example, it holds $\mathcal{X} = \mathcal{C}_{\mathbb{X}}[0, \infty)$ if the index set is given by $\mathbb{T} = [0, \infty)$ and the process has continuous sample paths.[2]

The reaction jump processes defined in Chap. 1, however, do not have continuous sample paths, but are piecewise constant on $\mathbb{X} = \mathbb{N}^L$ with jumps occurring at random times. A suitable path space, which includes such piecewise constant trajectories of Markov jump processes, is given by the space of *càdlàg-functions* (from French *continue à droite, limite à gauche*), i.e., functions that are everywhere right-continuous with existing left limits (see Fig. A.1). We denote the space of càdlàg-functions on $\mathbb{T} = [0, T]$ by $\mathcal{D}_{\mathbb{X}}[0, T]$; for $\mathbb{T} = [0, \infty)$, we analogously write $\mathcal{D}_{\mathbb{X}}[0, \infty)$. In applications,

---

[2] $\mathcal{C}[0, \infty)$ denotes the set of all continuous function $f : [0, \infty) \to \mathbb{X}$.

most stochastic processes have sample paths in $\mathcal{D}_{\mathbb{X}}[0, \infty)$, which motivates to consider this space. Note that it holds $\mathcal{C}_{\mathbb{X}}[0, \infty) \subset \mathcal{D}_{\mathbb{X}}[0, \infty)$.

Given a topology on the space $\mathcal{X}$ of sample paths, the stochastic process $X = (X(t))_{t \geq 0}$ can be interpreted as a random variable

$$X : (\Omega, \mathcal{E}, \mathbb{P}) \to (\mathcal{X}, \mathcal{B})$$

with values in $\mathcal{X}$, where $\mathcal{B} = \mathcal{B}(\mathcal{X})$ is the Borel σ-algebra on $\mathcal{X}$. The common approach is to define a metric $d_{\mathcal{X}}$ on $\mathcal{X}$ and to choose the topology induced by this metric. Then, given a sequence $(X_n)_{n \in \mathbb{N}}$ of stochastic processes $X_n = (X_n(t))_{t \geq 0}$, as well as a limit process $X = (X(t))_{t \geq 0}$, each of them having sample paths in $\mathcal{X}$, there are two main types of convergence:

- *Point-wise convergence*: It converges $X_n(t)$ to $X(t)$ for each $t \in \mathbb{T}$, i.e., it holds $d_{\mathbb{X}}(X_n(t), X(t)) \overset{n \to \infty}{\longrightarrow} 0$ in some stochastic convergence sense[3] for each $t \in \mathbb{T}$, where $d_{\mathbb{X}}$ is the metric on $\mathbb{X}$.

- *Path-wise convergence*: It holds $d_{\mathcal{X}}(X_n, X) \overset{n \to \infty}{\longrightarrow} 0$ in some stochastic convergence sense, where $d_{\mathcal{X}}$ is the metric on $\mathcal{X}$.

In order to consider path-wise convergence, one therefore has to define a suitable metric $d_{\mathcal{X}}$ on the path space $\mathcal{X}$. For the set of càdlàg-functions $\mathcal{X} = \mathcal{D}_{\mathbb{X}}[0, T]$ (or $\mathcal{X} = \mathcal{D}_{\mathbb{X}}[0, \infty)$), a reasonable choice is given by the *Skorokhod metric*, which will be considered in more detail in the following.

A.3.1   Skorokhod Space

We first consider the case $\mathbb{T} = [0, T]$ for some $T > 0$ and set $\mathbb{X} = \mathbb{R}$. Denote by $\mathcal{D}[0, T]$ the space of càdlàg-functions from $\mathbb{T} = [0, T]$ to $\mathbb{R}$. These functions are right-continuous at each point in $[0, T)$ with left limits existing for all points in $(0, T]$. Let further $\mathcal{I}$ be the set of strictly increasing, continuous bijections from $[0, T]$ to $[0, T]$ (which implies $I(0) = 0$ and $I(T) = T$ for $I \in \mathcal{I}$). Let

$$\|f\| := \sup_{t \leq T} |f(t)|$$

denote the uniform norm for functions $f : [0, T] \to \mathbb{R}$. For $f, g \in \mathcal{D}[0, T]$, the Skorokhod metric $d_{\mathcal{D}}$ is then defined by

$$d_{\mathcal{D}}(f, g) := \inf_{I \in \mathcal{I}} \max\{\|I - \mathrm{id}\|, \|f - g \circ I\|\}$$

---

[3]For example, convergence in distribution, convergence in probability, convergence almost surely, or convergence in mean.

where id : $[0, T] \to [0, T]$ is the identity function. It can be shown that $d_{\mathcal{D}}$ is indeed a metric [21]. Note that it holds $\|f\| < \infty$ for $f \in \mathcal{D}[0, T]$ because f can have at most finitely many time points at which the jump size exceeds a given positive number.

Remark A.11. The definition of the Skorokhod metric can be extended to the more general setting where $(\mathbb{X}, d_{\mathbb{X}})$ is a complete, separable metric space and $\mathcal{D}_{\mathbb{X}}[0, T]$ is the space of càdlàg-functions from $\mathbb{T} = [0, T]$ to $\mathbb{X}$, by setting

$$d_{\mathcal{D}}(f, g) := \inf_{I \in \mathcal{I}} \max \left\{ \|I - \mathrm{id}\|, \sup_{t \leq T} d_{\mathbb{X}}(f(t), g \circ I(t)) \right\}. \tag{A.28}$$

Intuitively, the topology induced by the Skorokhod metric allows functions to slightly deviate in time and space for still being "close". The functions $I \in \mathcal{I}$ hereby serve as time perturbations. In contrast, the traditional topology induced by the uniform norm only allows to slightly deviate in space. This difference is important for the discontinuity points of the limit function: For a sequence $(f_n)_{n \in \mathbb{N}}$ of functions $f_n \in \mathcal{D}_{\mathbb{X}}[0, T]$ to converge uniformly to $f \in \mathcal{D}_{\mathbb{X}}[0, T]$ with f having a jump at t^*, all f_n for large n must have a jump of almost the same size precisely at t^*. In contrast, for convergence with respect to the Skorokhod topology (i.e., $d_{\mathcal{D}}(f, f_n) \to 0$), these jumps of almost the same size can be located in small distance to t^* (see Example A.12 for an illustration). In case of a continuous limit function $f \in \mathcal{C}_{\mathbb{X}}[0, t]$, on the other hand, the convergence with respect to the Skorokhod topology is equivalent to uniform convergence.

Example A.12. Consider the sequence $(f_n)_{n \in \mathbb{N}}$ of functions $f_n : [0, T] \to \mathbb{R}$ defined by

$$f_n(t) := \begin{cases} 0 & \text{if } t < t^* + \frac{1}{n}, \\ 1 & \text{if } t \geq t^* + \frac{1}{n}. \end{cases} \tag{A.29}$$

for some $0 < t^* < T$. It holds $f_n \to f$ with respect to the Skorokhod topology, where f is defined by

$$f(t) := \begin{cases} 0 & \text{if } t < t^*, \\ 1 & \text{if } t \geq t^*. \end{cases}$$

This follows by choosing, for example, $I_n \in \mathcal{I}$ with

$$I_n(t) = \begin{cases} t + \frac{t}{nt^*} & \text{if } t < t^*, \\ t + \frac{1}{n} - \frac{t - t^*}{n(T - t^*)} & \text{if } t \geq t^*, \end{cases}$$

which implies $f_n \circ I_n = f$ for all n. In contrast, uniform convergence is not given, as it holds $\|f_n - f\| = |f_n(t^*) - f(t^*)| = 1$ for all n (see Fig. A.4 for an illustration) ◇

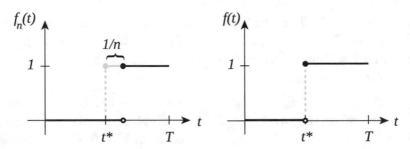

Figure A.4. Convergence with respect to Skorokhod metric. The functions f_n defined by (A.29) converge to f with respect to the Skorokhod metric, i.e., $d_\mathcal{D}(f, f_n) \to 0$ for $n \to \infty$; uniform convergence, on the other hand, is not given

Extension to $\mathcal{D}_\mathbb{X}[0, \infty)$

There exist different ways to extend the definition of the Skorokhod metric/topology to $\mathcal{D}_\mathbb{X}[0, \infty)$ for the infinite time interval $\mathbb{T} = [0, \infty)$. The basic idea is to consider the different restrictions of a function $f \in \mathcal{D}_\mathbb{X}[0, \infty)$ to each finite interval $[0, T]$ and to apply the definition given in (A.28) to these restrictions. This leads to the following formulation of convergence of a sequence of functions $f_n \in \mathcal{D}_\mathbb{X}[0, \infty)$ to the function $f \in \mathcal{D}_\mathbb{X}[0, \infty)$ [10]: It holds $f_n \to f$ with respect to the Skorokhod topology if and only if there exists a sequence of strictly increasing, continuous bijections $I_n : [0, \infty) \to [0, \infty)$ such that for each $T > 0$

$$\lim_{n \to \infty} \sup_{t \leq T} |I_n(t) - t| = 0 \quad \text{and} \quad \lim_{n \to \infty} \sup_{t \leq T} d_\mathbb{X}(f(t), f_n \circ I_n(t)) = 0.$$

The formulation in terms of restrictions to finite time intervals $[0, T]$ can cause problems for values of T at which the limit process has a positive probability for jumping. It may therefore be more convenient to define the Skorokhod topology on $\mathcal{D}_\mathbb{X}[0, \infty)$ directly, instead of deriving its properties from those of $\mathcal{D}_\mathbb{X}[0, T]$ [192]. Following [70, 85], let $\mathcal{I}_\infty$ denote the set of strictly increasing, continuous bijections from $[0, \infty)$ to $[0, \infty)$ and define

$$\gamma(I) := \sup_{s > t \geq 0} \left| \log \frac{I(s) - I(t)}{s - t} \right|$$

for $I \in \mathcal{I}_\infty$. For $f, g \in \mathcal{D}_\mathbb{X}[0, \infty)$, $I \in \mathcal{I}_\infty$, $u \geq 0$, set

$$d_0(f, g, I, u) := \sup_{t \geq 0} \frac{d_\mathbb{X}(f(t \wedge u), g(I(t) \wedge u))}{1 + d_\mathbb{X}(f(t \wedge u), g(I(t) \wedge u))}$$

where $t \wedge u := \min\{t, u\}$. Then the Skorokhod distance between f and g can be defined by

$$d_{\mathcal{D}}(f, g) := \inf_{I \in \mathcal{I}_{\infty}} \max \left\{ \gamma(I), \int_0^{\infty} \exp(-u) d_0(f, g, I, u) \, du \right\}.$$

Remark A.13. If the limit function f is continuous, i.e., $f \in \mathcal{C}_{\mathbb{X}}[0, \infty)$, then convergence of f_n to f with respect to the Skorokhod metric is equivalent to uniform convergence on compacta. That is, it holds $\lim_{n \to \infty} d_{\mathcal{D}}(f, f_n) = 0$ for $f \in \mathcal{C}_{\mathbb{X}}[0, \infty)$ if and only if

$$\lim_{n \to \infty} \sup_{t \le T} d_{\mathbb{X}}(f_n(t), f(t)) = 0 \quad \forall T > 0,$$

where $d_{\mathbb{X}}$ is the given metric on $\mathbb{X}$.

A.3.2 Types of Path-Wise Convergence

After having specified what the convergence $f_n \to f$ means for a sequence of càdlàg-functions f_n with general limit $f \in \mathcal{D}_{\mathbb{X}}[0, \infty)$, we now consider a sequence $(X_n)_{n \in \mathbb{N}}$ of stochastic processes $X_n = (X_n(t))_{t \ge 0}$ with sample paths in $\mathcal{D}_{\mathbb{X}}[0, \infty)$ and define different types of stochastic convergence. Given the metric space $(\mathcal{D}_{\mathbb{X}}[0, \infty), d_{\mathcal{D}})$, the common definitions of convergence in distribution, convergence in probability, and almost sure convergence can directly be formulated for stochastic processes.

Convergence in Distribution

One possible way to define convergence in distribution for a sequence $(X_n)_{n \in \mathbb{N}}$ of $\mathcal{X}$-valued random variables, given the metric space $(\mathcal{X}, d)$, is the following: Let $\mathcal{C}_b(\mathcal{X})$ denote the space of real-valued bounded, continuous functions on $(\mathcal{X}, d)$. Then $(X_n)_{n \in \mathbb{N}}$ converges in distribution to $X \in \mathcal{X}$ if it holds

$$\lim_{n \to \infty} \mathbb{E}(f(X_n)) = \mathbb{E}(f(X))$$

for each $f \in \mathcal{C}_b(\mathcal{X})$. This definition can directly be applied to stochastic processes with sample paths in $\mathcal{D}_{\mathbb{X}}[0, \infty)$ by setting $(\mathcal{X}, d) = (\mathcal{D}_{\mathbb{X}}[0, \infty), d_{\mathcal{D}})$ where $d_{\mathcal{D}}$ is the Skorokhod metric. We denote convergence in distribution by $X_n \Rightarrow X$.

The convergence in distribution $X_n \Rightarrow X$ is equivalent to the weak convergence of the induced probability measures $(P_{X_n} \overset{w}{\to} P_X)$, which are defined on the measurable space $(\mathcal{D}_{\mathbb{X}}[0, \infty), \mathcal{B}_S)$ with $\mathcal{B}_S$ denoting the Borel σ-algebra induced by the Skorokhod topology.

Convergence in Probability

The sequence of stochastic processes $(X_n)_{n \in \mathbb{N}}$ with sample paths in $\mathcal{D}_{\mathbb{X}}[0, \infty)$ converges in probability to $X = (X(t))_{t \geq 0}$ if it holds

$$\lim_{n \to \infty} \mathbb{P}\left(d_{\mathcal{D}}(X_n, X) > \varepsilon\right) = 0$$

for all $\varepsilon > 0$. If the limit process is almost surely globally continuous, i.e., $X \in \mathcal{C}_{\mathbb{X}}[0, \infty)$ almost surely (a.s.), we can replace the Skorokhod metric by the uniform metric. Then, convergence in probability requires that

$$\lim_{n \to \infty} \mathbb{P}\left(\sup_{t \leq T} |X_n(t) - X(t)| > \varepsilon\right) = 0$$

for all $\varepsilon > 0, T > 0$.

Convergence Almost Surely

It holds $X_n \to X$ almost surely in the Skorokhod topology if

$$\mathbb{P}\left(d_{\mathcal{D}}(X_n, X) \overset{n \to \infty}{\longrightarrow} 0\right) = 1. \tag{A.30}$$

Again, given that the limit process is a.s. globally continuous ($X \in \mathcal{C}_{\mathbb{X}}[0, \infty)$ a.s.), almost sure convergence in the Skorokhod topology is equivalent to uniform convergence a.s.

$$\mathbb{P}\left(\lim_{n \to \infty} \sup_{t \leq T} |X_n(t) - X(t)| = 0 \quad \forall T > 0\right) = 1.$$

If, instead, the limit process is not globally continuous ($X \notin \mathcal{C}_{\mathbb{X}}[0, \infty)$), but it is almost surely continuous at fixed $t \geq 0$, the Skorokhod convergence (A.30) implies

$$\mathbb{P}\left(X_n(t) \overset{n \to \infty}{\longrightarrow} X(t)\right) = 1$$

for all t, which is referred to as *point-wise convergence almost surely* and means that it holds $\lim_{n \to \infty} d_{\mathbb{X}}(X_n(t), X(t)) = 0$ a.s. for all t [85].

A.3.3 Large Population Limit

Given a unit-rate Poisson process $(\mathcal{U}(t))_{t \geq 0}$, we consider the rescaled process $\frac{1}{V}\mathcal{U}(Vt)$ for $V > 0$. It is well-known that in the large volume limit $V \to \infty$

this rescaled process can be approximated by its first-order moment given by

$$\mathbb{E}\left(\frac{1}{V}\mathcal{U}(Vt)\right) = t.$$

More precisely, it holds

$$\lim_{V\to\infty}\sup_{s\le t}\left|\frac{1}{V}\mathcal{U}(Vs) - s\right| = 0 \quad \text{a.s.}$$

for every $t > 0$, which follows from the strong law of large numbers (see [160] or Theorem 1.2 in [9]). For the centered process $\tilde{\mathcal{U}}(t) := \mathcal{U}(t) - t$, this implies

$$\lim_{V\to\infty}\sup_{s\le t}\left|\frac{1}{V}\tilde{\mathcal{U}}(Vs)\right| = 0, \quad \text{a.s.} \tag{A.31}$$

for every $t > 0$.

This fact can be used to prove the statement about the large volume limit given by Eq. (2.7) in Sect. 2.1. There, we started with the concentration version of the volume-dependent reaction jump process

$$\boldsymbol{C}^V(t) = \boldsymbol{C}^V(0) + \sum_{k=1}^{K}\frac{1}{V}\mathcal{U}_k\left(V\int_0^t \frac{1}{V}\tilde{\alpha}_k^V\left(\boldsymbol{C}^V(s)\right)ds\right)\boldsymbol{\nu}_k,$$

where the $\mathcal{U}_k$ denote independent standard Poisson processes, V is the volume, $\tilde{\alpha}_k^V$ are the volume-dependent propensity functions, and $\boldsymbol{\nu}_k$ are the state-change vectors of the reactions under consideration. We know that, for $V \to \infty$,

$$\frac{1}{V}\tilde{\alpha}_k^V \to \tilde{\alpha}_k,$$

(see Assumption 2.1). Therefore, we make the following simplification which is not necessary for the following theorem and its proof but makes all the subsequent arguments much less involved: We consider the concentration-based reaction jump process $\boldsymbol{C}^V$ given by

$$\boldsymbol{C}^V(t) = \boldsymbol{C}^V(0) + \sum_{k=1}^{K}\frac{1}{V}\mathcal{U}_k\left(V\int_0^t \tilde{\alpha}_k\left(\boldsymbol{C}^V(s)\right)ds\right)\boldsymbol{\nu}_k, \tag{A.32}$$

and want to prove the following statement

Theorem A.14. *Let the propensity functions $\tilde{\alpha}_k$ be globally Lipschitz-continuous, and let them additionally satisfy*

$$a_k := \sup_{\boldsymbol{c}} \tilde{\alpha}_k(\boldsymbol{c}) < \infty, \tag{A.33}$$

*where the supremum runs over the entire state space of the process C^V.
Then, whenever $C^V(0) \to c_0$ for $V \to \infty$, we have, for every fixed $t > 0$,*

$$\lim_{V \to \infty} \sup_{s \le t} \|C^V(s) - C(s)\| = 0 \quad \text{a.s.,}$$

*where $\| \cdot \|$ denotes a vector norm in $\mathbb{R}^L$ and the process $C = (C(t))_{t \ge 0}$ is
given by*

$$C(t) = c_0 + \int_0^t F(C(s))ds,$$

with

$$F(c) := \sum_{k=1}^K \tilde{\alpha}_k(c)\nu_k.$$

Proof: First, we center the Poisson processes at their mean by setting
$\tilde{\mathcal{U}}_k(t) = \mathcal{U}_k(t) - t$. This leads to

$$C^V(t) = C^V(0) + \sum_{k=1}^K \frac{1}{V}\tilde{\mathcal{U}}_k\left(V\int_0^t \tilde{\alpha}_k\left(C^V(s)\right)ds\right)\nu_k + \int_0^t F\left(C^V(s)\right)ds.$$
$$(A.34)$$

Next, we denote the middle term of the right-hand side of this equation by

$$\delta^V(t) := \sum_{k=1}^K \frac{1}{V}\tilde{\mathcal{U}}_k\left(V\int_0^t \tilde{\alpha}_k\left(C^V(s)\right)ds\right)\nu_k.$$

With these preparations we see that

$$\begin{aligned}
\varepsilon^V(t) &:= \sup_{u \le t}\left\|C^V(u) - C^V(0) - \int_0^u F\left(C^V(s)\right)ds\right\| \\
&= \sup_{u \le t}\|\delta^V(u)\| \\
&\le \sum_{k=1}^K \|\nu_k\| \sup_{u \le t}\left|\frac{1}{V}\tilde{\mathcal{U}}_k(Vua_k)\right| \quad \text{a.s.,}
\end{aligned}$$

with a_k defined in (A.33). The above form (A.31) of the law of large numbers
for $\tilde{\mathcal{U}}_k$ implies

$$\lim_{V \to \infty}\varepsilon^V(t) \le \sum_{k=1}^K \|\nu_k\| \lim_{V \to \infty}\sup_{u \le t}\left|\frac{1}{V}\tilde{\mathcal{U}}_k(Vua_k)\right| = 0 \quad \text{a.s.}$$

Therefore

$$\|\boldsymbol{C}^V(t) - \boldsymbol{C}(t)\|$$
$$= \left\|\boldsymbol{C}^V(0) - \boldsymbol{c}_0 + \delta^V(t) + \int_0^t F(\boldsymbol{C}^V(s)) - F(\boldsymbol{C}(s))ds\right\|$$
$$\leq \|\boldsymbol{C}^V(0) - \boldsymbol{c}_0\| + \|\delta^V(t)\| + \int_0^t \|F(\boldsymbol{C}^V(s)) - F(\boldsymbol{C}(s))\|ds$$
$$\leq \|\boldsymbol{C}^V(0) - \boldsymbol{c}_0\| + \varepsilon^V(t) + L\int_0^t \|\boldsymbol{C}^V(s) - \boldsymbol{C}(s)\|ds,$$

where L denotes the Lipschitz constant of F (which is Lipschitz, because all $\tilde{\alpha}_k$ are). The last inequality has the form

$$g(t) \leq f(t) + L\int_0^t g(s)ds,$$

with $L \geq 0$, a non-negative function g, and a non-negative and monotonically increasing function f. Then, the well-known Gronwall lemma states that

$$g(t) \leq f(t)e^{tL},$$

which, in our case, means

$$\|\boldsymbol{C}^V(t) - \boldsymbol{C}(t)\| \leq \left(\|\boldsymbol{C}^V(0) - \boldsymbol{c}_0\| + \varepsilon^V(t)\right)\exp(Lt),$$

from which the assertion follows directly.   □

In [70] (Chapter 11), the authors show that the assumption on the propensity functions made in Theorem A.14 can be weakened considerably. They show that the upper bound condition (A.33) and the Lipschitz continuity just have to be satisfied in a compact set around the trajectory of $\boldsymbol{C}(s)$, $s \in [0, t]$. These weaker conditions are satisfied by the classical mass action propensities.

A.4 Hybrid Stochastic Processes

The large population limit considered in Sect. A.3.3 states that a certain class of dynamical systems given by ODEs (viz., RREs) is obtained as the limit of scaled Markov jump processes. This approach fails if the scaling cannot be carried out equally across all entities. In this case, an unequal scaling of the dynamics can lead to limit processes which are not purely deterministic but combine deterministic motion (or diffusion processes) with random jumps. We here summarize some facts about these hybrid stochastic processes as a background for Sect. 2.2.

A.4.1   Piecewise-Deterministic Markov Processes

Piecewise-deterministic Markov processes (PDMPs) were introduced by Mark Davis in 1984 [44]. We here give a short description, for more general settings (see [44]).

The state of a PDMP at some time $t \geq 0$ is given by a couple $\mathbf{\Theta}(t) = (\mathbf{Y}(t), \mathbf{C}(t))$ containing a discrete variable $\mathbf{Y}(t) \in \mathbb{Y}$ for some discrete set $\mathbb{Y}$ and a continuous variable $\mathbf{C}(t) \in \mathbb{R}^n$. As the name suggests, the process $\mathbf{\Theta} = (\mathbf{\Theta}(t))_{t \geq 0}$ follows a deterministic motion which is interrupted by random jumps. The lengths of the time intervals between two successive jumps are random variables that are called *waiting times* τ_j. The *jump times* $(t_j)_{j \in \mathbb{N}_0}$ are recursively defined by $t_0 := 0$ and $t_{j+1} := t_j + \tau_j$. Within each of the time intervals $[t_j, t_{j+1})$, the discrete variable remains constant, while the continuous variable evolves according to a given ODE. More precisely, the dynamics of the PDMP $(\mathbf{\Theta}(t))_{t \geq 0}$ within the state space

$$\mathbb{S} = \mathbb{Y} \times \mathbb{R}^n,$$

provided with the Borel σ-algebra $\mathcal{B}(\mathbb{S})$, are determined by the following characteristics.

Characteristics of a PDMP:

1. The *deterministic flow*: For each $\mathbf{y} \in \mathbb{Y}$, there is a vector field given by a locally Lipschitz continuous map $\mu_{\mathbf{y}} : \mathbb{R}^n \to \mathbb{R}^n$ which defines the deterministic flow $\Phi_{\mathbf{y}}^t$ by

$$\frac{d}{dt}\Phi_{\mathbf{y}}^t \mathbf{c} = \mu_{\mathbf{y}}(\Phi_{\mathbf{y}}^t \mathbf{c}), \quad \Phi_{\mathbf{y}}^0 \mathbf{c} = \mathbf{c} \quad \forall \mathbf{c} \in \mathbb{R}^n. \qquad (A.35)$$

   The flow defines the evolution of the continuous variable over time $t \geq 0$, given its initial value at time $t = 0$.

2. The *jump rate function* $\lambda : \mathbb{S} \to [0, \infty)$ which determines the distribution of waiting times of the process within a branch $\{\mathbf{y}\} \times \mathbb{R}^n$ of the state space.

3. The *transition kernel* $\mathcal{K} : \mathbb{S} \times \mathcal{B}(\mathbb{S}) \to [0, 1]$ specifying the distribution of the process after a jump.

Typically, it is assumed that for all vector fields $\mu_{\mathbf{y}}$ there is no finite time blow up (where, vaguely explained, "blow up" means that the solution of (A.35) converges to infinity in finite time), such that the corresponding flow Φ_y^t is uniquely determined by (A.35) for all $t \geq 0$.

With these ingredients, a right-continuous sample path $(\Theta(t))_{t \geq 0} = (Y(t), C(t))_{t \geq 0}$ of the PDMP starting at time $t_0 = 0$ in $(y, c) \in \mathbb{S}$ is recursively constructed as follows. Define

$$\Theta(t) = (y, \Phi_y^t c) \quad \text{for } t \in [0, \tau_1)$$

where τ_1 is the waiting time until the first jump, having the distribution

$$\mathbb{P}(\tau_1 > t) = \exp\left(-\int_0^t \lambda(y, \Phi_y^s c)\, ds\right), \quad t \geq 0.$$

Then, the post-jump state at the jump time $t_1 = t_0 + \tau_1$ is selected independently according to the distribution

$$\mathbb{P}((Y(t_1), C(t_1)) \in A) = \mathcal{K}\left((y, \Phi_y^{\tau_1} c), A\right), \quad A \in \mathcal{B}(\mathbb{S}).$$

Up to the first jump time, the process is thus defined by

$$\Theta(t) := (Y(t), C(t)) = \begin{cases} (y, \Phi_y^s c) & \text{for } t \in [0, t_1), \\ (Y(t_1), C(t_1)) & \text{for } t = t_1. \end{cases}$$

Restarting the process at time t_1 in $\Theta(t_1)$ and proceeding recursively according to the same scheme, one obtains a sequence of jump-time realizations $t_1, t_2, \ldots$ In between two consecutive random jump times, the process Θ follows a deterministic motion determined by the vector fields μ_y. That is, within a time period (t_j, t_{j+1}), the discrete component Y is constant, while the continuous component C follows the ODE

$$\frac{d}{dt} C(t) = \mu_y(C(t)). \tag{A.36}$$

At the jump times t_j, both the discrete and the continuous variables can instantaneously change their values by the transitions

$$Y(t_j) \to Y(t_{j+1}), \quad \Phi_{Y(t_j)}^{\tau_j} C(t_j) \to C(t_{j+1}).$$

If it holds

$$C(t_{j+1}) = \Phi_{Y(t_j)}^{\tau_j} C(t_j) \quad \forall j \in \mathbb{N}_0, \tag{A.37}$$

then the trajectories of the continuous variable are globally continuous, and the only effect of a jump is a change in the regime by a new value of the discrete variable. Otherwise the trajectories of the continuous variable are only piecewise continuous.

A.4.2 Convergence of the Partially Scaled Reaction Jump Process to a PDMP

In Sect. 2.2.1 the piecewise-deterministic reaction process (PDRP)

$$\boldsymbol{\Theta}(t) = \boldsymbol{\Theta}(0) + \sum_{k \in \mathbb{K}_0} \mathcal{U}_k \left(\int_0^t \tilde{\alpha}_k(\boldsymbol{\Theta}(s)) \, ds \right) \tilde{\boldsymbol{\nu}}_k + \sum_{k \in \mathbb{K}_1} \int_0^t \tilde{\alpha}_k(\boldsymbol{\Theta}(s)) \tilde{\boldsymbol{\nu}}_k \, ds,$$

(see (2.33)) is considered as a special type of PDMP, arising as the limit of a partially scaled reaction jump process $(\boldsymbol{\Theta}^V(t))_{t \geq 0}$ defined in (2.28). For the definition of $\mathbb{K}_0$ and $\mathbb{K}_1$, see page 60. The relation between this special limit process and the general PDMP can be summarized as follows. It holds $\mathbb{Y} = \mathbb{N}_0^{l'}$ with the discrete process $(\boldsymbol{Y}(t))_{t \geq 0}$ given by the first l' (unscaled, integer-valued) components $(\Theta_1(t), \ldots, \Theta_{l'}(t))_{t \geq 0}$ of the limit process $\boldsymbol{\Theta}$. The continuous process $(\boldsymbol{C}(t))_{t \geq 0}$ is given by $\boldsymbol{C}(t) = (\Theta_{l'+1}(t), \ldots, \Theta_L(t)) \in \mathbb{R}^n$ with $n = L - l'$, such that

$$\mathbb{S} = \mathbb{N}_0^{l'} \times \mathbb{R}^{L-l'}.$$

Both $\boldsymbol{\Theta}^V$ and $\boldsymbol{\Theta}$ have sample paths in $\mathcal{D}_{\mathbb{S}}[0, \infty)$ (see Sect. A.3). The vector field $\mu_{\boldsymbol{y}}$ which works on the subspace of the continuous component is related to the vector field μ defined in (2.35) via

$$\mu(\boldsymbol{\theta}) = (0, \mu_{\boldsymbol{y}}(\boldsymbol{c})) \tag{A.38}$$

for $\boldsymbol{\theta} = (\boldsymbol{y}, \boldsymbol{c})$. The jump rate function λ is determined by the reaction propensities $\tilde{\alpha}$ via

$$\lambda(\boldsymbol{\theta}) = \sum_{k \in \mathbb{K}_0} \tilde{\alpha}_k(\boldsymbol{\theta}) \tag{A.39}$$

and the transition kernel $\mathcal{K}$ is given by

$$\mathcal{K}(\boldsymbol{\theta}, \{\boldsymbol{\theta} + \tilde{\boldsymbol{\nu}}_k\}) = \frac{\tilde{\alpha}_k(\boldsymbol{\theta})}{\sum_{k' \in \mathbb{K}_0} \tilde{\alpha}_{k'}(\boldsymbol{\theta})} \tag{A.40}$$

for each $k \in \mathbb{K}_0$.

There exist several statements about the convergence of the partially scaled process $\boldsymbol{\Theta}^V$ given in (2.28) to a PDMP (see [40, 85, 140]). The proofs are based on Assumption 2.6 and the following standard conditions for a PDMP.

Assumption A.15 (Standard Conditions for PDMP).

(a) Each vector field $\mu_{\boldsymbol{y}}$ is a C^1-function. (More general: Locally Lipschitz continuous (see Remark 2.4 in [40]).)

(b) For each $\boldsymbol{\theta} = (\boldsymbol{y}, \boldsymbol{c}) \in \mathbb{S}$, there exists $\varepsilon(\boldsymbol{\theta}) > 0$ such that $\int_0^{\varepsilon(\boldsymbol{\theta})} \lambda(\boldsymbol{y}, \Phi_{\boldsymbol{y}}^t \boldsymbol{c}) \, dt < \infty$.

(c) It holds $\mathcal{K}(\boldsymbol{\theta}, \{\boldsymbol{\theta}\}) = 0$ for all $\boldsymbol{\theta} \in \mathbb{S}$ which means that only "proper" jumps are considered as jumps.

(d) For every starting point $(\boldsymbol{y}, \boldsymbol{c}) \in \mathbb{S}$ and all times $t \geq 0$, it holds $\mathbb{E}(N(t)) < \infty$ where $N(t)$ is the number of jumps that occur within the time interval $[0, t]$.

It has been shown in [40] that in the partial thermodynamic limit (i.e., when the number of low-abundant species tends to infinity) the partially scaled Markov jump processes converge in distribution to a PDMP. This is stated in the following theorem.

Theorem A.16 (Convergence to PDMP). *Consider the partially scaled processes $(\boldsymbol{\Theta}^V(t))_{t \geq 0}$ defined in (2.28) with adapted reaction propensities $\tilde{\alpha}_k^V$ and state-change vectors $\tilde{\boldsymbol{\nu}}_k^V$. Let $(\boldsymbol{\Theta}(t))_{t \geq 0}$ be a PDMP starting at $\boldsymbol{\theta}_0 = (\boldsymbol{y}_0, \boldsymbol{c}_0)$ whose deterministic flow is given by the vector fields $\mu_{\boldsymbol{y}}$ defined by (A.38), while the jump rate function is given by (A.39), and the transition kernel is defined by (A.40). Assume that $(\boldsymbol{\Theta}(t))_{t \geq 0}$ fulfills the standard conditions of Assumption A.15 and $\boldsymbol{\Theta}^V(0)$ converges in distribution to $\boldsymbol{\theta}_0 = (\boldsymbol{y}_0, \boldsymbol{c}_0)$. Then, for $V \to \infty$, the partially scaled process $\boldsymbol{\Theta}^V = (\boldsymbol{\Theta}^V(t))_{t \geq 0}$ converges in distribution to $\boldsymbol{\Theta} = (\boldsymbol{\Theta}(t))_{t \geq 0}$, that is, it holds $\boldsymbol{\Theta}^V \Rightarrow \boldsymbol{\Theta}$ as defined in Sect. A.3.2.*

For the proof and technical details, we refer to [40].

At the same time, it has been proven in [85] that it holds $\boldsymbol{\Theta}^V \to \boldsymbol{\Theta}$ a.s. in the Skorokhod topology. Moreover, it can be shown that, fixing a time t, the limit process is a.s. continuous at t, which implies that it also holds $\boldsymbol{\Theta}^V(t) \to \boldsymbol{\Theta}(t)$ a.s. for each $t \geq 0$.

A.4.3 Hybrid Diffusion Processes

Between two consecutive jump times t_j and t_{j+1}, the continuous variable $(C(t))_{t \geq 0}$ of a PDMP is given by the deterministic flow $C(t) = \Phi_{\boldsymbol{y}}^t \boldsymbol{c}$ with $\boldsymbol{Y}(t_j) = \boldsymbol{y}$ and $C(t_j) = \boldsymbol{c}$ and satisfies the ODE $dC(t) = \mu_{\boldsymbol{y}}(C(t)) \, dt$, (see Eq. (A.36)). In [41], it is suggested to consider an Itô stochastic differential equation instead of this ODE, thus adding a diffusion term to produce noise in the flow of the continuous variable, which leads to processes called *hybrid diffusion*. Similar to a PDMP, such a hybrid diffusion is given by a process $(\boldsymbol{\Theta}(t))_{t \geq 0}$ of the form $\boldsymbol{\Theta}(t) = (\boldsymbol{Y}(t), C(t))$ which consists of a discrete component $\boldsymbol{Y}(t) \in \mathbb{Y}$ for some discrete set $\mathbb{Y}$ and a continuous component $\boldsymbol{C}(t) \in \mathbb{R}^n$. Both components have piecewise-continuous trajectories interrupted by jumps at jump times t_j, $j \in \mathbb{N}_0$. Again, the jump dynamics are specified by a rate function λ and a transition kernel $\mathcal{K}$. Instead of the deterministic flow (A.35) between two jump times, however, we consider the randomized flow $\Phi_{\boldsymbol{y}}^t \boldsymbol{c}$ defined by

$$d\Phi_{\boldsymbol{y}}^t \boldsymbol{c} = \mu_{\boldsymbol{y}}(\Phi_{\boldsymbol{y}}^t \boldsymbol{c})\, dt + \sigma_{\boldsymbol{y}}(\Phi_{\boldsymbol{y}}^t \boldsymbol{c})\, d\boldsymbol{W}(t), \quad \Phi_{\boldsymbol{y}}^0 \boldsymbol{c} = \boldsymbol{c}, \tag{A.41}$$

where $(\boldsymbol{W}(t))_{t \geq 0}$ is an m-dimensional Wiener process for some $m \in \mathbb{N}$, $\mu_{\boldsymbol{y}} : \mathbb{R}^n \to \mathbb{R}^n$ is a continuous vector field, and $\sigma_{\boldsymbol{y}} : \mathbb{R}^n \to \mathbb{R}^{n,m}$ are the noise terms, i.e., $\sigma_{\boldsymbol{y}}(\Phi_{\boldsymbol{y}}^t \boldsymbol{c})$ is a diffusion matrix for every $\boldsymbol{y} \in \mathbb{Y}$.

A sample path of the right-continuous hybrid diffusion process $(\boldsymbol{\Theta}(t))_{t \geq 0}$ follows the same recursive construction scheme as the PDMP with the deterministic flow (A.35) replaced by the randomized flow (A.41). More precisely, given that $\boldsymbol{\Theta}(t_j) = (\boldsymbol{y}, \boldsymbol{c})$ at a jump time t_j, it holds

$$\boldsymbol{\Theta}(t_j + s) = (\boldsymbol{Y}(t_j + s), \boldsymbol{C}(t_j + s)) = (\boldsymbol{y}, \Phi_{\boldsymbol{y}}^s \boldsymbol{c}), \quad s \in [0, \tau_j),$$

where τ_j is the waiting time for the next jump. Instead of the ODE (A.36), the dynamics of the continuous component in between the jump times are now given by the SDE

$$d\boldsymbol{C}(t) = \mu_{\boldsymbol{y}}(\boldsymbol{C}(t))\, dt + \sigma_{\boldsymbol{y}}(\boldsymbol{C}(t))\, d\boldsymbol{W}(t). \tag{A.42}$$

In Sect. 2.2.2, such a hybrid diffusion process appears in the form of the piecewise chemical Langevin dynamics (2.42), where drift and diffusion terms are determined by the reaction propensities and state-change vectors of the continuous reactions $\mathcal{R}_k$, $k \in \mathbb{K}_1$.

A.5 Multiscale Asymptotics of the CME

In the following, we will show how the CME can be reduced to a simpler system of equations by using methods of multiscale asymptotic expansion. In Sect. A.5.1, we consider the case of a population scaling which is uniform over all species, leading to an approximation of the CME by the Liouville equation. The approach is extended to multiple population scales in Sect. A.5.2, resulting in the hybrid CME as a reduced system of equations. Finally, multiple time scales will be considered in Sect. A.5.3.

A.5.1 Large Population Limit

In Sect. 2.1.1, we considered how to derive the reaction rate equation from the reaction jump process representation of the CME in the large copy number limit (see also Sect. A.3.3). However, we did not give details on what happens to the CME itself in this limit. In order to study the behavior of the solution of the CME in the large copy number limit, we have to deal with some relatively simple form of multiscale asymptotics. For this purpose, let us start with the form of the CME in terms of the volume-scaled quantities considered in Sect. 2.1.1. To this end, we first rewrite the CME (1.10) in terms of the concentration-based probability distribution

$$\rho^V(\boldsymbol{c}, t) := p(\boldsymbol{c}V, t),$$

together with the volume-dependent propensities

$$\tilde{\alpha}_k^V(\boldsymbol{c}) = \alpha_k^V(\boldsymbol{c}V),$$

of which we assume convergence for $V \to \infty$

$$V^{-1}\tilde{\alpha}_k^V(\boldsymbol{c}) \to \tilde{\alpha}_k(\boldsymbol{c}),$$

as in Sect. 2.1.1 in Assumption 2.1. Then, the CME reads

$$\frac{\partial \rho^V(\boldsymbol{c}, t)}{\partial t} = \sum_{k=1}^{K} [\tilde{\alpha}_k^V(\boldsymbol{c} - V^{-1}\boldsymbol{\nu}_k)\rho^V(\boldsymbol{c} - V^{-1}\boldsymbol{\nu}_k, t) - \tilde{\alpha}_k^V(\boldsymbol{c})\rho^V(\boldsymbol{c}, t)].$$

In order to adopt the notation typically used in multiscale asymptotics, we now introduce the smallness parameter $\varepsilon = V^{-1}$ and consider the limit $\varepsilon \to 0$. By redefining the notation in analogy, we thus have the CME

$$\frac{\partial \rho^\varepsilon(\boldsymbol{c}, t)}{\partial t} = \sum_{k=1}^{K} [\tilde{\alpha}_k^\varepsilon(\boldsymbol{c} - \varepsilon\boldsymbol{\nu}_k)\rho^\varepsilon(\boldsymbol{c} - \varepsilon\boldsymbol{\nu}_k, t) - \tilde{\alpha}_k^\varepsilon(\boldsymbol{c})\rho^\varepsilon(\boldsymbol{c}, t)]. \qquad (A.43)$$

In rewritten form, the convergence of the volume-dependent propensities reads

$$\varepsilon\tilde{\alpha}_k^\varepsilon(\boldsymbol{c}) \to \tilde{\alpha}_k(\boldsymbol{c}), \qquad \varepsilon \to 0,$$

which in terms of the Landau calculus means

$$\varepsilon\tilde{\alpha}_k^\varepsilon(\boldsymbol{c}) = \tilde{\alpha}_k(\boldsymbol{c}) + b_k^\varepsilon(\boldsymbol{c}),$$

with $b_k^\varepsilon(\boldsymbol{c}) = o(1)$. For the sake of simplicity, we will subsequently use the stronger assumption

$$b_k^\varepsilon(\boldsymbol{c}) = \varepsilon\tilde{b}_k^\varepsilon(\boldsymbol{c}), \qquad \tilde{b}_k^\varepsilon(\boldsymbol{c}) = \mathcal{O}(1),$$

not without emphasizing that the following considerations remain valid with the weaker form, too. Thus, we get

$$\frac{\partial \rho^\varepsilon(c, t)}{\partial t} = \frac{1}{\varepsilon} \sum_{k=1}^{K} \left[\tilde{\alpha}_k(c - \varepsilon \nu_k) \rho^\varepsilon(c - \varepsilon \nu_k, t) - \tilde{\alpha}_k(c) \rho^\varepsilon(c, t) \right]$$

$$+ \sum_{k=1}^{K} \left[\tilde{b}_k^\varepsilon(c - \varepsilon \nu_k) \rho^\varepsilon(c - \varepsilon \nu_k, t) - \tilde{b}_k^\varepsilon(c) \rho^\varepsilon(c, t) \right]. \tag{A.44}$$

Non-rigorously, we can now take the limit $\varepsilon \to 0$ and get that $\rho = \lim_{\varepsilon \to 0} \rho^\varepsilon$ satisfies[4]

$$\frac{\partial \rho(c, t)}{\partial t} = -\nabla_c \cdot \left[\sum_{k=1}^{K} \nu_k \tilde{\alpha}_k(c) \rho(c, t) \right] = -\mathrm{div}_c \left[\sum_{k=1}^{K} \nu_k \tilde{\alpha}_k(c) \rho(c, t) \right]. \tag{A.45}$$

Equation (A.45) is the so-called *generalized Liouville equation* (see page 221 for details). It appears in the main text in the form of (2.20) and is associated with the system of ordinary equations

$$\frac{d}{dt} C(t) = \sum_{k=1}^{K} \nu_k \tilde{\alpha}_k(C(t)), \tag{A.46}$$

that is, of the reaction rate equations, in the following sense: Let $C(t) = \Phi^t(C_0)$ denote the solution of (A.46) for initial conditions $C(0) = C_0$, i.e., let Φ^t denote the flow operator defined by (A.46). Then, the solution of the generalized Liouville equation (A.45) can be written as

$$\rho(c, t) = \frac{1}{\varsigma(c, t)} \rho(\Phi^{-t} c, 0), \tag{A.47}$$

with a prefactor resulting from the distortion of volume by the flow generated by the reaction rate equations

$$\varsigma(c, t) := \exp \left(\int_0^t \sum_{k=1}^{K} \nu_k^\mathsf{T} \nabla \tilde{\alpha}_k(\Phi^{-s} c) ds \right), \tag{A.48}$$

(see pages 221–224 for a derivation of (A.47) including the distortion factor). That is, the probability distribution at time t results from transporting the probability distribution at time $t = 0$ along the trajectories of the system of reaction rate equations plus a volume distortion factor.

---

[4]In (A.45), every summand refers to the directional derivative along the vector ν_k. Note that it holds in general $\nabla_\nu f(c) = \nu \cdot \nabla f(c) = \nabla_c \cdot (f(c)\nu)$ if the function f is differentiable at c.

Evolution of Expectation Values

This result has the following consequence: When we are interested in the expectation values of an observable function f of the concentrations in the system

$$\mathbb{E}^\varepsilon(f,t) := \int_{\mathbb{R}_+^L} f(c)\rho^\varepsilon(c,t)\,dc,$$

and ε is small enough, we can use the fact that $\rho^\varepsilon \to \rho$ for $\varepsilon \to 0$ to get an approximation of the expectation values via

$$\mathbb{E}^\varepsilon(f,t) \approx \int f(c)\rho(c,t)\,dc =: \mathbb{E}^0(f,t).$$

Now the above representation of $\rho(\cdot,t)$ via the initial distribution $\rho(\cdot,t=0)$ reveals, by means of substitution $\Phi^t u = c$,

$$
\begin{aligned}
\int f(c)\rho(c,t)dc &= \int f(c)\frac{1}{\varsigma(c,t)}\rho(\Phi^{-t}c,0)dc \\
&= \int f(\Phi^t u)\frac{1}{\varsigma(\Phi^t u,t)}\rho(u,0)|\det(D\Phi^t u)|\,du \\
&= \int f(\Phi^t u)\rho(u,0)\,du,
\end{aligned}
$$

where we used that the volume distortion factor ς is identical to the flow map variation $|\det(D\Phi^t u)|$. This yields

$$\mathbb{E}^\varepsilon(f,t) \approx \mathbb{E}^0(f,t) = \mathbb{E}^0(f \circ \Phi^t, t=0).$$

This means: If ε is only small enough, we can approximately compute the evolution of expectation values under the CME by starting trajectories of the reaction rate equation system from the initial distribution $\rho(\cdot,t=0)$.

Concentration of the Initial Distribution

The previous kind of formal derivation does work only if the initial distribution $\rho(\cdot,t=0)$ is continuously differentiable in c, i.e., if the sequence of probability distributions $\rho^\varepsilon(\cdot,0)$ converges to a smooth probability distribution $\rho(\cdot,0)$, because we have to be able to take the derivative with respect to c in (A.45). The situation we discussed in Sect. 2.1.1 is different. There we were interested in a probability distribution that concentrates around a single trajectory for $V \to \infty$ or for $\varepsilon \to 0$. In this case, we have to proceed with (A.43) differently and assume that the distributions ρ^ε concentrate for $\varepsilon \to 0$. That is, we have to assume that they are of the form

$$\rho^\varepsilon(\boldsymbol{c}, t) = \frac{1}{Z_\varepsilon} \exp\left(\frac{1}{\varepsilon} S^\varepsilon(\boldsymbol{c}, t)\right),$$

where Z_ε is a normalization constant and $S^\varepsilon = \mathcal{O}(1)$ is a function that goes to $-\infty$ for $\|\boldsymbol{c}\| \to \infty$ fast enough. Expanding S^ε asymptotically in ε results in the following asymptotic expansion/asymptotic ansatz for ρ^ε

$$\rho^\varepsilon(\boldsymbol{c}, t) = \frac{1}{\sqrt{\varepsilon}} \exp\left(\frac{1}{\varepsilon} S_0(\boldsymbol{c}, t)\right)\left(U_0(\boldsymbol{c}, t) + \varepsilon U_1(\boldsymbol{c}, t) + \dots\right)$$

for suitable function S_0 and $U_0, U_1, \dots$, where the factor $\sqrt{\varepsilon}$ results from normalization. This has the following consequences

$$\frac{\partial}{\partial t}\rho^\varepsilon = \frac{1}{\varepsilon}\left(\frac{\partial}{\partial t}S_0\right)\rho^\varepsilon + \frac{1}{\sqrt{\varepsilon}}\mathcal{O}(1),$$

$$\rho^\varepsilon(\boldsymbol{c} - \varepsilon\boldsymbol{\nu}_k, t) = \exp(-\nabla S_0(\boldsymbol{c}, t) \cdot \boldsymbol{\nu}_k^\mathsf{T})\rho^\varepsilon(\boldsymbol{c}, t) + \mathcal{O}(\sqrt{\varepsilon}),$$

$$\tilde{\alpha}_k(\boldsymbol{c} - \varepsilon\boldsymbol{\nu}_k) = \tilde{\alpha}_k(\boldsymbol{c}) + \mathcal{O}(\varepsilon),$$

where the second and third lines follow from a Taylor expansion of $\rho^\varepsilon(\boldsymbol{c} - \varepsilon\boldsymbol{\nu}_k, t)$ and $\tilde{\alpha}_k(\boldsymbol{c} - \varepsilon\boldsymbol{\nu}_k)$ in ε around $\varepsilon = 0$. This implies that

$$\tilde{\alpha}_k(\boldsymbol{c} - \varepsilon\boldsymbol{\nu}_k)\rho^\varepsilon(\boldsymbol{c} - \varepsilon\boldsymbol{\nu}_k, t) - \tilde{\alpha}_k(\boldsymbol{c})\rho^\varepsilon(\boldsymbol{c}, t)$$
$$= \left(\exp(-\nabla S_0(\boldsymbol{c}, t) \cdot \boldsymbol{\nu}_k^\mathsf{T}) - 1\right)\tilde{\alpha}_k(\boldsymbol{c})\rho^\varepsilon(\boldsymbol{c}, t) + \mathcal{O}(\sqrt{\varepsilon}),$$

Putting this into (A.44) and multiplying the entire equation with $\sqrt{\varepsilon}$ now yield

$$\frac{1}{\varepsilon}\left(\frac{\partial}{\partial t}S_0(\boldsymbol{c}, t)\right)\sqrt{\varepsilon}\rho^\varepsilon = \frac{1}{\varepsilon}\sum_{k=1}^K\left(\exp(-\nabla S_0(\boldsymbol{c}, t) \cdot \boldsymbol{\nu}_k^\mathsf{T}) - 1\right)\tilde{\alpha}_k(\boldsymbol{c})\sqrt{\varepsilon}\rho^\varepsilon + \mathcal{O}(1),$$

where $\sqrt{\varepsilon}\rho^\varepsilon = \mathcal{O}(1)$, and thus the only terms on the scale ε^{-1} are

$$\frac{\partial}{\partial t}S_0(\boldsymbol{c}, t) = \sum_{k=1}^K\left(\exp(-\nabla S_0(\boldsymbol{c}, t) \cdot \boldsymbol{\nu}_k^\mathsf{T}) - 1\right)\tilde{\alpha}_k(\boldsymbol{c}).$$

This is a Hamilton-Jacobi equation for the eikonal function S_0, which can be rewritten in the typical form

$$\frac{\partial}{\partial t}S_0(\boldsymbol{c}, t) + H(\boldsymbol{c}, \boldsymbol{\xi}) = 0$$

with $\boldsymbol{\xi} := \nabla S_0$ and the Hamiltonian function

$$H(\boldsymbol{c}, \boldsymbol{\xi}) := \sum_{k=1}^{K} \left(1 - \exp(-\boldsymbol{\xi} \cdot \boldsymbol{\nu}_k^\mathsf{T})\right) \tilde{\alpha}_k(\boldsymbol{c}).$$

The evolution equations for the corresponding Hamiltonian system are

$$\frac{d}{dt}\boldsymbol{c} = \nabla_p H = \sum_{k=1}^{K} \boldsymbol{\nu}_k \exp(-\boldsymbol{\xi} \cdot \boldsymbol{\nu}_k^\mathsf{T}) \tilde{\alpha}_k(\boldsymbol{c}),$$

$$\frac{d}{dt}\boldsymbol{\xi} = -\nabla_c H = -\sum_{k=1}^{K} \left(1 - \exp(-\boldsymbol{\xi} \cdot \boldsymbol{\nu}_k^\mathsf{T})\right) \nabla \tilde{\alpha}_k(\boldsymbol{c}).$$

Since we are interested in the solutions of (A.44) that concentrate around one point, we can assume that S_0 has a unique maximum (so that the ρ^ε have a unique maximum for small enough ε). If we start with $\boldsymbol{\xi} = \boldsymbol{0}$ in the second characteristic equation, we stay at $\boldsymbol{\xi} = \boldsymbol{0}$ for all times. That is, the unique maximum belongs to the solution of the characteristic equations for $\boldsymbol{\xi}(t) = \boldsymbol{0}$, meaning that it travels in time along the trajectories of the reaction rate equations

$$\frac{d}{dt}\boldsymbol{c} = \sum_{k=1}^{K} \boldsymbol{\nu}_k \tilde{\alpha}_k(\boldsymbol{c}), \qquad \boldsymbol{c}(0) = \boldsymbol{c}_0,$$

where $\boldsymbol{c}_0$ is the concentration point given by the unique maximum of S_0 at $t = 0$, i.e., by the unique maximum of the distributions ρ^ε at $t = 0$.

By further analysis using the Laplace integral method (see [181]), one can show that the expectation values of an observable function f of the concentrations in the system can be computed via

$$\mathbb{E}^\varepsilon(f, t) = \int f(\boldsymbol{c}) \rho^\varepsilon(\boldsymbol{c}, t) d\boldsymbol{c} = f(\boldsymbol{C}(t)) + \mathcal{O}(\varepsilon), \qquad (A.49)$$

where $\boldsymbol{C}(t) = \Phi^t \boldsymbol{c}_0$ is the (unique) solution of the above reaction rate equations if the initial distributions $\rho^\varepsilon(\cdot, t = 0)$ concentrate around $\boldsymbol{c}_0$.

Solution Formula for the Generalized Liouville Equation

The aim of this subsection solely is to give a derivation of the solution formula (A.47) for the Liouville equation (A.45). To this end we start with the dynamical system

$$\frac{d}{dt}\boldsymbol{c} = f(\boldsymbol{c})$$

for a function $f : \mathbb{R}^L \to \mathbb{R}^L$, for which we are mainly interested in the particular case $f(\boldsymbol{c}) := \sum_{k=1}^{K} \boldsymbol{\nu}_k^{(z)} \alpha_k(\boldsymbol{c})$.

We start with the classical Liouville equation for a Hamiltonian system with an arbitrary Hamiltonian $H = H(\boldsymbol{c}, \boldsymbol{\xi})$ (a smooth scalar function), where the Liouville equation reads

$$\frac{\partial}{\partial t}\rho + \left(\nabla_{\boldsymbol{c}}\rho \cdot \nabla_{\boldsymbol{\xi}}H - \nabla_{\boldsymbol{\xi}}\rho \cdot \nabla_{\boldsymbol{c}}H\right) = 0, \qquad \rho(\boldsymbol{c}, \boldsymbol{\xi}, t = 0) = \rho_0(\boldsymbol{c}, \boldsymbol{\xi}),$$

and its solution can be expressed by

$$\rho(\boldsymbol{c}, \boldsymbol{\xi}, t) = \rho_0(\Psi^{-t}(\boldsymbol{c}, \boldsymbol{\xi})),$$

where Ψ^t is the solution operator of the Hamiltonian equations of motion associated with H

$$\frac{d}{dt}\boldsymbol{c} = \nabla_{\boldsymbol{\xi}}H,$$

$$\frac{d}{dt}\boldsymbol{\xi} = -\nabla_{\boldsymbol{c}}H.$$

From now on we consider the Hamiltonian

$$H(\boldsymbol{c}, \boldsymbol{\xi}) := f(\boldsymbol{c})^{\mathsf{T}}\boldsymbol{\xi},$$

such that the Hamiltonian equations of motion get the form

$$\frac{d}{dt}\boldsymbol{c} = f(\boldsymbol{c}), \tag{A.50}$$

$$\frac{d}{dt}\boldsymbol{\xi} = -(\nabla_{\boldsymbol{c}} \cdot f)\boldsymbol{\xi}. \tag{A.51}$$

Let us denote the solution operator of the first equation (A.50) by Φ^t, i.e., let $\boldsymbol{c}(t) = \Phi^t\boldsymbol{c}(0)$ denote the solution to (A.50). Then, the second equation (A.51) is of the form

$$\frac{d}{dt}\boldsymbol{\xi}(t) = A(t)\boldsymbol{\xi}(t), \qquad \boldsymbol{\xi}(0) = \boldsymbol{\xi}_0,$$

with $A(t) := -\nabla_{\boldsymbol{c}} \cdot f(\Phi^t\boldsymbol{c}_0)$. Now, let $\mathcal{S}(t, s)$ denote the evolution semigroup of this last equation for $\boldsymbol{\xi}$ such that $\boldsymbol{\xi}(t) = \mathcal{S}(t, 0)\boldsymbol{\xi}(0)$. Then the evolution of the density ρ for this system takes the form

$$\rho(\boldsymbol{c}, \boldsymbol{\xi}, t) = \rho_0(\Phi^{-t}\boldsymbol{c}, \mathcal{S}(0, t)\boldsymbol{\xi}),$$

with $\mathcal{S}(0,t) = \mathcal{S}(t,0)^{-1}$. The associated marginal density $\hat{\rho}(\boldsymbol{c},t) := \int \rho(\boldsymbol{c},\boldsymbol{\xi},t)d\boldsymbol{\xi}$ thus satisfies

$$\hat{\rho}(\boldsymbol{c},t) = \int \rho_0(\Phi^{-t}\boldsymbol{c}, \mathcal{S}(0,t)\boldsymbol{\xi})d\boldsymbol{\xi}$$

$$= \int \rho_0(\Phi^{-t}\boldsymbol{c}, \boldsymbol{v}) \, |\det(\mathcal{S}(t,0))| \, d\boldsymbol{v}$$

$$= |\det(\mathcal{S}(t,0))| \cdot \hat{\rho}(\Phi^{-t}\boldsymbol{c}, 0),$$

where we used the substitution $\boldsymbol{\xi} = \mathcal{S}(t,0)\boldsymbol{v}$ in the middle step. The fact that $\mathcal{S}(t,0)$ is invertible allows us to show that[5]

$$|\det(\mathcal{S}(t,0))| = \exp\left(-\int_0^t \mathrm{tr}(\nabla_{\boldsymbol{c}} f(\Phi^s \boldsymbol{c}_0))ds\right) = \exp\left(-\int_0^t \mathrm{div}_{\boldsymbol{c}} f(\Phi^s \boldsymbol{c}_0)ds\right),$$

but with $\boldsymbol{c}_0 = \Phi^{-t}\boldsymbol{c}$ such that we get

$$\hat{\rho}(\boldsymbol{c},t) = \exp\left(-\int_0^t \mathrm{div}_{\boldsymbol{c}} f(\Phi^{s-t}\boldsymbol{c})ds\right)\hat{\rho}(\Phi^{-t}\boldsymbol{c}, 0),$$

or, finally,

$$\hat{\rho}(\boldsymbol{c},t) = \exp\left(-\int_0^t \mathrm{div}_{\boldsymbol{c}} f(\Phi^{-s}\boldsymbol{c})ds\right)\hat{\rho}(\Phi^{-t}\boldsymbol{c}, 0).$$

Comparing with Eq. (A.47) on page 218 and setting $f(\boldsymbol{c}) = \sum_{k=1}^K \boldsymbol{\nu}_k \tilde{\alpha}_k(\boldsymbol{c})$, we see that the distortion factor is given by

$$\varsigma(\boldsymbol{c},t) = \exp\left(\int_0^t \mathrm{div}_{\boldsymbol{c}} f(\Phi^{-s}\boldsymbol{c})ds\right)$$

$$= \exp\left(\int_0^t \nabla \cdot \sum_{k=1}^K \boldsymbol{\nu}_k \tilde{\alpha}_k(\Phi^{-s}\boldsymbol{c})ds\right)$$

$$= \exp\left(\int_0^t \sum_{k=1}^K \boldsymbol{\nu}_k^\mathsf{T} \nabla \tilde{\alpha}_k(\Phi^{-s}\boldsymbol{c})ds\right)$$

as defined in (A.48).

The generalized Liouville equation for $\hat{\rho}$ can be computed from the Liouville equation for ρ

$$\frac{\partial}{\partial t}\rho + \left(\nabla_{\boldsymbol{c}}\rho \cdot f - \nabla_{\boldsymbol{\xi}}\rho \cdot (\nabla_{\boldsymbol{c}} \cdot f)\boldsymbol{\xi}\right) = 0, \qquad \rho(\boldsymbol{c},\boldsymbol{\xi},t=0) = \rho_0(\boldsymbol{c},\boldsymbol{\xi}),$$

---

[5]We use the Leibniz formula $\frac{d}{dt}\det(B(t)) = \det(B(t)) \cdot \mathrm{tr}[B(t)^{-1}\frac{d}{dt}B(t)]$ for invertible square matrices $B = B(t)$, together with $\frac{d}{dt}\mathcal{S}(t,0) = A(t)\mathcal{S}(t,0) = \mathcal{S}(t,0)A(t)$.

by integration in $\boldsymbol{\xi}$

$$
\begin{aligned}
0 &= \frac{\partial}{\partial t}\hat{\rho} + \left(\nabla_c\hat{\rho}\cdot f - (\nabla_c\cdot f)\int \boldsymbol{\xi}\cdot\nabla_{\boldsymbol{\xi}}\rho\,d\boldsymbol{\xi}\right) \\
&= \frac{\partial}{\partial t}\hat{\rho} + \left(\nabla_c\hat{\rho}\cdot f + (\nabla_c\cdot f)\hat{\rho}\right) \\
&= \frac{\partial}{\partial t}\hat{\rho} + \operatorname{div}_c\left[f\hat{\rho}\right],
\end{aligned}
$$
$$
\hat{\rho}(\boldsymbol{c},t=0) = \int \rho_0(\boldsymbol{c},\boldsymbol{\xi})d\boldsymbol{\xi},
$$

where the second line results from partial integration. Setting again $f(\boldsymbol{c}) = \sum_{k=1}^K \boldsymbol{\nu}_k\tilde{\alpha}_k(\boldsymbol{c})$ gives (A.45).

A.5.2   Multiple Population Scales

We will now analyze the case considered in Sect. 2.2.4 of a CME with high-abundance species, described by concentrations $\boldsymbol{c}$, and low-abundance ones, described by (discrete) particle numbers $\boldsymbol{y}$. We again assume that the propensities fall into two categories, volume-dependent propensities $\tilde{\alpha}_k^\varepsilon$, $k \in \mathbb{K}_c$, scaling like above, i.e.,

$$
\varepsilon\tilde{\alpha}_k^\varepsilon(\boldsymbol{y},\boldsymbol{c}) = \tilde{\alpha}_k(\boldsymbol{y},\boldsymbol{c}),
$$

for which the state-change vectors do *not* act on the particle numbers $\boldsymbol{y}$, and the rest, $\tilde{\alpha}_k^\varepsilon$, $k \in \mathbb{K}_d$, with

$$
\tilde{\alpha}_k^\varepsilon(\boldsymbol{y},\boldsymbol{c}) = \tilde{\alpha}_k(\boldsymbol{y},\boldsymbol{c}),
$$

(see also page 62 for a definition of the sets $\mathbb{K}_c$ and $\mathbb{K}_d$).

We consider the joint CME for $p^\varepsilon(\boldsymbol{y},\boldsymbol{c},t) := p(\boldsymbol{y},\boldsymbol{c}/\varepsilon,t)$

$$
\begin{aligned}
\frac{\partial p^\varepsilon(\boldsymbol{y},\boldsymbol{c},t)}{\partial t} &= \frac{1}{\varepsilon}\sum_{k\in\mathbb{K}_c}\left[\tilde{\alpha}_k(\boldsymbol{y},\boldsymbol{c}-\varepsilon\boldsymbol{\nu}_k^{(z)})p^\varepsilon(\boldsymbol{y},\boldsymbol{c}-\varepsilon\boldsymbol{\nu}_k^{(z)},t) - \tilde{\alpha}_k(\boldsymbol{y},\boldsymbol{c})p^\varepsilon(\boldsymbol{c},t)\right] \\
&\quad + \sum_{k\in\mathbb{K}_d}\Big[\tilde{\alpha}_k(\boldsymbol{y}-\boldsymbol{\nu}_k^{(y)},\boldsymbol{c}-\varepsilon\boldsymbol{\nu}_k^{(z)})p^\varepsilon(\boldsymbol{y}-\boldsymbol{\nu}_k^{(y)},\boldsymbol{c}-\varepsilon\boldsymbol{\nu}_k^{(z)},t) \\
&\qquad\qquad - \tilde{\alpha}_k(\boldsymbol{y},\boldsymbol{c})p^\varepsilon(\boldsymbol{y},\boldsymbol{c},t)\Big]
\end{aligned}
$$

$$
\text{(A.52)}
$$

for state-change vectors $\boldsymbol{\nu}_k^{(y)} \in \mathbb{N}_0^{l'}$ and $\boldsymbol{\nu}_k^{(z)} \in \mathbb{N}_0^{L-l'}$, where l' is the number of low-abundant species given w.l.o.g. by $\mathcal{S}_1,\ldots,\mathcal{S}_{l'}$. Furthermore, let the probability distribution p^ε have the Bayesian decomposition

$$
p^\varepsilon(\boldsymbol{y},\boldsymbol{c},t) = \rho^\varepsilon(\boldsymbol{c},t|\boldsymbol{y})p^\varepsilon(\boldsymbol{y},t),
$$

where $p^\varepsilon(\boldsymbol{y}, t) := \int_{\mathbb{R}_+^{L-l'}} p^\varepsilon(\boldsymbol{y}, \boldsymbol{c}, t) \, d\boldsymbol{c}$.

Partial Population Scaling via Multiscale Expansion

Now we assume again that the initial distribution $p^\varepsilon(\boldsymbol{y}, \boldsymbol{c}, t = 0)$ concentrates on a single initial point $\boldsymbol{c}_0$ for all $\boldsymbol{y}$. Then, in analogy to the above multiscale analysis, we utilize the ansatz

$$\rho^\varepsilon(\boldsymbol{c}, t | \boldsymbol{y}) = \frac{1}{\sqrt{\varepsilon}} \exp\left(\frac{1}{\varepsilon} S_0(\boldsymbol{c}, t | \boldsymbol{y})\right) \left(U_0(\boldsymbol{c}, t | \boldsymbol{y}) + \varepsilon U_1(\boldsymbol{c}, t | \boldsymbol{y}) + \ldots\right),$$

$$p^\varepsilon(\boldsymbol{y}, t) = p^0(\boldsymbol{y}, t) + \varepsilon p^1(\boldsymbol{y}, t) + \ldots$$

for suitable functions S_0 and $U_0, U_1, \ldots$. As above, the leading order in ε yields a Hamilton-Jacobi equation for $S_0(\cdot | \boldsymbol{y})$ and to the insight that for all smooth F (as observables of $\boldsymbol{c}$ and $\boldsymbol{y}$), the partial expectation

$$\mathbb{E}_{\boldsymbol{y}}^\varepsilon(F, t) := \int F(\boldsymbol{y}, \boldsymbol{c}) p^\varepsilon(\boldsymbol{y}, \boldsymbol{c}, t) d\boldsymbol{c}$$

fulfills

$$\mathbb{E}_{\boldsymbol{y}}^\varepsilon(F, t) = \mathbb{E}_{\boldsymbol{y}}^0(F, t) + \mathcal{O}(\varepsilon)$$

where $\mathbb{E}_{\boldsymbol{y}}^0(F, t)$ is defined via

$$\mathbb{E}_{\boldsymbol{y}}^0(F, t) := F(\boldsymbol{y}, \Phi_{\boldsymbol{y}}^t \boldsymbol{c}_0) p^0(\boldsymbol{y}, t) \qquad (A.53)$$

and $\Phi_{\boldsymbol{y}}^t$ is the flow operator of the $\boldsymbol{y}$-dependent RRE system

$$\frac{d}{dt} \boldsymbol{C}(t) = \sum_{k \in \mathbb{K}_c} \boldsymbol{\nu}_k^{(z)} \tilde{\alpha}_k(\boldsymbol{y}, \boldsymbol{C}(t)). \qquad (A.54)$$

Partial Population Scaling via Splitting Approach

In order to get more insight into the form of the underlying system of equations that has to be solved in order to realize the above $\mathcal{O}(\varepsilon)$-approximation of expectation values, we consider the asymptotic expansion

$$p^\varepsilon(\boldsymbol{y}, \boldsymbol{c}, t) = p^0(\boldsymbol{y}, \boldsymbol{c}, t) + \varepsilon p^1(\boldsymbol{y}, \boldsymbol{c}, t) + \ldots$$

Putting this into (A.52) and collecting the highest-order terms in ε, we get

$$\frac{\partial p^0(\boldsymbol{y}, \boldsymbol{c}, t)}{\partial t} = -\sum_{k \in \mathbb{K}_c} \operatorname{div}_{\boldsymbol{c}} \left[\boldsymbol{\nu}_k^{(z)} \tilde{\alpha}_k(\boldsymbol{y}, \boldsymbol{c}) p^0(\boldsymbol{y}, \boldsymbol{c}, t)\right]$$

$$+ \sum_{k \in \mathbb{K}_d} \left[\tilde{\alpha}_k(\boldsymbol{y} - \boldsymbol{\nu}_k^{(y)}, \boldsymbol{c}) p^0(\boldsymbol{y} - \boldsymbol{\nu}_k^{(y)}, \boldsymbol{c}, t) - \tilde{\alpha}_k(\boldsymbol{y}, \boldsymbol{c}) p^0(\boldsymbol{y}, \boldsymbol{c}, t)\right].$$

$$(A.55)$$

This hybrid system can be rewritten in the following abstract form:

$$\frac{\partial p^0}{\partial t} = (\mathcal{L} + \mathcal{G})p^0, \tag{A.56}$$

where the full operator on the right-hand side is split into the Liouville operator appearing in the first line of (A.55)

$$\mathcal{L}p^0(\boldsymbol{y}, \boldsymbol{c}, t) := -\sum_{k \in \mathbb{K}_c} \mathrm{div}_c \left[\boldsymbol{\nu}_k^{(z)} \tilde{\alpha}_k(\boldsymbol{y}, \boldsymbol{c}) p^0(\boldsymbol{y}, \boldsymbol{c}, t) \right],$$

and the CME operator given by the second line of (A.55)

$$\mathcal{G}p^0(\boldsymbol{y}, \boldsymbol{c}, t) := \sum_{k \in \mathbb{K}_d} \left[\tilde{\alpha}_k(\boldsymbol{y} - \boldsymbol{\nu}_k^{(y)}, \boldsymbol{c}) p^0(\boldsymbol{y} - \boldsymbol{\nu}_k^{(y)}, \boldsymbol{c}, t) - \tilde{\alpha}_k(\boldsymbol{y}, \boldsymbol{c}) p^0(\boldsymbol{y}, \boldsymbol{c}, t) \right].$$

The Liouville operator $\mathcal{L}$ acts on $\boldsymbol{c}$ and depends parametrically on $\boldsymbol{y}$, while $\mathcal{G}$ acts on $\boldsymbol{y}$ and depends parametrically on $\boldsymbol{c}$. Our above results for the Liouville case tell us that the evolution semigroup $\exp(t\mathcal{L})$, acting on distributions ρ, can be represented as

$$\exp(t\mathcal{L})\rho(\boldsymbol{y}, \boldsymbol{c}) = \frac{1}{\varsigma(\boldsymbol{y}, \boldsymbol{c}, t)} \, \rho(\boldsymbol{y}, \Phi_{\boldsymbol{y}}^{-t}\boldsymbol{c})$$

where $\Phi_{\boldsymbol{y}}^t$ again denotes the solution operator of the $\boldsymbol{y}$-dependent RRE system (A.54) and

$$\varsigma(\boldsymbol{y}, \boldsymbol{c}, t) := \exp\left(\int_0^t \sum_{k \in \mathbb{K}_c} (\boldsymbol{\nu}_k^{(z)})^\mathsf{T} \nabla \tilde{\alpha}_k(\boldsymbol{y}, \Phi_{\boldsymbol{y}}^{-s}\boldsymbol{c}) ds \right),$$

in analogy to the definition of the volume distortion factor ς in (A.48). Moreover, the above results yield by means of substitution $\Phi_{\boldsymbol{y}}^t \boldsymbol{u} = \boldsymbol{c}$ for smooth bounded observables F and densities ρ

$$\int F(\boldsymbol{y}, \boldsymbol{c}) \exp(t\mathcal{L})\rho(\boldsymbol{y}, \boldsymbol{c}) d\boldsymbol{c} = \int F(\boldsymbol{y}, \boldsymbol{c}) \frac{1}{\varsigma(\boldsymbol{y}, \boldsymbol{c}, t)} \rho(\boldsymbol{y}, \Phi_{\boldsymbol{y}}^{-t}\boldsymbol{c}) \, d\boldsymbol{c}$$

$$= \int F(\boldsymbol{y}, \Phi_{\boldsymbol{y}}^t \boldsymbol{u}) \rho(\boldsymbol{y}, \boldsymbol{u}) \, d\boldsymbol{u},$$

where we again used the equality $\varsigma(\boldsymbol{y}, \Phi_{\boldsymbol{y}}^t \boldsymbol{u}, t) = |\det(D\Phi_{\boldsymbol{y}}^t \boldsymbol{u})|$ as in Sect. A.5.1. In terms of the duality bracket $\langle \cdot, \cdot \rangle$ defined by

$$\langle F, \rho \rangle := \int F(\boldsymbol{y}, \boldsymbol{c}) \rho(\boldsymbol{y}, \boldsymbol{c}) \, d\boldsymbol{c}$$

this means that

$$\langle F, \exp(t\mathcal{L})\rho \rangle = \langle \exp(t\mathcal{L}^*)F, \rho \rangle,$$

with the semigroup of the adjoint operator $\mathcal{L}^*$ being defined as

$$\exp(t\mathcal{L}^*)F(\boldsymbol{y}, \boldsymbol{c}) := F(\boldsymbol{y}, \Phi_{\boldsymbol{y}}^t \boldsymbol{c}).$$

Next, we want to use this representation of the Liouville semigroup for solving the hybrid CME (A.55) by introducing the time-dependent density

$$\rho^0(\boldsymbol{y}, \boldsymbol{c}, t) := \exp(-t\mathcal{L})p^0(\boldsymbol{y}, \boldsymbol{c}, t).$$

With ρ^0 all partial expectation values of observables F with respect to p^0 can be rewritten as

$$\mathbb{E}_{\boldsymbol{y}}^0(F, t) := \int F(\boldsymbol{y}, \boldsymbol{c})p^0(\boldsymbol{y}, \boldsymbol{c}, t)d\boldsymbol{c} = \langle F, p^0(\cdot, \boldsymbol{y}, t) \rangle = \langle \exp(t\mathcal{L}^*)F, \rho^0 \rangle.$$
(A.57)

The abstract form (A.56) of the CME yields for the evolution of ρ^0

$$\frac{\partial \rho^0}{\partial t} = \exp(-t\mathcal{L})\,\mathcal{G}\,\exp(t\mathcal{L})\rho^0.$$

This implies that the partial expectation value $\mathbb{E}_{\boldsymbol{y}}^0(f, t)$ satisfies the equation

$$\frac{\partial}{\partial t}\mathbb{E}_{\boldsymbol{y}}^0(F, t) = \left\langle \frac{\partial}{\partial t}\exp(t\mathcal{L}^*)F, \rho^0 \right\rangle + \left\langle \exp(t\mathcal{L}^*)F, \frac{\partial}{\partial t}\rho^0 \right\rangle,$$

which gives

$$\frac{\partial}{\partial t}\mathbb{E}_{\boldsymbol{y}}^0(F, t) - \left\langle \frac{\partial}{\partial t}\exp(t\mathcal{L}^*)F, \rho^0 \right\rangle \qquad (A.58)$$
$$= \langle \exp(t\mathcal{L}^*)F, \exp(-t\mathcal{L})\,\mathcal{G}\,\exp(t\mathcal{L})\rho^0 \rangle$$
$$= \langle F, \mathcal{G}\,\exp(t\mathcal{L})\rho^0 \rangle = \langle \exp(t\mathcal{L}^*)\mathcal{G}F, \rho^0 \rangle,$$

where we used $\mathcal{G} = \mathcal{G}^*$ in the last step which follows from the fact that $\mathcal{G}$ acts on $\boldsymbol{y}$ and not on $\boldsymbol{c}$.

When comparing (A.53) with (A.57), we see that the case in which the initial probability distribution concentrates at a point $\boldsymbol{c}_0$ with respect to $\boldsymbol{c}$ formally fits into the latter case when

$$\rho^0(\boldsymbol{y}, \boldsymbol{c}, t) = \delta(\boldsymbol{c} - \boldsymbol{c}_0)\,p^0(\boldsymbol{y}, t), \qquad (A.59)$$

where $\delta(\cdot - \boldsymbol{c}_0)$ denotes the Dirac distribution concentrated on $\boldsymbol{c}_0$ and $p^0(\boldsymbol{y}, t)$ being defined by

$$p^0(\boldsymbol{y}, t) := \int \rho^0(\boldsymbol{y}, \boldsymbol{c}, t)d\boldsymbol{c}.$$

With this form (A.59) of ρ^0 and $F(\boldsymbol{y}, \boldsymbol{c}) = \mathbb{1}$, the constant 1-function, we have

$$\mathbb{E}_{\boldsymbol{y}}^0(\mathbb{1}, t) = \int p^0(\boldsymbol{y}, \boldsymbol{c}, t)d\boldsymbol{c} = p^0(\boldsymbol{y}, t).$$

Putting this into (A.58) and using the specific forms of $\mathcal{L}$ and $\mathcal{G}$ yield

$$\frac{\partial}{\partial t}p^0(\boldsymbol{y}, t) \tag{A.60}$$

$$= \sum_{k \in \mathbb{K}_d} \left[\tilde{\alpha}_k(\boldsymbol{y} - \boldsymbol{\nu}_k^{(y)}, \Phi_{\boldsymbol{y}-\boldsymbol{\nu}_k^{(y)}}^t \boldsymbol{c}_0)p^0(\boldsymbol{y} - \boldsymbol{\nu}_k^{(y)}, t) - \tilde{\alpha}_k(\boldsymbol{y}, \Phi_{\boldsymbol{y}}^t \boldsymbol{c}_0)p^0(\boldsymbol{y}, t) \right].$$

Similarly, putting $F(\boldsymbol{y}, \boldsymbol{c}) = \boldsymbol{c}$ and the specific form (A.59) of ρ^0 into (A.57), we get

$$\mathbb{E}_{\boldsymbol{y}}^0(\boldsymbol{c}, t) = p^0(\boldsymbol{y}, t)\Phi_{\boldsymbol{y}}^t \boldsymbol{c}_0.$$

Inserting this into (A.58), and using the specific forms of $\mathcal{L}$ and $\mathcal{G}$ again, yields

$$\frac{\partial}{\partial t}\mathbb{E}_{\boldsymbol{y}}^0(\boldsymbol{c}, t) = \left(\frac{d}{dt}\Phi_{\boldsymbol{y}}^t \boldsymbol{c}_0 \right) p^0(\boldsymbol{y}, t)$$

$$+ \sum_{k \in \mathbb{K}_d} \left[\tilde{\alpha}_k(\boldsymbol{y} - \boldsymbol{\nu}_k^{(y)}, \Phi_{\boldsymbol{y}-\boldsymbol{\nu}_k^{(y)}}^t \boldsymbol{c}_0)\mathbb{E}_{\boldsymbol{y}-\boldsymbol{\nu}_k^{(y)}}^0(\boldsymbol{c}, t) - \tilde{\alpha}_k(\boldsymbol{y}, \Phi_{\boldsymbol{y}}^t \boldsymbol{c}_0)\mathbb{E}_{\boldsymbol{y}}^0(\boldsymbol{c}, t) \right],$$

which finally, by using the ODE that defines the flow $\Phi_{\boldsymbol{y}}^t \boldsymbol{c}_0$, results in

$$\frac{\partial}{\partial t}\mathbb{E}_{\boldsymbol{y}}^0(\boldsymbol{c}, t) = \sum_{k \in \mathbb{K}_c} \boldsymbol{\nu}_k^{(z)} \tilde{\alpha}_k(\boldsymbol{y}, \Phi_{\boldsymbol{y}}^t \boldsymbol{c}_0)p^0(\boldsymbol{y}, t) \tag{A.61}$$

$$+ \sum_{k \in \mathbb{K}_d} \left[\tilde{\alpha}_k(\boldsymbol{y} - \boldsymbol{\nu}_k^{(y)}, \Phi_{\boldsymbol{y}-\boldsymbol{\nu}_k^{(y)}}^t \boldsymbol{c}_0)\mathbb{E}_{\boldsymbol{y}-\boldsymbol{\nu}_k^{(y)}}^0(\boldsymbol{c}, t) - \tilde{\alpha}_k(\boldsymbol{y}, \Phi_{\boldsymbol{y}}^t \boldsymbol{c}_0)\mathbb{E}_{\boldsymbol{y}}^0(\boldsymbol{c}, t) \right].$$

Remark A.17. This derivation shows that in leading order the full CME reduces to a hybrid model (A.55) composed of a Liouville equation for the continuous concentration part and a CME for the remaining discrete states. For an initial condition that concentrates at a single point with respect to the continuous concentrations, this hybrid model can be decomposed, leading to a individual reduced CME (A.60) for the discrete states coupled to a system of ordinary differential equations (A.61) for the expected concentrations of the continuous states, where these two sets of equations both depend on RRE trajectories in concentration space (one trajectory for each of the discrete states).

Remark A.18. The above kind of derivation can be repeated for all higher-order moments $\mathbb{E}_{\boldsymbol{y}}^0(\boldsymbol{c}^k, t)$ and always lead to a closed set of equations.

A.5.3 Multiple Time Scales

We now consider the problem of multiple time scales for the CME as it appears in Sects. 3.1 and 3.2. The starting point is the CME

$$\frac{dp}{dt} = \left(\frac{1}{\varepsilon}\mathcal{G}_f + \mathcal{G}_s\right)p \tag{A.62}$$

with ε again denoting a smallness parameter, $p = p(\boldsymbol{x}, \boldsymbol{y}, t)$ a probability distribution, and $\mathcal{G}_f$ and $\mathcal{G}_s$ generators of which $\mathcal{G}_s$ acts on $\boldsymbol{x}$ and depends on $\boldsymbol{y}$ parametrically only and $\mathcal{G}_f$ acts only on $\boldsymbol{y}$ and depends on $\boldsymbol{x}$ parametrically only. We here denote the state by $(\boldsymbol{x}, \boldsymbol{y})$ as a placeholder for the state $(\boldsymbol{r}^s, \boldsymbol{r}^f)$ of slow and fast reaction extents given in Sect. 3.1 and for the state $(\boldsymbol{v}, \boldsymbol{w})$ of slow and fast variables given in Sect. 3.2, respectively. One example for such a pair of generators is discussed in Sect. 3.1.3

$$\mathcal{G}_f u(\boldsymbol{x}, \boldsymbol{y}) := \sum_{k \in \mathbb{K}_f} \Big(b_k(\boldsymbol{x}, \boldsymbol{y} - \boldsymbol{\nu}_k^y) u(\boldsymbol{x}, \boldsymbol{y} - \boldsymbol{\nu}_k^y) - b_k(\boldsymbol{x}, \boldsymbol{y}) u(\boldsymbol{x}, \boldsymbol{y}) \Big),$$

$$\mathcal{G}_s u(\boldsymbol{x}, \boldsymbol{y}) := \sum_{k \in \mathbb{K}_s} \Big(\beta_k(\boldsymbol{x} - \boldsymbol{\nu}_k^x, \boldsymbol{y}) u(\boldsymbol{x} - \boldsymbol{\nu}_k^x, \boldsymbol{y}) - \beta_k(\boldsymbol{x}, \boldsymbol{y}) u(\boldsymbol{x}, \boldsymbol{y}) \Big), \tag{A.63}$$

for probability density functions $u = u(\boldsymbol{x}, \boldsymbol{y})$ and given propensity functions β_k, $k = 1, \dots, K$, with $b_k = \varepsilon \beta_k$ for $k \in \mathbb{K}_f$ (see Eq. (3.13)). As in Chap. 3, $\mathbb{K}_f$ and $\mathbb{K}_s$ are disjoint index sets for the fast and the slow reactions, respectively (see page 109). The properties of $\mathcal{G}_f$ as a generator imply that the summation operator S_y defined by

$$S_y u(\boldsymbol{x}) := \sum_y u(\boldsymbol{x}, \boldsymbol{y})$$

satisfies

$$S_y \mathcal{G}_f = 0. \tag{A.64}$$

Moreover, $\mathcal{G}_f$ generates an evolution semigroup $\exp(t\mathcal{G}_f)$ that denotes the formal solution operator of the CME solely generated by $\mathcal{G}_f$: Given that $p(0) = p_0$ and $\frac{dp}{dt} = \mathcal{G}_f p$, it follows

$$p(t) = \exp(t\mathcal{G}_f)p_0.$$

Letting S_y act on this semigroup yields

$$S_y \exp(t\mathcal{G}_f) = S_y. \tag{A.65}$$

Method of Multiple Time Scales

Our CME exhibits two disparate time scales, t and $\tau = t/\varepsilon$. We will deal with this by using a method called *two-timing*, also called *the method of multiple (time)scales* (see [170] or [155] (Section 9.8)), where multiple time scales are used in the very same expansion as independent variables. According to this approach, we use the following ansatz for the solution of the CME

$$p^\varepsilon(\boldsymbol{x}, \boldsymbol{y}, t, t/\varepsilon) = p^\varepsilon(\boldsymbol{x}, \boldsymbol{y}, t, \tau).$$

With these two temporal variables t and τ, the time derivative now takes the following form

$$\frac{dp^\varepsilon}{dt} = \frac{\partial p^\varepsilon}{\partial t} + \frac{d\tau}{dt}\frac{\partial p^\varepsilon}{\partial \tau}$$
$$= \frac{\partial p^\varepsilon}{\partial t} + \frac{1}{\varepsilon}\frac{\partial p^\varepsilon}{\partial \tau}.$$

Next, we expand p^ε asymptotically in terms of ε

$$p^\varepsilon(\boldsymbol{x}, \boldsymbol{y}, t, \tau) = p^0(\boldsymbol{x}, \boldsymbol{y}, t, \tau) + \varepsilon p^1(\boldsymbol{x}, \boldsymbol{y}, t, \tau) + \mathcal{O}(\varepsilon^2),$$

where all terms p^i must stay bounded in t and τ. Inserting this ansatz into the CME (A.62) results in

$$\frac{1}{\varepsilon}\frac{\partial p^0}{\partial \tau} + \frac{\partial p^0}{\partial t} + \frac{\partial p^1}{\partial \tau} + \mathcal{O}(\varepsilon) = \frac{1}{\varepsilon}\mathcal{G}_f p^0 + \mathcal{G}_f p^1 + \mathcal{G}_s p^0 + \mathcal{O}(\varepsilon)$$

so that the two leading orders in ε yield

$$\frac{\partial p^0}{\partial \tau} = \mathcal{G}_f p^0 \tag{A.66}$$

$$\frac{\partial p^0}{\partial t} + \frac{\partial p^1}{\partial \tau} = \mathcal{G}_f p^1 + \mathcal{G}_s p^0. \tag{A.67}$$

Now we define the marginal distributions

$$\hat{p}^i := S_y p^i, \tag{A.68}$$

i.e., $\hat{p}^i(\boldsymbol{x}, t, \tau) := \sum_{\boldsymbol{y}} p^i(\boldsymbol{x}, \boldsymbol{y}, t, \tau)$ for $i = 0, 1$, and introduce the conditional probability $\rho^0(\boldsymbol{y}, t, \tau|\boldsymbol{x})$ of $\boldsymbol{y}$ for given $\boldsymbol{x}$, such that the probability distribution p^0 can be decomposed into

$$p^0(\boldsymbol{x}, \boldsymbol{y}, t, \tau) := \rho^0(\boldsymbol{y}, t, \tau|\boldsymbol{x}) \cdot \hat{p}^0(\boldsymbol{x}, t, \tau).$$

By letting S_y act on both sides of (A.66) and using (A.64), we immediately see that

$$\frac{\partial \hat{p}^0}{\partial \tau} = S_y \mathcal{G}_f p^0 = 0,$$

so that $\hat{p}^0$ cannot depend on τ, i.e., only depends on x and t, $\hat{p}^0 = \hat{p}^0(x, t)$. This implies

$$p^0(x, y, t, \tau) = \rho^0(y, t, \tau | x) \cdot \hat{p}^0(x, t)$$

for a τ-independent function $\hat{p}^0$ and a distribution ρ^0 that depends on x only parametrically. Inserting this once again into (A.66) reveals that, because $\mathcal{G}_f$ acts on the y-component of p only,

$$p^0(x, \cdot, t, \tau) = \hat{p}^0(x, t) \exp(\tau \mathcal{G}_f) \rho^0(\cdot, t, 0 | x), \tag{A.69}$$

which we will write in shorthand as

$$p^0(\tau) = \exp(\tau \mathcal{G}_f) p^0(\tau = 0), \tag{A.70}$$

implying

$$\frac{\partial p^0}{\partial t} = \exp(\tau \mathcal{G}_f) \frac{\partial p^0}{\partial t}(\tau = 0).$$

This can now be inserted into (A.67) to give

$$\frac{\partial p^1}{\partial \tau} = \mathcal{G}_f p^1 + \mathcal{G}_s p^0 - \exp(\tau \mathcal{G}_f) \frac{\partial p^0}{\partial t}(\tau = 0).$$

Integrating over τ and using the variation of constants method as well as our shorthand notation (A.70), this now leads to

$$p^1(\tau) = \exp(\tau \mathcal{G}_f) p^1(\tau = 0)$$
$$+ \exp(\tau \mathcal{G}_f) \int_0^\tau \exp(-\sigma \mathcal{G}_f) \mathcal{G}_s \exp(\sigma \mathcal{G}_f) p^0(\tau = 0) d\sigma$$
$$- \tau \exp(\tau \mathcal{G}_f) \frac{\partial p^0}{\partial t}(\tau = 0).$$

Letting S_y act on this equation and using (A.65) and (A.68), we get

$$\hat{p}^1(\tau) = \hat{p}^1(\tau = 0)$$
$$+ \int_0^\tau S_y \mathcal{G}_s \exp(\sigma \mathcal{G}_f) p^0(\tau = 0) d\sigma \tag{A.71}$$
$$- \tau \frac{\partial \hat{p}^0}{\partial t}.$$

From our initial assumption that p^1 stays bounded in τ, we can follow

$$\lim_{\tau \to \infty} \frac{1}{\tau} \hat{p}^1(\tau) = 0.$$

Thus, by dividing (A.71) by τ and using (A.69) (i.e., replacing $p^0(\tau = 0)$ by $\hat{p}^0(\boldsymbol{x}, t)\rho^0(\cdot, t, 0|\boldsymbol{x}))$, we get

$$\frac{\partial \hat{p}^0}{\partial t}(\boldsymbol{x}, t) = \lim_{\tau \to \infty} \frac{1}{\tau} \int_0^\tau S_{\boldsymbol{y}} \mathcal{G}_s \hat{p}^0(\boldsymbol{x}, t) \exp(\sigma \mathcal{G}_f) \rho^0(\boldsymbol{y}, t, 0|\boldsymbol{x}) d\sigma$$

$$= S_{\boldsymbol{y}} \mathcal{G}_s \hat{p}^0(\boldsymbol{x}, t) \lim_{\tau \to \infty} \frac{1}{\tau} \int_0^\tau \exp(\sigma \mathcal{G}_f) \rho^0(\boldsymbol{y}, t, 0|\boldsymbol{x}) d\sigma.$$

That is, the limit

$$\bar{\rho}^0(\boldsymbol{y}, t|\boldsymbol{x}) := \lim_{\tau \to \infty} \frac{1}{\tau} \int_0^\tau \rho^0(\boldsymbol{y}, t, \sigma|\boldsymbol{x}) d\sigma \tag{A.72}$$

$$= \lim_{\tau \to \infty} \frac{1}{\tau} \int_0^\tau \exp(\sigma \mathcal{G}_f) \rho^0(\boldsymbol{y}, t, 0|\boldsymbol{x}) d\sigma.$$

has to exist, and we get an evolution equation for $\hat{p}^0$

$$\frac{\partial \hat{p}^0}{\partial t}(\boldsymbol{x}, t) = \left(S_{\boldsymbol{y}} \mathcal{G}_s \bar{\rho}^0(\boldsymbol{y}, t|\boldsymbol{x}) \right) \hat{p}^0(\boldsymbol{x}, t).$$

For the specific form of $\mathcal{G}_s$ given in (A.63), this equation takes the form

$$\frac{\partial \hat{p}^0}{\partial t}(\boldsymbol{x}, t) = \sum_{k \in \mathbb{K}_s} \left(\bar{\beta}_k(\boldsymbol{x} - \boldsymbol{\nu}_k^x, t) \hat{p}^0(\boldsymbol{x} - \boldsymbol{\nu}_k^x, t) - \bar{\beta}_k(\boldsymbol{x}, t) \hat{p}^0(\boldsymbol{x}, t) \right) \tag{A.73}$$

with

$$\bar{\beta}_k(\boldsymbol{x}, t) := \sum_{\boldsymbol{y}} \beta_k(\boldsymbol{x}, \boldsymbol{y}) \bar{\rho}^0(\boldsymbol{y}, t|\boldsymbol{x}). \tag{A.74}$$

In order to see how $\bar{\rho}^0$ as defined in (A.72) can be computed, we distinguish between tow cases.

The Fast Process Is Geometrically Ergodic

As a first case, we assume that $\mathcal{G}_f$ has a stationary distribution.

Assumption A.19. The process defined by $\mathcal{G}_f$ is geometrically ergodic on the state space spanned by $\boldsymbol{y}$, that is, for every $\boldsymbol{x}$ there exists a (unique) stationary probability distribution $\pi(\cdot|\boldsymbol{x})$ such that for any probability density function $u = u(\boldsymbol{x}, \boldsymbol{y})$ and for some $r > 0$ that does not depend on $\boldsymbol{x}$,

$$\exp(t\mathcal{G}_f)u(\boldsymbol{x}, \boldsymbol{y}) = \pi(\boldsymbol{y}|\boldsymbol{x}) + \mathcal{O}(e^{-rt}).$$

In this case, we obviously have, for all $t \geq 0$,

$$\bar{p}^0(\boldsymbol{y}, t | \boldsymbol{x}) = \pi(\boldsymbol{y}|\boldsymbol{x}),$$

and the evolution equation for $\hat{p}^0$ takes the form

$$\frac{\partial \hat{p}^0}{\partial t} = \mathcal{L}_s \hat{p}^0(t),$$

where

$$\mathcal{L}_s = S_{\boldsymbol{y}} \mathcal{G}_s \pi(\boldsymbol{y}|\boldsymbol{x})$$

is π-averaged with respect to $\boldsymbol{y}$ but still acts on $\boldsymbol{x}$. For the specific form of $\mathcal{G}_s$ given in (A.63), $\mathcal{L}_s$ takes the form

$$\mathcal{L}_s u(\boldsymbol{x}, t) = \sum_{k \in \mathbb{K}_s} \left(\bar{\beta}_k(\boldsymbol{x} - \boldsymbol{\nu}_k^x) u(\boldsymbol{x} - \boldsymbol{\nu}_k^x, t) - \bar{\beta}_k(\boldsymbol{x}) u(\boldsymbol{x}, t) \right),$$

with the averaged propensity functions

$$\bar{\beta}_k(\boldsymbol{x}) = \sum_{\boldsymbol{y}} \beta_k(\boldsymbol{x}, \boldsymbol{y}) \pi(\boldsymbol{y}|\boldsymbol{x})$$

for $k \in \mathbb{K}_s$ and $\pi(\cdot|\boldsymbol{x})$ satisfying

$$\sum_{k \in \mathbb{K}_f} \left(b_k(\boldsymbol{x}, \boldsymbol{y} - \boldsymbol{\nu}_k^y) \pi(\boldsymbol{y} - \boldsymbol{\nu}_k^y | \boldsymbol{x}) - b_k(\boldsymbol{x}, \boldsymbol{y}) \pi(\boldsymbol{y}|\boldsymbol{x}) \right) = 0.$$

In this case, solving (A.73) for $\hat{p}^0$ with the above-averaged propensity functions $\bar{\beta}_k$ yields an approximation of the full solution

$$p^\varepsilon(\boldsymbol{x}, \boldsymbol{y}, t) = \pi(\boldsymbol{y}|\boldsymbol{x}) \hat{p}^0(\boldsymbol{x}, t) + \mathcal{O}(\varepsilon) + \mathcal{O}(e^{-rt/\varepsilon}).$$

Remark A.20. In [243] another multiscale asymptotic approach to the temporal multiscale CME (A.62) has been published. Therein the time scale $\tau = t/\varepsilon$ and its effect on the evolution of p^ε are ignored. The above result can be seen as a justification of the simpler approach in [243] on time scale of order $\mathcal{O}(1)$.

The Fast Process Is Increasing But Bounded

Lastly, let us consider the case of the fast-slow reaction extent CME (3.13) considered in Sect. 3.1.3. In this case, the slow reactions $\boldsymbol{r}^s$ are represented by our $\boldsymbol{x}$-variables and the fast reactions $\boldsymbol{r}^f$ by $\boldsymbol{y}$. For fixed $\boldsymbol{r}^s = \boldsymbol{x}$, the fast reaction process $Y(t|\boldsymbol{x}) = \boldsymbol{r}^f(t|\boldsymbol{r}^s)$ generated by $\mathcal{G}_f$ is monotonically increasing. Let us assume that it cannot increase beyond a maximal value

$$Y(t|\boldsymbol{x}) \leq Y_{\max}(t|\boldsymbol{x}),$$

e.g., because the number of reactant molecules is limited. Then

$$\bar{\rho}^0(\boldsymbol{y}, t|\boldsymbol{x}) = \delta_{Y_{\max}(t|\boldsymbol{x})}(\boldsymbol{y}),$$

and we get a closed form of (A.73) for $\hat{p}^0$ with

$$\bar{\beta}_k(\boldsymbol{x}, t) = \beta_k(\boldsymbol{x}, Y_{\max}(t|\boldsymbol{x})).$$

Hereby we complete the appendix.

Symbol Directory

$a(t, x, y)$	Transition function 172, 186
$A(t)$	Transition matrix 172, 186
$\mathcal{A}^t$	Transfer operator 189
$\boldsymbol{b}$	Vector defining a variable 114
$B_k = (B_k(t))_{t \geq 0}$	Standard Brownian motion in $\mathbb{R}$ 47
B	Matrix of basis vectors 114
B^+	Pseudoinverse of B 114
$B(v, q)$	Binomial distribution 29, 128
$\mathcal{B}$	Borel σ-algebra on $\mathbb{R}^L$ 52
$\mathcal{B}(\mathbb{X})$	Borel σ-algebra on $\mathbb{X}$ 212
$\mathcal{B}_S$	Borel σ-algebra induced by the Skorokhod topology 207
c_l	Concentration of species $\mathcal{S}_l$ 40
$\boldsymbol{c}$	$= (c_1, \ldots, c_L)^\mathsf{T}$ 40
$\boldsymbol{C}^V = (\boldsymbol{C}^V(t))_{t \geq 0}$	Scaled process 40
$\boldsymbol{C}^V(t)$	$= (C_1^V(t), \ldots, C_L^V(t))^\mathsf{T}$ 40
$C_l(t)$	Concentration of species $\mathcal{S}_l$ at time t 43
$\boldsymbol{C} = (\boldsymbol{C}(t))_{t \geq 0}$	$= (C_1(t), \ldots, C_L(t))_{t \geq 0}^\mathsf{T}$ limit process of concentrations 43
	Continuous component of the PDMP 212
$\tilde{\boldsymbol{C}}^V = (\tilde{\boldsymbol{C}}^V(t))_{t \geq 0}$	Second-order approximation of $\boldsymbol{C}^V$ 47
$\bar{C}(t \vert \boldsymbol{y})$	Maximum of eikonal function S_0 81
$\mathcal{C}_\mathbb{X}[0, \infty)$	Space of continuous functions $f : [0, \infty) \to \mathbb{X}$ 203

© The Editor(s) (if applicable) and The Author(s), under exclusive license to Springer Nature Switzerland AG 2020
S. Winkelmann, C. Schütte, *Stochastic Dynamics in Computational Biology*, Frontiers in Applied Dynamical Systems: Reviews and Tutorials 8, https://doi.org/10.1007/978-3-030-62387-6

$\mathcal{C}_b(\mathcal{X})$	Space of real-valued bounded, continuous functions on $(\mathcal{X}, d)$ 207
d	Dimension of spatial environment 132
$d_{\mathcal{D}} : \mathcal{D}[0,T]^2 \to \mathbb{R}_{\geq 0}$	Skorokhod metric on $\mathcal{D}[0,T]$ 204
$d_{\mathbb{X}} : \mathbb{X}^2 \to [0, \infty)$	Metric on $\mathbb{X}$ 205
D_n	Diffusion coefficient of particle n 137
D_l	Diffusion coefficient of species $\mathcal{S}_l$ 149
D	Diffusion coefficient 134
$\boldsymbol{D}$	Diffusion coefficient matrix 149
$\mathcal{D}[0,T]$	Space of càdlàg-functions $f : [0,T] \to \mathbb{R}$ 204
$\mathcal{D}_{\mathbb{X}}[0, \infty)$	Space of càdlàg-functions $f : [0, \infty) \to \mathbb{X}$ 203
$\mathbb{D}$	Space of motion, domain for the particles' movement 132
$\mathbb{D}_i$	Compartment/subdomain of $\mathbb{D}$ 159
$\boldsymbol{e}_{li}$	Matrix whose elements are all zero except the entry (l, i) which is one 160
E	Enzyme in enzyme kinetics example 107
ES	Complex in enzyme kinetics example 107
$E_n(t)$	Running mean for time t 23
$\mathcal{E}$	σ-Algebra 185
$\mathbb{E}$	Expectation/first-order moment 14
$F : \mathbb{X} \to \mathbb{R}$	Observable of $\boldsymbol{x}$ 17
$g \in \mathcal{D}[0,T]$	Càdlàg-function 205
$g(\tau|t)$	Integrated rate function 67
G	Generator matrix 19
$\mathcal{G} : \ell_1^1 \to \ell_1^1$	CME operator 13
$\mathcal{G}_f : \ell_1^1 \to \ell_1^1$	Fast-scale CME operator 112
$\mathcal{G}_s : \ell_1^1 \to \ell_1^1$	Slow-scale CME operator 112
h	Width of compartments 163
$h_k : \mathbb{X} \to \mathbb{N}_0$	Combinatorial function 4
H	Hamiltonian function 221
id: $\mathbb{X} \to \mathbb{X}$	Identity function 17
Id	Identity matrix 187
$I : [0,T] \to [0,T]$	Bijection in $\mathcal{I}$ 204
$I : [0, \infty] \to [0, \infty]$	Bijection in $\mathcal{I}_\infty$ 204
$\mathcal{I}$	Set of strictly increasing, continuous bijections on $[0,T]$ 204
$\mathcal{I}_\infty$	Set of strictly increasing, continuous bijections on $[0, \infty]$ 206

$\mathbb{I}$	Index set of spatial compartments/voxels 159
$\mathbb{I}_{\text{int}}$	Index set 200
$|\mathbb{I}|$	Number of voxels/compartments of spatial discretization 159
J	Number of basis vectors $\boldsymbol{b}_j$ 114
$\boldsymbol{J} = (J_1, \ldots, J_L)$	Rank vector 21
k	Index of reaction 2
k_B	Boltzmann constant 135
K	Number of reactions 2
$\mathbb{K} = \{1, \ldots, K\}$	Index set of reactions 60
$\mathbb{K}_0$, $\mathbb{K}_1$, $\mathbb{K}_2$	Index subset of reactions 60, 75
$\mathbb{K}_c$, $\mathbb{K}_d$	Index subset of continuous and discrete reactions 62, 79
$\mathbb{K}_f$, $\mathbb{K}_s$	Index subset of fast and slow reactions 109
$\mathcal{K} : \mathbb{S} \times \mathcal{B}(\mathbb{S}) \to [0,1]$	Transition kernel of a PDMP 88, 212
l	Index of species 2
ℓ^1	Space of functions u on the discrete set $\mathbb{X}$ with $\sum_{x \in \mathbb{X}} |u(x)| < \infty$ 189
ℓ_1^1	Set of non-negative functions on $\mathbb{X}$ that sum up to one 12
$\ell^\infty(\mathbb{X})$	Space of bounded functions on $\mathbb{X}$ 17
L	Number of species 2
$L^2(\mathbb{R}^d)$	$= \left\{ u : \mathbb{R}^d \to \mathbb{R} \,\middle|\, \int_{\mathbb{R}^d} u(q)^2 \, dq < \infty \right\}$ 141
$\mathcal{L}$	Liouville operator 226
$\mathcal{L}^{(0)}, \mathcal{L}^{(1)}, \mathcal{L}^{(2)}$	Operators of system size expansion 55
$\mathcal{L}^{(D)}$	Diffusion operator 148
$\mathcal{L}^{(R)}$	Reaction operator 148
m	Order of a reaction 3
m_0	Mass of particle 134
M	Number of elements of the finite state space 19
$\mathcal{M}$	Ansatz manifold 22
n	Index of individual particle 133
n	Number of Monte Carlo simulations 23
n_0	Number of discretization steps in each spatial dimension 200
N	Number of particles 137
$N(t)$	Number of jumps that occurred within $[0, t]$ 215
$\mathcal{N}(\mu, \sigma^2)$	Normal distribution with mean μ and variance σ^2
$\mathbb{N}_0$	Set of natural numbers including zero 2

$o(\ldots)$	Asymptotic notation 3
$\mathcal{O}(\ldots)$	Asymptotic notation 40
	Order of convergence 43
$p(\boldsymbol{x}, t)$	Probability for $\boldsymbol{X}(t) = \boldsymbol{x}$ 10
p_0	Initial distribution of the reaction jump process 10
$\boldsymbol{p}(t)$	Probability row vector 19, 188
$\boldsymbol{p}_0$	Initial distribution as a row vector 188
$\hat{p}(\boldsymbol{x}, \boldsymbol{y})$	Transition probabilities of embedded Markov chain 24
$p(\boldsymbol{y}, t)$	Marginal distribution 79
$p(\boldsymbol{z}, t\|\boldsymbol{y})$	Conditional distribution 79
$p^i(\boldsymbol{y}, t)$	Basis functions in multiscale expansion 81
$p(\boldsymbol{r}, t)$	Probability for $\boldsymbol{R}(t) = \boldsymbol{r}$ 107
$p^0(\boldsymbol{v}, t)$	Slow-scale distribution 117
P	Product in enzyme kinetics example 107
P	Protein in gene expression example 177
$(\mathcal{P}_v(t))_{t \geq 0}$	Poisson process 192
$\boldsymbol{q}$	$= (q_1, \ldots, q_N)$, configuration state 137
$Q_n(t)$	Spatial position of particle n at time t 133
r_k	Number of reactions R_k 107
$\boldsymbol{r}$	$= (r_1, \ldots, r_K)^\mathsf{T}$ 107
$\boldsymbol{r}^s$	$= (r_1, \ldots, r_m)^\mathsf{T}$ 109
$\boldsymbol{r}^f$	$= (r_{m+1}, \ldots, r_K)^\mathsf{T}$ 109
$\mathcal{R}_k$	The kth reaction 2
$R_k(t)$	Number of times the reaction R_k has occurred by time t 6
$R_k(\boldsymbol{x}, \tau)$	Number of R_k-reactions during $[t, t + \tau]$ given $\boldsymbol{X}(t) = \boldsymbol{x}$ 27
$\boldsymbol{R}(t)$	$= (R_1(t), \ldots, R_K(t))$ 106
$\boldsymbol{R}^s(t)$	Extent/advancement of slow reactions 109
$\boldsymbol{R}^f(t)$	Extent/advancement of fast reactions 109
$\mathbb{R}_+$	$= \{x \in \mathbb{R} \| x \geq 0\}$ 14
$\mathbb{R}_{>0}$	$= \{x \in \mathbb{R} \| x > 0\}$ 193
s_{kl}, s'_{kl}	Stoichiometric coefficients 2
S	Substrate in enzyme kinetics example 107
$\mathcal{S}_l$	The lth chemical species 2
$\mathcal{S}(t, s)$	Evolution semigroup 222
$S_0(\boldsymbol{c}, t\|\boldsymbol{y})$	Eikonal function 81

S_x	Summation operator 229
$\mathbb{S}$	State space of the PDMP 86,212
t	Continuous time index 6
T_0	Temperature 135
T	Time horizon 28
T_j	Jump times of a Markov jump process 190
T^f	Time length for the inner SSA 119
$\mathcal{T}^t$	Backward transfer operators $\mathcal{T}^t$ 189
$\mathbb{T}$	Index set of time 203
$u_{j_l}^{(l)}$	Basis functions in Galerkin ansatz 21
$U : \mathcal{D} \to \mathbb{R}$	Potential energy function 134
$U_n : \mathcal{D}^N \to \mathbb{R}$	Interaction potential 137
U_i	Basis functions in multiscale ansatz 81, 220
$U(0,1)$	Uniform distribution on the interval $[0,1]$ 24
$\mathcal{U}_k$	Unit-rate Poisson process 8
$\mathcal{U}_{ijk}^{(D)}, \mathcal{U}_{ik}^{(R)}$	Independent unit-rate Poisson processes 162
v_j	$= \boldsymbol{b}_j^{\mathsf{T}} \boldsymbol{x}$ slow variable 114
$\boldsymbol{v}$	$= B\boldsymbol{x}$ vector of slow variables 114
$V > 0$	Volume 38
$\mathbb{V}$	Variance/second-order moment 50
$\boldsymbol{w}$	Orthogonal complement of the slow variables $\boldsymbol{v}$ 114
$(W(t))_{t \geq 0}, (W_k(t))_{t \geq 0}$	Standard Brownian motion/Wiener process in $\mathbb{R}$ 46
$(W_n(t))_{t \geq 0}$	Wiener process in $\mathbb{R}^d$ 133
$(\boldsymbol{W}(t))_{t \geq 0}$	Wiener process in $\mathbb{R}^m$ 216
x_l	Number of particles of species $\mathcal{S}_l$ 3
$\boldsymbol{x}$	$= (x_1, \ldots, x_L)^{\mathsf{T}}$ 3
$X_l(t)$	Number of particles of species $\mathcal{S}_l$ at time t 6
$\boldsymbol{X} = (\boldsymbol{X}(t))_{t \geq 0}$	$= (X_1(t), \ldots, X_L(t))_{t \geq 0}^{\mathsf{T}}$ reaction jump process 6
$X_l^V(t)$	Number of particles of species $\mathcal{S}_l$ at time t given the volume V 38
$\boldsymbol{X}^V = (\boldsymbol{X}^V(t))_{t \geq 0}$	$= (X_1^V(t), \ldots, X_L^V(t))_{t \geq 0}^{\mathsf{T}}$ reaction jump process given the volume V 38
$\tilde{\boldsymbol{X}}^V = (\tilde{\boldsymbol{X}}^V(t))_{t \geq 0}$	CLE-process 47
$\tilde{\boldsymbol{X}}^V = (\tilde{\boldsymbol{X}}^V(t))_{t \geq 0}$	Three-level adaptive hybrid process 98
$X = (X(t))_{t \geq 0}$	Stochastic process, Markov jump process 185

$\mathcal{X}$ Set of sample paths of a stochastic process $(X(t))_{t\geq 0}$ 203

$\mathbb{X} \subset \mathbb{N}_0^L$ State space of the reaction jump process $(\boldsymbol{X}(t))_{t\geq 0}$ 7

$\boldsymbol{y}$ Value of $\boldsymbol{Y}(t)$ 62

$\boldsymbol{Y}^V = (\boldsymbol{Y}^V(t))_{t\geq 0}$ Low-abundant component of the multiscale process $\boldsymbol{X}^V$ 78

$(\boldsymbol{Y}(t))_{t\geq 0}$ Discrete component of the PDMP $(\boldsymbol{\Theta}(t))_{t\geq 0}$ 62, 212

$\mathbb{Y}$ Range of the discrete component of a PDMP 212

$\boldsymbol{Z}^V = (\boldsymbol{Z}^V(t))_{t\geq 0}$ High-abundant component of the multiscale process $\boldsymbol{X}^V$ 78

$Z_{l,j}^{(D)}(y,t)$ Gaussian space-time white noise for diffusion dynamics 152

$\boldsymbol{Z}_l^{(D)}(y,t)$ $= (Z_{l,1}^{(D)}(y,t), \ldots, Z_{l,d}^{(D)}(y,t))$ 152

$Z_k^{(R)}(y,t)$ Gaussian space-time white noise for reaction dynamics 152

$\alpha_k(\boldsymbol{x})$ Propensity for reaction R_k in terms of the discrete state $\boldsymbol{x}$ 4

$\alpha_k^V(\boldsymbol{x})$ Volume-scaled propensity 39

$\tilde{\alpha}_k^V(\boldsymbol{c})$ Propensity in terms of concentrations 40

$\tilde{\alpha}_k(\boldsymbol{c})$ Limit propensity in terms of concentrations 40

$\tilde{\alpha}_k^V(\theta)$ Propensity in terms of the combined state θ 59

$\tilde{\alpha}_k(\boldsymbol{x})$ Approximative averaged propensity 119

$\alpha_k^{(t)}(\boldsymbol{y})$ Conditional propensity 80

$\tilde{\alpha}_k^\varepsilon(\boldsymbol{y}, \boldsymbol{c})$ Propensity in the hybrid setting 80

$\hat{\alpha}_k(\boldsymbol{w}, \boldsymbol{v})$ Propensity in terms of fast and slow variables 115

$\bar{\alpha}_k(\boldsymbol{v})$ Averaged propensity 117

$\beta_k(\boldsymbol{r})$ Propensity in terms of reaction extents 107

$\bar{\beta}_k(\boldsymbol{r}^s, t)$ Conditional propensity in terms of reaction extents 110

γ_k Reaction rate constant for stochastic reaction kinetics 2

$\tilde{\gamma}_k$ Reaction rate constant for classical mass-action kinetics 40

γ_d Dissociation constant 124

$\gamma_1^{\mathrm{micro}}$ Microscopic reaction rate constant 139

$\Gamma(N_\mathcal{R}, 1/\lambda(\boldsymbol{x}))$ Gamma distribution 30

$\delta(\cdot)$	Dirac delta function 55, 134, 152
δ_{xy}	Kronecker delta with $\delta_{xy} = 1$ for $x = y$ and zero otherwise 152, 186
Δt	Discrete time step 68, 145
$\boldsymbol{\epsilon} = (\boldsymbol{\epsilon}(t))_{t\geq 0}$	$= (\epsilon_1(t), \dots, \epsilon_L(t))_{t\geq 0}$, difference process 55
ε	$= \frac{1}{V}$, smallness parameter 217
ζ	Friction 134
η	Outward pointing normal to a given surface 174
θ	State of the partially scaled process or of the PDMP 59, 61
$\boldsymbol{\Theta}^V = (\boldsymbol{\Theta}^V(t))_{t\geq 0}$	Partially scaled process 59
$\boldsymbol{\Theta} = (\boldsymbol{\Theta}(t))_{t\geq 0}$	Piecewise-deterministic reaction process 61
	Three-level hybrid process 76
ϑ_1	Reaction radius 138
ϑ	Relative distance between two particles 138
ι	Radius of disk 173
ι_l	Radius of spherical particles 134
κ	Permeability of membrane 174
λ_{ij}^l	Jump rates for transitions between compartments 151
$\Lambda = (\lambda_{ij})_{i,j=1,\dots,|\mathbb{0}|}$	Rate matrix 172
$\lambda : \mathbb{X} \to [0, \infty)$	Jump rate function of the reaction jump process 23
$\tilde{\lambda} : \mathbb{S} \to [0, \infty)$	Jump rate function of the piecewise-deterministic reaction process 61
$\lambda : \mathbb{X} \to [0, \infty)$	Jump rate function of a Markov jump process 190
$\lambda : \mathbb{S} \to [0, \infty)$	Jump rate function of a PDMP 212
$\mu : \mathbb{R}^L \to \mathbb{R}^L$	Drift vector field for the CLE/CFPE 49
$\tilde{\mu} : \mathbb{R}^L \to \mathbb{R}^L$	Drift vector field for the RRE/Liouville equation 52
$\mu_y : \mathbb{R}^n \to \mathbb{R}^n$	Drift vector field of a PDMP or PCLE 62, 71, 212
$\mu : \mathbb{R}^d \to \mathbb{R}^d$	Drift vector field of the diffusion process $(\Psi(t))_{t\geq 0}$ 198
$\boldsymbol{\nu}_k$	Stoichiometric vector/state-change vector 2
$\boldsymbol{\nu}_k^{(y)}, \boldsymbol{\nu}_k^{(z)}$	Components of $\boldsymbol{\nu}_k$ 79
$\tilde{\boldsymbol{\nu}}_k^V$	Partially scaled state-change vector 59
ν	Stoichiometric matrix 107
ξ	Standard Gaussian random variable in $\mathbb{R}^d$ 145
$(\xi(t))_{t\geq 0}, (\xi_n(t))_{t\geq 0}$	White noise process 134, 154

$\xi(\mu, \sigma^2)$	Normal random variable with mean μ and variance σ^2 56
$\boldsymbol{\xi}$	$= \nabla S_0$ 221
$\Pi(\boldsymbol{\varepsilon}, t)$	Probability density of $\boldsymbol{\varepsilon}(t)$ 55
$\pi(\boldsymbol{x})$	Equilibrium distribution, stationary distribution 19, 190
$\pi(\boldsymbol{w}|\boldsymbol{v})$	Quasi-equilibrium distribution 116
$\pi(x)$	Stationary distribution for the total number x of $\mathcal{S}_1$-particles 165
π	Stationary distribution on $\mathbb{D}$ 172
$\rho(\boldsymbol{c}, t)$	Density in terms of concentration 52, 222
ρ_0	Initial distribution 53
$\rho^V(\boldsymbol{c}, t), \rho^\varepsilon(\boldsymbol{c}, t)$	Density of the scaled process 217
$\rho^V(\boldsymbol{x}, t)$	Density in terms of particle numbers 51
ρ_0^V	Initial distribution 52
$\rho^\varepsilon(\boldsymbol{c}, t|\boldsymbol{y})$	Conditional density 80
$\rho(\vartheta, t)$	Probability density for the relative distance ϑ 138
$\rho_n(q, t)$	Probability density for position of particle in space $\mathbb{D}$ 135
$\varsigma(\boldsymbol{c}, t)$	Volume distortion factor 218
$\sigma : \mathbb{R}^L \to \mathbb{R}^{L,K}$	Noise function 49
$\sigma_{\boldsymbol{y}} : \mathbb{R}^{L-l'} \to \mathbb{R}^{L-l',m}$	Noise function 71
$\sigma_{\boldsymbol{y}} : \mathbb{R}^n \to \mathbb{R}^{n,m}$	Noise function 216
$\boldsymbol{\Sigma}$	Diffusion matrix 52
$\tau(t)$	Residual life time 23
τ	Random jump time 24
	Fixed lag time 27
τ_k	Random jump time for reaction k 26
$\upsilon > 0$	Rate constant of a Poisson process 192
$\Upsilon : \mathbb{R}_+ \to \mathbb{R}_{>0}$	Intensity function of a Poisson process 193
$\Upsilon_{\text{PDRP}} : \mathbb{R}_+ \to \mathbb{R}_+$	Rate function 95
$\Upsilon_{\text{PCLE}} : \mathbb{R}_+ \to \mathbb{R}_+$	Rate function 95
$\Upsilon_{\text{CME}} : \mathbb{R}_+ \to \mathbb{R}_+$	Rate function 95
φ	Orthogonal complement of $B\boldsymbol{\nu}_k$ 116
$\phi_n(q, t)$	Dirac delta density 153
Φ^t	Flow operator 54, 218
	Solution operator 222
$\Phi_{\boldsymbol{y}}^t \boldsymbol{c}$	Deterministic flow of a PDMP 212
	Randomized flow of a hybrid diffusion process 216

Φ_y^t	Flow operator 81, 225
$\chi(y,t)$	Noise field 154
$\tilde{\chi}(y,t)$	Noise field 154
ψ_1, ψ_2	Random numbers for Gillespie simulation 24
$\omega \in \Omega$	Event in Ω 185
Ω	Sample space 185
$\|\cdot\|$	Euclidean norm on $\mathbb{R}^d$ 138
$|\mathbb{D}_j|$	Size of compartment $\mathbb{D}_j$ 151
$\langle \cdot, \cdot \rangle$	Scalar product on Hilbert space 141
$\langle \cdot, \cdot \rangle$	Duality bracket 226
$\mathbb{1}_{\boldsymbol{x}}$	Indicator function of the state $\boldsymbol{x}$ 17
$\mathbb{1}_A$	Indicator function of the set A 67
$\mathbb{1}_{r_1}$	Indicator distance function 139
$\mathbf{1}_k$	kth column of the $K \times K$-identity matrix 107
$\mathbf{1}_i'$	Row vector with the value 1 at entry i and zeros elsewhere, i.e., ith row of the $|\mathbb{I}| \times |\mathbb{I}|$-identity matrix 160
∇	Nabla/del operator 134
∇^2	Laplace operator 139

Bibliography

1. N. Agmon, Diffusion with back reaction. J. Chem. Phys. **81**(6), 2811–2817 (1984)

2. A.H. Al-Mohy, N.J. Higham, Computing the action of the matrix exponential, with an application to exponential integrators. SIAM J. Sci. Comput. **33**(2), 488–511 (2011)

3. A. Alfonsi, E. Cancès, G. Turinici, B. Di Ventura, W. Huisinga, Adaptive simulation of hybrid stochastic and deterministic models for biochemical systems, in *ESAIM: Proceedings*, vol. 14, pp. 1–13 (EDP Sciences, 2005)

4. D. Altintan, A. Ganguly, H. Koeppl, Efficient simulation of multiscale reaction networks: A multilevel partitioning approach, in *2016 American Control Conference* (IEEE, 2016), pp. 6073–6078

5. M. Ander, P. Beltrao, B. Di Ventura, J. Ferkinghoff-Borg, MAFM Foglierini, C. Lemerle, I. Tomas-Oliveira, L. Serrano, Smartcell, a framework to simulate cellular processes that combines stochastic approximation with diffusion and localisation: Analysis of simple networks. Systems Biology **1**(1), 129–138 (2004)

6. D.F. Anderson, A modified next reaction method for simulating chemical systems with time dependent propensities and delays. J. Chem. Phys. **127**(21), 214107 (2007)

S. Winkelmann, C. Schütte, *Stochastic Dynamics in Computational Biology*, Frontiers in Applied Dynamical Systems: Reviews and Tutorials 8, https://doi.org/10.1007/978-3-030-62387-6

7. D.F. Anderson, Incorporating postleap checks in tau-leaping. J. Chem. Phys. **128**(5), 054103 (2008)

8. D.F. Anderson, A. Ganguly, T.G. Kurtz, Error analysis of tau-leap simulation methods. Ann. Appl. Probab. **21**(6), 2226–2262 (2011)

9. D.F. Anderson, T.G. Kurtz, Continuous time Markov chain models for chemical reaction networks, in *Design and Analysis of Biomolecular Circuits* (Springer, Berlin, 2011), pp. 3–42

10. D.F. Anderson, T.G. Kurtz, *Stochastic Analysis of Biochemical Systems* (Springer, Berlin, 2015)

11. S. Andrews, N. Addy, R. Brent, A. Arkin, Detailed simulations of cell biology with smoldyn 2.1. PLoS Comput. Biol. **6**(3), e1000705 (2010)

12. S. Andrews, D. Bray, Stochastic simulation of chemical reactions with spatial resolution and single molecule detail. Physical Biology **1**(3), 137 (2004)

13. P. Arányi, J. Tóth, A full stochastic description of the Michaelis-Menten reaction for small systems. Acta Biochim. Biophys. Acad. Sci. Hung. **12**(4), 375–388 (1977)

14. M. Arcak, Certifying spatially uniform behavior in reaction–diffusion PDE and compartmental ODE systems. Automatica **47**(6), 1219–1229 (2011)

15. S. Asmussen, P.W. Glynn, *Stochastic Simulation: Algorithms and Analysis*, vol. 57 (Springer Science & Business Media, 2007)

16. A. Auger, P. Chatelain, P. Koumoutsakos, R-leaping: Accelerating the stochastic simulation algorithm by reaction leaps. J. Chem. Phys. **125**(8), 084103 (2006)

17. B. Bayati, P. Chatelain, P. Koumoutsakos, Adaptive mesh refinement for stochastic reaction–diffusion processes. J. Comput. Phys. **230**(1), 13–26 (2011)

18. D. Ben-Avraham, S. Havlin, *Diffusion and Reactions in Fractals and Disordered Systems* (Cambridge university press, Cambridge, 2000)

19. H. Berry, Monte Carlo simulations of enzyme reactions in two dimensions: Fractal kinetics and spatial segregation. Biophys. J. **83**(4), 1891–1901 (2002)

20. A.K. Bhattacharjee, K. Balakrishnan, A.L. Garcia, J.B. Bell, A. Donev, Fluctuating hydrodynamics of multi-species reactive mixtures. J. Chem. Phys. **142**(22), 224107 (2015)

21. P. Billingsley, *Convergence of Probability Measures* (Wiley, 2013)

22. P. Bokes, J.R. King, A. Wood, M. Loose, Multiscale stochastic modelling of gene expression. J. Math. Biol. **65**(3), 493–520 (2012)

23. G.R. Bowman, X. Huang, V.S. Pande, Using generalized ensemble simulations and Markov state models to identify conformational states. Methods **49**(2), 197–201 (2009)

24. G.R. Bowman, V.S. Pande, F. Noé, *An Introduction to Markov State Models and Their Application to Long Timescale Molecular Simulation.* Advances in Experimental Medicine and Biology (Springer, Berlin, 2014)

25. P. Brémaud, *Markov Chains: Gibbs Fields, Monte Carlo Simulation, and Queues*, vol. 31 (Springer Science & Business Media, 2013)

26. J.C. Butcher, *Numerical Methods for Ordinary Differential Equations* (Wiley, 2016)

27. Y. Cao, D.T. Gillespie, L.R. Petzold, Avoiding negative populations in explicit poisson tau-leaping. J. Chem. Phys. **123**(5), 054104 (2005)

28. Y. Cao, D.T. Gillespie, L.R. Petzold, Multiscale stochastic simulation algorithm with stochastic partial equilibrium assumption for chemically reacting systems. J. Comput. Phys. **206**(2), 395–411 (2005)

29. Y. Cao, D.T. Gillespie, L.R. Petzold, The slow-scale stochastic simulation algorithm. J. Chem. Phys. **122**(1), 014116 (2005)

30. Y. Cao, D.T. Gillespie, L.R. Petzold, Efficient step size selection for the tau-leaping simulation method. J. Chem. Phys. **124**(4), 044109 (2006)

31. Y. Cao, D.T. Gillespie, L.R. Petzold, Adaptive explicit-implicit tau-leaping method with automatic tau selection. J. Chem. Phys. **126**(22), 224101 (2007)

32. Y. Cao, H. Li, L.R. Petzold, Efficient formulation of the stochastic simulation algorithm for chemically reacting systems. J. Chem. Phys. **121**(9), 4059–4067 (2004)

33. L. Cardelli, M. Kwiatkowska, L. Laurenti, A stochastic hybrid approximation for chemical kinetics based on the linear noise approximation, in *International Conference on Computational Methods in Systems Biology* (Springer, Berlin, 2016), pp. 147–167

34. A. Chatterjee, D.G. Vlachos, M.A. Katsoulakis, Binomial distribution based τ-leap accelerated stochastic simulation. J. Chem. Phys. **122**(2), 024112 (2005)

35. J.D. Chodera, K.A. Dill, N. Singhal, V.S. Pande, W.C. Swope, J.W. Pitera, Automatic discovery of metastable states for the construction of Markov models of macromolecular conformational dynamics. J. Chem. Phys. **126**, 155101 (2007)

36. J.D. Chodera, P.J. Elms, W.C. Swope, J.-H. Prinz, S. Marqusee, C. Bustamante, F. Noé, V.S. Pande, A robust approach to estimating rates from time-correlation functions (2011). arXiv:1108.2304

37. C. Cianci, S. Smith, R. Grima, Molecular finite-size effects in stochastic models of equilibrium chemical systems. J. Chem. Phys. **144**(8), 084101 (2016)

38. F.C. Collins, G.E. Kimball, Diffusion-controlled reaction rates. J. Colloid Sci. **4**(4), 425–437 (1949)

39. F. Cornalba, T. Shardlow, J. Zimmer, A regularized Dean–Kawasaki model: Derivation and analysis. SIAM J. Math. Anal. **51**(2), 1137–1187 (2019)

40. A. Crudu, A. Debussche, A. Muller, O. Radulescu, Convergence of stochastic gene networks to hybrid piecewise deterministic processes. Ann. Appl. Probab. **22**(5), 1822–1859 (2012)

41. A. Crudu, A. Debussche, O. Radulescu, Hybrid stochastic simplifications for multiscale gene networks. BMC Syst. Biol. **3**(1), 89 (2009)

42. J. Cullhed, S. Engblom, A. Hellander, The URDME manual version 1.0. Technical report, Department of Information Technology, Uppsala University, Sweden, 2008

43. I.G. Darvey, P.J. Staff, Stochastic approach to first-order chemical reaction kinetics. J. Chem. Phys. **44**(3), 990–997 (1966)

44. M.H.A. Davis, Piecewise-deterministic Markov processes: A general class of non-diffusion stochastic models. J. R. Stat. Soc. B (Methodol.) **46**(3), 353–388 (1984)

45. D.S. Dean, Langevin equation for the density of a system of interacting Langevin processes. J. Phys. A Math. Gen. **29**(24), L613 (1996)

46. M.J. Del Razo, H. Qian, A discrete stochastic formulation for reversible bimolecular reactions via diffusion encounter. Preprint (2015). arXiv:1511.08798

47. M. Delbrück, Statistical fluctuations in autocatalytic reactions. J. Chem. Phys. **8**(1), 120–124 (1940)

48. P. Deuflhard, W. Huisinga, T. Jahnke, M. Wulkow, Adaptive discrete Galerkin methods applied to the chemical master equation. SIAM J. Sci. Comput. **30**(6), 2990–3011 (2008)

49. P. Deuflhard, S. Roeblitz, *A Guide to Numerical Modelling in Systems Biology*, vol. 12 of *Texts in Computational Science and Engineering* (Springer, Berlin, 2015)

50. P. Deuflhard, M. Wulkow, Computational treatment of polyreaction kinetics by orthogonal polynomials of a discrete variable. IMPACT Comput. Sci. Eng. **1**(3), 269–301 (1989)

51. M. Dibak, M.J. del Razo, D. De Sancho, C. Schütte, F. Noé, MSM/RD: Coupling Markov state models of molecular kinetics with reaction-diffusion simulations. J. Chem. Phys. **148**(21), 214107 (2018)

52. K.N. Dinh, R.B. Sidje, Understanding the finite state projection and related methods for solving the chemical master equation. Phys. Biol. **13**(3), 035003 (2016)

53. N. Djurdjevac-Conrad, L. Helfmann, J. Zonker, S. Winkelmann, C. Schütte, Human mobility and innovation spreading in ancient times: A stochastic agent-based simulation approach. EPJ Data Sci. **7**(1), 24 (2018)

54. M. Dobrzyński, J.V. Rodríguez, J.A. Kaandorp, J.G. Blom, Computational methods for diffusion-influenced biochemical reactions. Bioinformatics **23**(15), 1969–1977 (2007)

55. M. Doi, Second quantization representation for classical many-particle system. J. Phys. A Math. Gen. **9**(9), 1465 (1976)

56. M. Doi, Stochastic theory of diffusion-controlled reaction. J. Phys. A Math. Gen. **9**(9), 1479 (1976)

57. A. Donev, A. Nonaka, A.K. Bhattacharjee, A.L. Garcia, J.B. Bell, Low mach number fluctuating hydrodynamics of multispecies liquid mixtures. Phys. Fluids **27**(3), 037103 (2015)

58. A. Donev, E. Vanden-Eijnden, A. Garcia, J. Bell, On the accuracy of finite-volume schemes for fluctuating hydrodynamics. Commun. Appl. Math. Comput. Sci. **5**(2), 149–197 (2010)

59. A. Duncan, R. Erban, K. Zygalakis, Hybrid framework for the simulation of stochastic chemical kinetics. J. Comput. Phys. **326**, 398–419 (2016)

60. S. Duwal, L. Dickinson, S. Khoo, M. von Kleist, Hybrid stochastic framework predicts efficacy of prophylaxis against HIV: An example with different dolutegravir prophylaxis schemes. PLoS Comput. Biol. **14**(6), e1006155 (2018)

61. H. El Samad, M. Khammash, L. Petzold, D. Gillespie, Stochastic modelling of gene regulatory networks. Int. J. Robust Nonlinear Control IFAC Affiliated J. **15**(15), 691–711 (2005)

62. J. Elf, M. Ehrenberg, Spontaneous separation of bi-stable biochemical systems into spatial domains of opposite phases. Syst. Biol. **1**(2), 230–236 (2004)

63. S. Engblom, A discrete spectral method for the chemical master equation. Technical Report 2006-036 (2006)

64. S. Engblom, L. Ferm, A. Hellander, P. Lötstedt, Simulation of stochastic reaction-diffusion processes on unstructured meshes. SIAM J. Sci. Comput. **31**(3), 1774–1797 (2009)

65. S. Engblom, P. Lötstedt, L. Meinecke, Mesoscopic modeling of random walk and reactions in crowded media. Phys. Rev. E **98**(3), 033304 (2018)

66. R. Erban, J. Chapman, P. Maini, A practical guide to stochastic simulations of reaction-diffusion processes. Preprint (2007). arXiv:0704.1908

67. R. Erban, S.J. Chapman, Stochastic modelling of reaction-diffusion processes: Algorithms for bimolecular reactions. Phys. Biol. **6**(4), 046001 (2009)

68. R. Erban, M.B. Flegg, G.A. Papoian, Multiscale stochastic reaction–diffusion modeling: Application to actin dynamics in filopodia. Bull. Math. Biol. **76**(4), 799–818 (2014)

69. D.L. Ermak, J.A. McCammon, Brownian dynamics with hydrodynamic interactions. J. Chem. Phys. **69**(4), 1352–1360 (1978)

70. S.N. Ethier, T.G. Kurtz, *Markov Processes: Characterization and Convergence*, vol. 282 (Wiley, 2009)

71. A. Faggionato, D. Gabrielli, M.R. Crivellari, Averaging and large deviation principles for fully-coupled piecewise deterministic Markov processes and applications to molecular motors. Markov Process. Relat. Fields **16**(3), 497–548 (2010)

72. D. Fanelli, A.J. McKane, Diffusion in a crowded environment. Phys. Rev. E **82**(2), 021113 (2010)

73. D. Fange, O.G. Berg, P. Sjöberg, J. Elf, Stochastic reaction-diffusion kinetics in the microscopic limit. Proc. Natl. Acad. Sci. **107**(46), 19820–19825 (2010)

74. B. Fehrman, B. Gess, Well-posedness of nonlinear diffusion equations with nonlinear, conservative noise. Arch. Ration. Mech. Anal., 1–74 (2019)

75. N. Fenichel, Geometric singular perturbation theory for ordinary differential equations. J. Differ. Equ. **31**, 53–98 (1979)

76. L. Ferm, A. Hellander, P. Lötstedt, An adaptive algorithm for simulation of stochastic reaction–diffusion processes. J. Comput. Phys. **229**(2), 343–360 (2010)

77. L. Ferm, P. Lötstedt, Numerical method for coupling the macro and meso scales in stochastic chemical kinetics. BIT Numer. Math. **47**(4), 735–762 (2007)

78. H. Flanders, Differentiation under the integral sign. Am. Math. Monthly **80**(6), 615–627 (1973)

79. M.B. Flegg, S.J. Chapman, R. Erban, The two-regime method for optimizing stochastic reaction–diffusion simulations. J. R. Soc. Interface **9**(70), 859–868 (2012)

80. M.B. Flegg, S.J. Chapman, L. Zheng, R. Erban, Analysis of the two-regime method on square meshes. SIAM J. Sci. Comput. **36**(3), B561–B588 (2014)

81. M.B. Flegg, S. Hellander, R. Erban, Convergence of methods for coupling of microscopic and mesoscopic reaction–diffusion simulations. J. Comput. Phys. **289**, 1–17 (2015)

82. M.B. Flegg, S. Rüdiger, R. Erban, Diffusive spatio-temporal noise in a first-passage time model for intracellular calcium release. J. Chem. Phys. **138**(15), 04B606 (2013)

83. E.G. Flekkøy, J. Feder, G. Wagner, Coupling particles and fields in a diffusive hybrid model. Phys. Rev. E **64**(6), 066302 (2001)

84. B. Franz, M.B. Flegg, S.J. Chapman, R. Erban, Multiscale reaction-diffusion algorithms: PDE-assisted Brownian dynamics. SIAM J. Appl. Math. **73**(3), 1224–1247 (2013)

85. U. Franz, V. Liebscher, S. Zeiser, Piecewise-deterministic Markov processes as limits of Markov jump processes. Adv. Appl. Probab. **44**(03), 729–748 (2012)

86. D. Frenkel, B. Smit, *Understanding Molecular Simulation: From Algorithms to Applications*, vol. 1 (Elsevier, 2001)

87. N. Friedman, L. Cai, X.S. Xie, Linking stochastic dynamics to population distribution: An analytical framework of gene expression. Phys. Rev. Lett. **97**(16), 168302 (2006)

88. C. Gadgil, C.H. Lee, H.G. Othmer, A stochastic analysis of first-order reaction networks. Bull. Math. Biol. **67**(5), 901–946 (2005)

89. A. Ganguly, D. Altintan, H. Koeppl, Jump-diffusion approximation of stochastic reaction dynamics: Error bounds and algorithms. Multiscale Model. Simul. **13**(4), 1390–1419 (2015)

90. C.W. Gardiner, *Handbook of Stochastic Methods*, vol. 3 (Springer, Berlin, 1985)

91. C.W. Gardiner, S. Chaturvedi, The poisson representation. i. a new technique for chemical master equations. J. Stat. Phys. **17**(6), 429–468 (1977)

92. C.W. Gardiner, K.J. McNeil, D.F. Walls, I.S. Matheson, Correlations in stochastic theories of chemical reactions. J. Stat. Phys. **14**(4), 307–331 (1976)

93. T.S. Gardner, C.R. Cantor, J.J. Collins, Construction of a genetic toggle switch in Escherichia coli. Nature **403**(6767), 339 (2000)

94. L. Gauckler, H. Yserentant, Regularity and approximability of the solutions to the chemical master equation. ESAIM Math. Model. Numer. Anal. **48**(6), 1757–1775 (2014)

95. T. Geyer, C. Gorba, V. Helms, Interfacing Brownian dynamics simulations. J. Chem. Phys. **120**(10), 4573–4580 (2004)

96. A. Ghosh, A. Leier, T.T. Marquez-Lago, The spatial chemical Langevin equation and reaction diffusion master equations: Moments and qualitative solutions. Theor. Biol. Med. Model. **12**(1), 5 (2015)

97. M. Giaquinta, S. Hildebrandt, *Calculus of Variations II*, vol. 311 (Springer Science & Business Media, 2013)

98. M.A. Gibson, J. Bruck, Efficient exact stochastic simulation of chemical systems with many species and many channels. J. Phys. Chem. A **104**(9), 1876–1889 (2000)

99. D.T. Gillespie, A general method for numerically simulating the stochastic time evolution of coupled chemical reactions. J. Comput. Phys. **22**(4), 403–434 (1976)

100. D.T. Gillespie, Exact stochastic simulation of coupled chemical reactions. J. Phys. Chem. **81**(25), 2340–2361 (1977)

101. D.T. Gillespie, *Markov Processes: An Introduction for Physical Scientists* (Elsevier, 1991)

102. D.T. Gillespie, A rigorous derivation of the chemical master equation. Phys. A Stat. Mech. Appl. **188**(1), 404–425 (1992)

103. D.T. Gillespie, The chemical Langevin equation. J. Chem. Phys. **113**(1), 297–306 (2000)

104. D.T. Gillespie, Approximate accelerated stochastic simulation of chemically reacting systems. J. Chem. Phys. **115**(4), 1716–1733 (2001)

105. D.T. Gillespie, Stochastic simulation of chemical kinetics. Annu. Rev. Phys. Chem. **58**, 35–55 (2007)

106. D.T. Gillespie, A. Hellander, L. Petzold, Perspective: Stochastic algorithms for chemical kinetics. J. Chem. Phys. **138**(17), 170901 (2013)

107. D.T. Gillespie, L.R. Petzold, Improved leap-size selection for accelerated stochastic simulation. J. Chem. Phys. **119**(16), 8229–8234 (2003)

108. A.N. Gorban, I.V. Karlin, *Invariant Manifolds for Physical and Chemical Kinetics* (Springer, Berlin, 2004)

109. J. Goutsias, Quasiequilibrium approximation of fast reaction kinetics in stochastic biochemical systems. J. Chem. Phys. **122**(18), 184102 (2005)

110. J. Goutsias, G. Jenkinson, Markovian dynamics on complex reaction networks. Phys. Rep. **529**(2), 199–264 (2013)

111. R. Grima, An effective rate equation approach to reaction kinetics in small volumes: Theory and application to biochemical reactions in nonequilibrium steady-state conditions. J. Chem. Phys. **133**(3), 07B604 (2010)

112. R. Grima, D.R. Schmidt, T.J. Newman, Steady-state fluctuations of a genetic feedback loop: An exact solution. J. Chem. Phys. **137**(3), 035104 (2012)

113. R. Grima, S. Schnell, A systematic investigation of the rate laws valid in intracellular environments. Biophys. Chem. **124**(1), 1–10 (2006)

114. R. Grima, P. Thomas, A.V. Straube, How accurate are the nonlinear chemical Fokker-Planck and chemical Langevin equations? J. Chem. Phys. **135**(8), 084103 (2011)

115. C.M. Guldberg, Concerning the laws of chemical affinity. CM Forhandlinger Videnskabs-Selskabet i Christiana **111**, 1864 (1864)

116. E.L. Haseltine, J.B. Rawlings, Approximate simulation of coupled fast and slow reactions for stochastic chemical kinetics. J. Chem. Phys. **117**(15), 6959–6969 (2002)

117. E.L. Haseltine, J.B. Rawlings, On the origins of approximations for stochastic chemical kinetics. J. Chem. Phys. **123**(16), 164115 (2005)

118. J. Hattne, D. Fange, J. Elf, Stochastic reaction-diffusion simulation with MesoRD. Bioinformatics **21**(12), 2923–2924 (2005)

119. S. Havlin, D. Ben-Avraham, Diffusion in disordered media. Adv. Phys. **36**(6), 695–798 (1987)

120. L. Helfmann, A. Djurdjevac, N. Djurdjevac-Conrad, S. Winkelmann, C. Schütte, From interacting agents to density-based modeling with stochastic PDEs Accepted for publication in Communications in Appl. Math. Comput Sci., to appear 2021

121. A. Hellander, S. Hellander, P. Lötstedt, Coupled mesoscopic and microscopic simulation of stochastic reaction-diffusion processes in mixed dimensions. Multiscale Model. Simul. **10**(2), 585–611 (2012)

122. S. Hellander, A. Hellander, L. Petzold, Reaction-diffusion master equation in the microscopic limit. Phys. Rev. E **85**(4), 042901 (2012)

123. S. Hellander, A. Hellander, L. Petzold, Reaction rates for mesoscopic reaction-diffusion kinetics. Phys. Rev. E **91**(2), 023312 (2015)

124. S. Hellander, L. Petzold, Reaction rates for a generalized reaction-diffusion master equation. Phys. Rev. E **93**(1), 013307 (2016)

125. T.A. Henzinger, L. Mikeev, M. Mateescu, V. Wolf, Hybrid numerical solution of the chemical master equation, in *Proceedings of the 8th International Conference on Computational Methods in Systems Biology* (ACM, 2010), pp. 55–65

126. D.J. Higham, An algorithmic introduction to numerical simulation of stochastic differential equations. SIAM Rev. **43**(3), 525–546 (2001)

127. D.J. Higham, Modeling and simulating chemical reactions. SIAM Rev. **50**(2), 347–368 (2008)

128. Y. Hu, T. Li, B. Min, A weak second order tau-leaping method for chemical kinetic systems. J. Chem. Phys. **135**(2), 024113 (2011)

129. G. Hummer, A. Szabo, Optimal dimensionality reduction of multistate kinetic and Markov state models. J. Phys. Chem. B **119**(29), 9029–9037 (2015). PMID: 25296279

130. S.A. Isaacson, The reaction-diffusion master equation as an asymptotic approximation of diffusion to a small target. SIAM J. Appl. Math. **70**(1), 77–111 (2009)

131. S.A. Isaacson, A convergent reaction-diffusion master equation. J. Chem. Phys. **139**(5), 054101 (2013)

132. S.A. Isaacson, D.M. McQueen, C.S. Peskin, The influence of volume exclusion by chromatin on the time required to find specific dna binding sites by diffusion. Proc. Natl. Acad. Sci. **108**(9), 3815–3820 (2011)

133. S.A. Isaacson, J. Newby, Uniform asymptotic approximation of diffusion to a small target. Phys. Rev. E **88**, 012820 (2013)

134. S.A. Isaacson, C.S. Peskin, Incorporating diffusion in complex geometries into stochastic chemical kinetics simulations. SIAM J. Sci. Comput. **28**(1), 47–74 (2006)

135. S.A. Isaacson, Y. Zhang, An unstructured mesh convergent reaction-diffusion master equation for reversible reactions. Preprint (2017). arXiv:1711.04220

136. K. Ishida, Stochastic model for bimolecular reaction. J. Chem. Phys. **41**(8), 2472–2478 (1964)

137. T. Jahnke, On reduced models for the chemical master equation. Multiscale Model. Simul. **9**(4), 1646–1676 (2011)

138. T. Jahnke, W. Huisinga, Solving the chemical master equation for monomolecular reaction systems analytically. J. Math. Biol. **54**(1), 1–26 (2007)

139. T. Jahnke, W. Huisinga, A dynamical low-rank approach to the chemical master equation. Bull. Math. Biol. **70**(8), 2283–2302 (2008)

140. T. Jahnke, M. Kreim, Error bound for piecewise deterministic processes modeling stochastic reaction systems. SIAM Multiscale Model. Simul. **10**(4), 1119–1147 (2012)

141. R. Kapral, Multiparticle collision dynamics: Simulation of complex systems on mesoscales. Adv. Chem. Phys. **140**, 89–146 (2008)

142. V. Kazeev, M. Khammash, M. Nip, C. Schwab, Direct solution of the chemical master equation using quantized tensor trains. PLoS Comput. Biol. **10**(3), e1003359 (2014)

143. I.G. Kevrekidis, C.W. Gear, G. Hummer, Equation-free: The computer-aided analysis of complex multiscale systems. AIChE J. **50**(7), 1346–1355 (2004)

144. I.G. Kevrekidis, G. Samaey, Equation-free multiscale computation: Algorithms and applications. Annu. Rev. Phys. Chem. **60**(1), 321–344 (2009)

145. C. Kim, A. Nonaka, J.B. Bell, A.L. Garcia, A. Donev, Stochastic simulation of reaction-diffusion systems: A fluctuating-hydrodynamics approach. J. Chem. Phys. **146**(12), 124110 (2017)

146. H. Kim, K.J. Shin, Exact solution of the reversible diffusion-influenced reaction for an isolated pair in three dimensions. Phys. Rev. Lett. **82**(7), 1578 (1999)

147. J.F.C. Kingman, *Poisson Processes* (Wiley Online Library, 1993)

148. M. Klann, A. Ganguly, H. Koeppl, Hybrid spatial Gillespie and particle tracking simulation. Bioinformatics **28**(18), i549–i555 (2012)

149. M. Klann, H. Koeppl, Spatial simulations in systems biology: From molecules to cells. Int. J. Mol. Sci. **13**(6), 7798–7827 (2012)

150. P.E. Kloeden, E. Platen, *Numerical Solution of Stochastic Differential Equations* (Springer, Berlin, 1992)

151. V. Kolokoltsov, *Nonlinear Markov Processes and Kinetic Equations*, vol. 182 of *Cambridge Tracts in Mathematics* (Cambridge University Press, Cambridge, 2010)

152. V. Konarovskyi, T. Lehmann, M.-K. von Renesse, Dean-Kawasaki dynamics: Ill-posedness vs. triviality. Electron. Commun. Probab. **24**, 9 pp. (2019)

153. H.A. Kramers, Brownian motion in a field of force and the diffusion model of chemical reactions. Physica **7**(4), 284–304 (1940)

154. I. Kryven, S. Roeblitz, C. Schütte, Solution of the chemical master equation by radial basis functions approximation with interface tracking. BMC Syst. Biol. **9**(67), 1–12 (2015)

155. C. Kuehn, *Multiple Time Scale Dynamics*, Vol. 191 (Springer, Berlin, 2015)

156. T.G. Kurtz, Solutions of ordinary differential equations as limits of pure jump Markov processes. J. Appl. Probab. **7**(1), 49–58 (1970)

157. T.G. Kurtz, Limit theorems for sequences of jump Markov processes approximating ordinary differential processes. J. Appl. Probab. **8**(2), 344–356 (1971)

158. T.G. Kurtz, The relationship between stochastic and deterministic models for chemical reactions. J. Chem. Phys. **57**(7), 2976–2978 (1972)

159. T.G. Kurtz, Limit theorems and diffusion approximations for density dependent Markov chains, in *Stochastic Systems: Modeling, Identification and Optimization, I* (Springer, Berlin, 1976), pp. 67–78

160. T.G. Kurtz, Strong approximation theorems for density dependent Markov chains. Stoch. Process. Appl. **6**(3), 223–240 (1978)

161. K.A. Landman, A.E. Fernando, Myopic random walkers and exclusion processes: Single and multispecies. Phys. A Stat. Mech. Appl. **390**(21–22), 3742–3753 (2011)

162. J. Lang, A. Walter, An adaptive Rothe method for nonlinear reaction-diffusion systems. Appl. Numer. Math. **13**, 135–146 (1993)

163. I.J. Laurenzi, An analytical solution of the stochastic master equation for reversible bimolecular reaction kinetics. J. Chem. Phys. **113**(8), 3315–3322 (2000)

164. G.F. Lawler, *Introduction to Stochastic Processes* (CRC Press, 2006)

165. W. Ledermann, G.E.H. Reuter, Spectral theory for the differential equations of simple birth and death processes. Phil. Trans. R. Soc. A **246**, 321–369 (1954)

166. A. Leier, T.T. Marquez-Lago, K. Burrage, Generalized binomial τ-leap method for biochemical kinetics incorporating both delay and intrinsic noises. J. Chem. Phys. **128**(20), 05B623 (2008)

167. H. Li, Y. Cao, L.R. Petzold, D.T. Gillespie, Algorithms and software for stochastic simulation of biochemical reacting systems. Biotechnol. Prog. **24**(1), 56–61 (2008)

168. H. Li, L. Petzold, Logarithmic direct method for discrete stochastic simulation of chemically reacting systems. J. Chem. Phys. **16**, 1–11 (2006)

169. T. Li, A. Abdulle, E. Weinan, Effectiveness of implicit methods for stiff stochastic differential equations, vol. 3(2) *Commun. Comput. Phys.* (Citeseer, 2008), pp. 295–307

170. W. Lick, Two-variable expansions and singluar perturbation problems. SIAM J. Appl. Math. **17**(4), 815–825 (1969)

171. J. Lipková, G. Arampatzis, P. Chatelain, B. Menze, P. Koumoutsakos, S-leaping: An adaptive, accelerated stochastic simulation algorithm, bridging τ-leaping and R-leaping. Bull. Math. Biol., 1–23 (2018)

172. J. Lipková, K.C. Zygalakis, S.J. Chapman, R. Erban, Analysis of Brownian dynamics simulations of reversible bimolecular reactions. SIAM J. Appl. Math. **71**(3), 714–730 (2011)

173. P. Lötstedt, L. Ferm, Dimensional reduction of the Fokker–Planck equation for stochastic chemical reactions. Multiscale Model. Simul. **5**(2), 593–614 (2006)

174. J. Maas, A. Mielke, Modeling of chemical reaction systems with detailed balance using gradient structures. Preprint (2020). arXiv:2004.02831

175. M. Martcheva, H.R. Thieme, T. Dhirasakdanon, Kolmogorov's differential equations and positive semigroups on first moment sequence spaces. J. Math. Biol. **53**, 642–671 (2006)

176. S. Mauch, M. Stalzer, Efficient formulations for exact stochastic simulation of chemical systems. IEEE/ACM Trans. Comput. Biol. Bioinform. **8**(1), 27–35 (2009)

177. J.M. McCollum, G.D. Peterson, C.D. Cox, M.L. Simpson, N.F. Samatova, The sorting direct method for stochastic simulation of biochemical systems with varying reaction execution behavior. Comput. Biol. Chem. **30**(1), 39–49 (2006)

178. D.A. McQuarrie, Stochastic approach to chemical kinetics. J. Appl. Probab. **4**(03), 413–478 (1967)

179. L. Meinecke, Multiscale modeling of diffusion in a crowded environment. Bull. Math. Biol. **79**(11), 2672–2695 (2017)

180. S. Menz, Hybrid stochastic-deterministic approaches for simulation and analysis of biochemical reaction networks. Ph.D. thesis, Freie Universität Berlin, 2013

181. S. Menz, J. Latorre, C. Schütte, W. Huisinga, Hybrid stochastic-deterministic solution of the chemical master equation. SIAM Interdiscip. J. Multiscale Model. Simul. (MMS) **10**(4), 1232–1262 (2012)

182. P. Metzner, C. Schütte, E. Vanden-Eijnden, Transition path theory for Markov jump processes. Multiscale Model. Simul. **7**(3), 1192–1219 (2009)

183. C. Moler, C. Van Loan, Nineteen dubious ways to compute the exponential of a matrix, twenty-five years later. SIAM Rev. **45**(1), 3–49 (2003)

184. E. Moro, Hybrid method for simulating front propagation in reaction-diffusion systems. Phys. Rev. E **69**(6), 060101 (2004)

185. J.E. Moyal, Stochastic processes and statistical physics. J. R. Stat. Soc. B (Methodol.) **11**(2), 150–210 (1949)

186. B. Munsky, M. Khammash, The finite state projection algorithm for the solution of the chemical master equation. J. Chem. Phys. **124**(4), 044104 (2006)

187. B. Øksendal, *Stochastic Differential Equations: An Introduction with Applications* (Springer, Berlin, 2003)

188. J. Pahle, Biochemical simulations: Stochastic, approximate stochastic and hybrid approaches. Brief. Bioinform. **10**(1), 53–64 (2009)

189. G. Pavliotis, A. Stuart, *Multiscale Methods: Averaging and Homogenization* (Springer Science & Business Media, 2008)

190. G.A. Pavliotis, *Stochastic Processes and Applications: Diffusion Processes, the Fokker-Planck and Langevin Equations*, vol. 60 (Springer, Berlin, 2014)

191. X. Peng, W. Zhou, Y. Wang, Efficient binomial leap method for simulating chemical kinetics. J. Chem. Phys. **126**(22), 224109 (2007)

192. D. Pollard, *Convergence of Stochastic Processes* (Springer Science & Business Media, 2012)

193. J.-H. Prinz, H. Wu, M. Sarich, B. Keller, M. Fischbach, M. Held, C. Schütte, J.D. Chodera, F. Noé, Markov models of molecular kinetics: Generation and validation. J. Chem. Phys. **134**(17), 174105 (2011)

194. T. Prüstel, M. Meier-Schellersheim, Exact green's function of the reversible diffusion-influenced reaction for an isolated pair in two dimensions. J. Chem. Phys. **137**(5), 054104 (2012)

195. M. Rathinam, L.R. Petzold, Y. Cao, D.T. Gillespie, Stiffness in stochastic chemically reacting systems: The implicit tau-leaping method. J. Chem. Phys. **119**(24), 12784–12794 (2003)

196. G.E.H. Reuter, W. Ledermann, On the differential equations for the transition probabilities of Markov processes with enumerably many states. Proc. Camb. Philos. Soc. **49**, 247–262 (1953)

197. D. Ridgway, G. Broderick, M.J. Ellison, Accommodating space, time and randomness in network simulation. Curr. Opin. Biotechnol. **17**(5), 493–498 (2006)

198. E. Roberts, J.E. Stone, Z. Luthey-Schulten, Lattice microbes: High-performance stochastic simulation method for the reaction-diffusion master equation. J. Comput. Chem. **34**(3), 245–255 (2013)

199. S. Röblitz, Statistical error estimation and grid-free hierarchical refinement in conformation dynamics. Ph.D. thesis, Freie Universität Berlin, 2009

200. D.M. Roma, R.A. O'Flanagan, A.E. Ruckenstein, A.M. Sengupta, R. Mukhopadhyay, Optimal path to epigenetic switching. Phys. Rev. E **71**(1), 011902 (2005)

201. H. Salis, Y. Kaznessis, Accurate hybrid stochastic simulation of a system of coupled chemical or biochemical reactions. J. Chem. Phys. **122**(5), 054103 (2005)

202. A. Samant, B.A. Ogunnaike, D.G. Vlachos, A hybrid multiscale Monte Carlo algorithm (HyMSMC) to cope with disparity in time scales and species populations in intracellular networks. BMC Bioinf. **8**(1), 175 (2007)

203. M. Sarich, R. Banisch, C. Hartmann, C. Schütte, Markov state models for rare events in molecular dynamics. Entropy **16**(1), 258 (2013)

204. M. Sarich, F. Noé, C. Schütte, On the approximation quality of Markov state models. Multiscale Model. Simul. **8**(4), 1154–1177 (2010)

205. M.J. Saxton, A biological interpretation of transient anomalous subd-iffusion. i. qualitative model. Biophys. J. **92**(4), 1178–1191 (2007)

206. W.E. Schiesser, *The Numerical Method of Lines: Integration of Partial Differential Equations* (Elsevier, 2012)

207. S. Schnell, T.E. Turner, Reaction kinetics in intracellular environments with macromolecular crowding: Simulations and rate laws. Prog. Bio-phys. Mol. Biol. **85**(2–3), 235–260 (2004)

208. D. Schnoerr, G. Sanguinetti, R. Grima, The complex chemical Langevin equation. J. Chem. Phys. **141**(2), 024103 (2014)

209. D. Schnoerr, G. Sanguinetti, R. Grima, Approximation and inference methods for stochastic biochemical kinetics – a tutorial review. J. Phys. A Math. Theor. **50**(9), 093001 (2017)

210. J. Schöneberg, F. Noé, Readdy – a software for particle-based reaction-diffusion dynamics in crowded cellular environments. PLoS ONE **8**(9), e74261 (2013)

211. J. Schöneberg, A. Ullrich, F. Noé, Simulation tools for particle-based reaction-diffusion dynamics in continuous space. BMC Biophys. **7**(1), 1 (2014)

212. C. Schütte, A. Fischer, W. Huisinga, P. Deuflhard, A direct approach to conformational dynamics based on hybrid Monte Carlo. J. Comput. Phys. **151**, 146–168 (1999)

213. C. Schütte, F. Noé, J. Lu, M. Sarich, E. Vanden-Eijnden, Markov state models based on milestoning. J. Chem. Phys. **134**(20), 204105 (2011)

214. C. Schütte, M. Sarich, *Metastability and Markov State Models in Molecular Dynamics: Modeling, Analysis, Algorithmic Approaches*, vol. 24 of *Courant Lecture Notes* (American Mathematical Soc., 2013)

215. D. Shoup, A. Szabo, Role of diffusion in ligand binding to macro-molecules and cell-bound receptors. Biophys. J. **40**(1), 33–39 (1982)

216. S. Smith, R. Grima, Breakdown of the reaction-diffusion master equa-tion with nonelementary rates. Phys. Rev. E **93**(5), 052135 (2016)

217. S. Smith, R. Grima, Spatial stochastic intracellular kinetics: A review of modelling approaches. Bull. Math. Biol., 1–50 (2018)

218. P.J. Staff, A stochastic development of the reversible Michaelis-Menten mechanism. J. Theor. Biol. **27**(2), 221–232 (1970)

219. D.J. Stekel, D.J. Jenkins, Strong negative self regulation of prokaryotic transcription factors increases the intrinsic noise of protein expression. BMC Syst. Biol. **2**(1), 6 (2008)

220. V. Sunkara, M. Hegland, An optimal finite state projection method. Procedia Comput. Sci. **1**(1), 1579–1586 (2010)

221. V. Sunkara, M. Hegland, Parallelising the finite state projection method. ANZIAM J. **52**, 853–865 (2010)

222. A. Szabo, K. Schulten, Z. Schulten, First passage time approach to diffusion controlled reactions. J. Chem. Phys. **72**(8), 4350–4357 (1980)

223. T. Székely Jr., K. Burrage, Stochastic simulation in systems biology. Comput. Struct. Biotechnol. J. **12**(20–21), 14–25 (2014)

224. P.R. Taylor, C.A. Yates, M.J. Simpson, R.E. Baker, Reconciling transport models across scales: The role of volume exclusion. Phys. Rev. E **92**(4), 040701 (2015)

225. V.H. Thanh, C. Priami, Simulation of biochemical reactions with time-dependent rates by the rejection-based algorithm. J. Chem. Phys. **143**(5), 08B601_1 (2015)

226. P. Thomas, H. Matuschek, R. Grima, Intrinsic noise analyzer: A software package for the exploration of stochastic biochemical kinetics using the system size expansion. PloS One **7**(6), e38518 (2012)

227. T. Tian, K. Burrage, Binomial leap methods for simulating stochastic chemical kinetics. J. Chem. Phys. **121**(21), 10356–10364 (2004)

228. P. Todorovic, *An Introduction to Stochastic Processes and Their Applications* (Springer Science & Business Media, 2012)

229. K. Tucci, R. Kapral, Mesoscopic model for diffusion-influenced reaction dynamics. J. Chem. Phys. **120**(17), 8262–8270 (2004)

230. T.E. Turner, S. Schnell, K. Burrage, Stochastic approaches for modelling in vivo reactions. Comput. Biol. Chem. **28**(3), 165–178 (2004)

231. M.v. Smoluchowski, Mathematical theory of the kinetics of the coagulation of colloidal solutions. Z. Phys. Chem. **92**, 129–168 (1917)

232. M.v. Smoluchowski, Versuch einer mathematischen Theorie der Koagulationskinetik kolloider Lösungen. Z. Physikal. Chemie **92**(1), 129–168 (1918)

233. N.G. van Kampen, A power series expansion of the master equation. Can. J. Phys. **39**(4), 551–567 (1961)

234. N.G. van Kampen, *Stochastic Processes in Physics and Chemistry*, 4th edn. (Elsevier, 2006)

235. J.S. van Zon, P.R. Ten Wolde, Green's-function reaction dynamics: A particle-based approach for simulating biochemical networks in time and space. J. Chem. Phys. **123**(23), 234910 (2005)

236. J.S. van Zon, P.R. Ten Wolde, Simulating biochemical networks at the particle level and in time and space: Green's function reaction dynamics. Phys. Rev. Lett. **94**(12), 128103 (2005)

237. C.L. Vestergaard, M. Génois, Temporal Gillespie algorithm: Fast simulation of contagion processes on time-varying networks. PLoS Comput. Biol. **11**(10), e1004579 (2015)

238. E.O. Voit, H.A. Martens, S.W. Omholt, 150 years of the mass action law. PLoS Comput. Biol. **11**(1), e1004012 (2015)

239. M. Voliotis, P. Thomas, R. Grima, C.G. Bowsher, Stochastic simulation of biomolecular networks in dynamic environments. PLoS Comput. Biol. **12**(6), e1004923 (2016)

240. A.F. Voter, Introduction to the kinetic Monte Carlo method. *Radiation Effects in Solids* (Springer, Berlin, 2007), pp. 1–23

241. P. Waage, Experiments for determining the affinity law. Forhandlinger Videnskabs-Selskabet i Christiana **92**, 1864 (1864)

242. P. Waage, C.M. Guldberg, Studies concerning affinity. J. Chem. Educ **63**(12), 1044 (1986)

243. E. Weinan, D. Liu, E. Vanden-Eijnden, Nested stochastic simulation algorithm for chemical kinetic systems with multiple time scales. J. Comput. Phys. **221**(1), 158–180 (2007)

244. E. Weinan, D. Liu, E. Vanden-Eijnden, Nested stochastic simulation algorithm for chemical kinetic systems with disparate rates. J. Chem. Phys. **123**, 194107 (2005)

245. M. Weiss, M. Elsner, F. Kartberg, T. Nilsson, Anomalous subdiffusion is a measure for cytoplasmic crowding in living cells. Biophys. J. **87**(5), 3518–3524 (2004)

246. S. Winkelmann, C. Schütte, The spatiotemporal master equation: Approximation of reaction-diffusion dynamics via Markov state modeling. J. Chem. Phys. **145**(21), 214107 (2016)

247. S. Winkelmann, C. Schütte, Hybrid models for chemical reaction networks: Multiscale theory and application to gene regulatory systems. J. Chem. Phys. **147**(11), 114115 (2017)

248. C.A. Yates, R.E. Baker, R. Erban, P.K. Maini, Going from microscopic to macroscopic on nonuniform growing domains. Phys. Rev. E **86**(2), 021921 (2012)

249. C.A. Yates, M.B. Flegg, The pseudo-compartment method for coupling partial differential equation and compartment-based models of diffusion. J. R. Soc. Interface **12**(106), 20150141 (2015)

250. S. Zeiser, U. Franz, V. Liebscher, Autocatalytic genetic networks modeled by piecewise-deterministic Markov processes. J. Math. Biol. **60**(2), 207–246 (2010)

251. S. Zeiser, U. Franz, O. Wittich, V. Liebscher, Simulation of genetic networks modelled by piecewise deterministic Markov processes. IET Syst. Biol. **2**(3), 113 (2008)

252. O.C. Zienkiewicz, R.L. Taylor, J.Z. Zhu, *The Finite Element Method: Its Basis and Fundamentals* (Elsevier, 2005)

Index

Printed in the United States
By Bookmasters